T0247727

MAKING SPACE ^{for} WOMEN

Pioneering Women: Leaders and Trailblazers
Sponsored by Nancy and Ted Paup
Texas Woman's University
Claire L. Sahlin, General Editor
With assistance from Kimberly C. Merenda

TEXAS WOMAN'S UNIVERSITY
JANE NELSON INSTITUTE
for WOMEN'S LEADERSHIP

MAKING SPACE
for WOMEN

Stories from Trailblazing Women
of NASA's Johnson Space Center

Edited by

JENNIFER M. ROSS-NAZZAL

TEXAS A&M UNIVERSITY PRESS ★ COLLEGE STATION

LIBRARY OF CONGRESS CATALOGING-IN-PUBLICATION DATA
Names: Ross-Nazzal, Jennifer M., editor.
Title: Making space for women: stories from trailblazing women of NASA's Johnson Space Center / edited by Jennifer M. Ross-Nazzal.
Other titles: Pioneering women, leaders and trailblazers.
Description: First edition. | College Station: Texas A&M University Press, [2021] | Series: Pioneering women: leaders and trailblazers | Includes bibliographical references and index.
Identifiers: LCCN 2021017015 | ISBN 9781623499938 (cloth) | ISBN 9781623499945 (ebook)
Subjects: LCSH: Lyndon B. Johnson Space Center—Officials and employees—Biography. | Women scientists—Texas—Houston—Biography. | Women in science—United States. | LCGFT: Biographies.
Classification: LCC Q141 .M242 2021 | DDC 629.4092/520973—dc23
LC record available at https://lccn.loc.gov/2021017015

 All images are courtesy of NASA unless otherwise indicated.

Contents

Have you ever thought about working for the National Aeronautics and Space Administration (NASA)? Did you worry that you wouldn't have "the right stuff," because you wouldn't have enough technical, scientific, or engineering training? Well, NASA is more than astronauts and rocket scientists. And space? Space is for everyone. So are the STEM fields: science, technology, engineering, and mathematics. Thousands of people with many different backgrounds keep NASA operating and exploring, including trainers, programmers, flight controllers, and—yes—photographers, writers, and scuba divers. All of them work together to support the wide range of missions within the nation's space agency.

More than thirty years ago, I learned of an opportunity with NASA while watching the evening news. President Ronald Reagan announced the Teacher in Space Program in 1984, at a time when my total focus was on teaching second graders in a little mountain town in Idaho. Many teachers applied for this opportunity because of all we could learn and bring back to our students. At the end of a terrific competition I was named as the backup, and I got to train with our Teacher in Space—social studies teacher Christa McAuliffe—and the crew of the Space Shuttle *Challenger*. Tragically, the *Challenger* accident in January 1986 changed the course of American history, and it changed the course of my life. I continued to teach in my classroom and to work with NASA until 1998, when I was selected to join a new class of astronaut candidates. Nine years later, in the summer of 2007, my crewmates and I served onboard the Space Shuttle *Endeavour* to help construct the International Space Station (ISS) and to engage with students and teachers here on Earth.

My experience demonstrates that there are no boundaries or limits to a career at NASA. I am not the only woman who is not an engineer but who has found herself happily at home in the business of space exploration.

Nonetheless, many girls and young women hesitate to take risks and to commit to an out-of-this-world career. As a teacher, I've seen many girls quietly sit by while boys willingly jump in. The women featured in this book demonstrate that the opportunities are there for those who choose to seize them.

For young women, parents, and teachers looking for inspirational stories about women working in STEM fields, *Making Space for Women* fills that gap. By sharing the oral history narratives of twenty-one women, this pathbreaking work looks at women's opportunities at NASA for more than fifty years. In particular, these stories highlight the changes witnessed by female employees over the years at the Johnson Space Center (JSC) in Texas. They reveal the challenges women have had to overcome, and the value of determination and stick-to-itiveness. Readers will learn about the women of NASA and the lessons they learned along the way to achieve the success they enjoyed.

Jennifer Ross-Nazzal has written a wonderful book. In these pages, you will find unique insights from women who, though their thoughts and contributions through the years have remained relatively unknown, will definitely inspire the next generation of young women to explore careers in technical fields with endless opportunities. As an educator and mother, I can attest to the value of these narratives. I encourage all those who are interested to take a risk and reach for the stars.

Barbara Morgan
Educator and NASA Astronaut, Retired

Preface

My office phone rang. I was always curious about who would be on the other end of the line. As the NASA Johnson Space Center historian, I frequently received requests for research assistance from media outlets, students, retirees, and NASA employees. This caller was from the prestigious Johns Hopkins University in Baltimore, Maryland, and as she explained her project, my heart sank. A student in the Department of History of Science and Technology, she hoped to write about female NASA engineers of the late 1960s and early 1970s. She asked if there were any archival collections at my center that she could mine. I agreed that it was a worthy subject, but one with scarce documentation beyond a handful of oral history interviews with women mathematicians. Because NASA is a scientific and technological agency, I said, very few of the records—even the personnel series at the National Archives—documented gender or sex in America's space agency.

Over the years I had received numerous requests on this very topic, and while my answer had disappointed many, I could speak with confidence, because for more than a decade I had combed through every possible location and source in Houston. A woman's historian myself, I came to NASA in 2000 as an intern with the JSC Oral History Project. Throughout that summer and during the following year I spent time gathering research on engineers, scientists, and astronauts who had worked on the lunar and follow-on programs. All of them were men. "What about the women?" I wondered. I saw them walking across the Houston campus; they served as flight directors and astronauts and worked in other highly visible positions. I even shared an office with some of them.

I began my search at the library, the Scientific and Technical Information Center, or STIC, which everyone pronounces as "stick" (NASA loves acronyms, as I learned), but found nothing academic or in-depth about gender, sex, and space history on the shelves. In the JSC Collection that houses the records of NASA's human spaceflight programs, I found little about any women. There was a marked absence of females in NASA history.

The writings about NASA had remained firmly male-centric until 2002, when authors started to document the efforts of Fellow Lady Astronaut Trainees, a group that was called the Mercury 13. Today, women astronauts continue to receive a majority of the attention, although they represent a tiny fraction of the female workforce. There were many more stories to tell, and as the JSC Historian, I realized I had a unique opportunity to fill that gap and give voice to other women who had made contributions to the nation's space program. I envisioned a book that would explore how opportunities changed over time as the workforce diversified at JSC, and fields that were once closed to women—particularly the astronaut corps and flight control—opened to their gender. As I worked, I came to see the book as a must-read for young women interested in pursuing a career in science, technology, engineering, or mathematics.

Since 1997 the JSC History Office had conducted hundreds of interviews, including some with women, but not all were suitable for publication. I drew up a manuscript proposal for *Making Space for Women* that would allow each woman to tell her own story through an oral history narrative. I reviewed the interviews already gathered but found obvious and significant gaps in occupations as well as in race and ethnicity. In consultation with Duane Ross, manager of the Astronaut Selection Office at the time, I worked to identify potential candidates to be interviewed specifically for this project to provide a more complete reflection of women's advancement at JSC. We included many of the "firsts" in their positions, such as the first female director of Human Resources and first female chief of the Astronaut Office.

In 2012, just months away from the birth of my son, I scheduled fourteen interviews to be completed over two months, and thus embarked on my own NASA mission to complete these sessions before I went on maternity leave. I almost succeeded. Test pilot and Space Shuttle Commander Eileen Collins no longer lived in Houston, so I wasn't able to meet her until she came to the JSC memorial service for Sally K. Ride, America's first woman in space. My growing belly stuck out as I extended my hand, explained my project, and asked her to participate. Months later, as we corresponded about a date and time during my maternity leave, Collins, a mother of two, offered to do the interview at my house so I could be close to my newborn.

What a difference between these targeted interviews and those I had conducted just a few months earlier in Florida for an oral history project focusing on the divers and crews working on the solid rocket booster recovery ships and the pilots and flight engineers of the Shuttle carrier aircraft. Not a single woman was on either of the NASA teams—and this was in the spring of 2012! By contrast, the people I interviewed over the span of two months later that year were all women. This was unusual for a NASA historian, but it was inspiring to me as a woman (especially as a pregnant one) to hear female voices speak about the challenges of balancing motherhood and careers, instead of designing and testing hardware. Their stories represent the undeniably softer side of NASA.

Consequently, no book is like this one. None have focused on the different occupations women have held at NASA since its creation more than fifty years ago. *Making Space for Women* is not intended, however, to fully explore all their positions or to serve as the definitive book about women in the space agency. This is just the beginning of my effort to document women's history within NASA. Many women have yet to be interviewed, and more sources need to be mined and analyzed. I hope that reading *Making Space for Women* will be an inspiration to any high school- and college-age women who are exploring career opportunities in science, technology, engineering, and mathematics fields, as well as those with a desire to learn more about women's contributions to America's space program.

Acknowledgments

Many years ago I envisioned this project, and I am happy that my vision has finally been published! I wish to thank the women who were interviewed and featured in this volume for their patience as I worked through the publication process.

Others made important contributions as well, and I would like to thank them here. This project would not have taken flight if not for Duane Ross's approval and the continued support and encouragement from my two colleagues Sandra Johnson and Rebecca Wright. Both have provided invaluable insight in their reviews of the manuscript over the years. JSC Engineer Amy Cassady helped me understand some complex concepts, and Fred Martin and Bob Reid reviewed Dottie Lee's chapter and introduction. Steve Garber at NASA Headquarters saw a value in publishing the work.

My current employer, MORI Associates, helped me navigate the final stages of publication, as did my colleagues in Knowledge Management and the JSC History Office. Thanks specifically to Janine Bolton, Tammy Hoke, Jeanne Newman, Jim Rostohar, and John Uri. And, of course, thanks to my husband and son, Jim and Julius, for their support.

MAKING SPACE for WOMEN

Introduction

Ask a girl what she wants to be when she grows up and it is unlikely you will hear that she wants to be an engineer, scientist, or astronaut.[1] Rarely do young girls express an interest in these careers. Although historically few women held positions in the fields of science, technology, engineering, and mathematics (STEM), women have made small but steady progress over the years in these nontraditional careers.

But why aren't even more girls gravitating toward STEM careers? One reason might be a lack of representation. Television shows and movies tend to feature men as engineers or scientists—almost always portrayed as "geeks"—working in federally funded research labs while playing with rockets and airplanes. Popular culture also encourages girls to aspire to so-called feminine strengths like shopping, music, and dancing—math, not so much. Amazingly, even presidents of prestigious universities feed these stereotypes. In 2005, Lawrence H. Summers, then president of Harvard University, explained women's underrepresentation in STEM fields as due to innate differences between the sexes. When girls have been bombarded by these images and attitudes during their childhood, it is not surprising that few girls pursue a professional STEM job or dream of working for the nation's space agency, NASA. Unless they have found a role model or live in a community where women in STEM are celebrated or prominent, they have no idea of the possibilities. However, research shows that sharing stories of women in these fields inspires girls.[2]

With only ten NASA centers spread across the country, women employed by the space agency cannot personally counsel all of the United States' girls on the exciting and distinct job opportunities related to outer space and the world beneath it. *Making Space for Women* offers a look at women who excelled in this dynamic environment and proved that cool jobs in aerospace are not just for men. I hope their stories will encourage more young women to consider careers in aerospace and other STEM-related fields. Even women who do not have technical training can pursue a career with NASA in a wide variety of support roles.

The women featured in these pages played leading roles in their chosen professions in spite of the social, cultural, and personal challenges they may have faced along the way. Some, for instance, grew up in an age when women were funneled into traditionally female careers. Others had to deal with family tragedies or overcome economic hurdles to achieve their goals. All made a difference after they arrived at NASA, and their contributions made noticeable impacts in the fields of space exploration, science, engineering, technology, and management.

Dorothy B. "Dottie" Lee was hired as a "computer," at the National Advisory Committee for Aeronautics (NACA) NASA's predecessor, by the Langley Memorial Aeronautical Laboratory in 1948. She demonstrated such an amazing aptitude that spacecraft designer Maxime A. "Max" Faget decided to train her to become an engineer in an age when few women stepped into that role.[3] In an era when nearly all of the women employed by NASA were secretaries, some men confused her for one because she wore a skirt and worked behind a desk. During the space race with the Soviet Union, Lee analyzed and interpreted data from the charred heat shields that protected astronauts from the high temperatures of reentry into Earth's atmosphere. One of Lee's greatest accomplishments came when NASA began to conceive its first reusable spacecraft, the Space Shuttle Orbiter, and she was one of two women selected to serve on the design team.

Ginger Kerrick first dreamed of becoming an astronaut at the age of five. Six years later her father died of a heart attack, leaving her mother to raise four small children. Undeterred by tragedy, Kerrick graduated second in her high school class and earned a scholarship to the University of Texas at El Paso. Hoping to work for NASA, she transferred in the fall of 1989 to Texas Tech University, where they had a cooperative education program with the space agency. When her grades fell, the opportunity slipped out of her grasp. An internship at the NASA Johnson Space Center (JSC) finally cracked open the door, but a hiring freeze nearly ended her chance to become a full-time employee. Doggedly determined to be one of the few selected, Kerrick called the co-op office every week for an update. Her persistence paid off and she joined NASA in 1994. Her plan seemed to be working, soon after joining NASA, she applied to be an astronaut. The Astronaut Selection Office called to schedule an interview and physical exam—but it was not to be. Kerrick had kidney stones, which permanently disqualified her from flight status. Although disappointed, she decided to stay at JSC and went on to become the first non-astronaut capsule communicator. Her selection for this Mission Control position changed NASA's longtime rule that only allowed astronauts to communicate with crews in space. Later, Kerrick directed Space Shuttle and International Space Station missions.[4]

Eileen M. Collins grew up poor, living in a housing project for low-income families in Elmira, New York; to make ends meet her family sometimes relied on food stamps. At age nineteen, Collins began working part time to pay for flying lessons and attending community college; she later graduated from Syracuse University. A trailblazer, she was one the first women selected to participate in flight training at Vance Air Force Base in 1978, and later became a flight instructor and test pilot for the United States Air Force. NASA included Collins in its 1990 astronaut class, and she made history by becoming the first female pilot in the corps. She served as commander on two of her four Space Shuttle flights; one of these was the 2005 Return to Flight mission that followed the tragic loss of the Space Shuttle *Columbia*.

Estella Hernández Gillette came to the Johnson Space Center in 1963 just out of high school. Only eighteen years old and a newly naturalized citizen—she was born in Mexico—she started as a secretary in the Engineering Directorate. Years later while working full time and juggling a family, Gillette earned her bachelor's in business administration in 1986 and her master's in human resource management in 1994. By the time she retired, she was second in charge of the External Relations Office, which handled NASA's relationships with schools, universities, the local community, Congress, and the media. She continued with her pursuit of education, completing her doctorate of education in 2012 from prestigious George Washington University.

It is my hope that the stories of these four women and the others included in this book will inspire girls to embrace the exciting challenges found in the fields of STEM and seek employment in space exploration. Within these pages, this diverse group of women share their passion for America's space program, explain the challenges they faced, and talk about how they overcame struggles in career advancement. They also offer lessons learned along the way, and through their narratives the evolution of opportunities for women at JSC unfolds. Their memories provide an account of more than fifty years of progress at this federal institution. It wasn't always easy—either for them or the center—but it was definitely rewarding.

The engaging stories in this book have been compiled from edited transcripts, largely from the Johnson Space Center Oral History Project. Initiated in 1996 by Center Director George W. S. Abbey, as of 2021 the JSC Oral History Project had recorded more than 1,400 interviews. I began working full-time on the project in 2002. On my first day I interviewed America's first woman in space, Sally K. Ride. What a memorable day! Since then I've gone on to talk with hundreds of individuals about the innumerable activities, people, and events associated with human spaceflight.

The twenty-one women I chose to feature in this book are an assorted lot with different backgrounds, education, training, and life experiences. Many of these women, just like the many girls and women who will read this book, had not considered that there could be opportunities for them with the space agency or in any STEM-related field. Vickie L. Kloeris, who grew up down the road from the Johnson Space Center, went on to manage the food systems for the Space Shuttle and International Space Station Programs. Albuquerque native Natalie V. Saiz became the chief of JSC Human Resources, though when she had visited the Houston center as a high school student she did not realize that NASA had opportunities for non-technical workers. Astronaut Joan E. Higginbotham had planned to work for business giant IBM after graduating from college when NASA offered her a job working on the Return to Flight after the 1986 *Challenger* accident. Some of the women featured had just graduated from high school before accepting entry-level positions with NASA, while two—Peggy A. Whitson and Carolyn L. Huntoon—were National Research Council fellows. A few joined NASA as cooperative education students or interns. Many held bachelor's degrees and went on to complete a master's while working for the space agency, while others held advanced degrees in science and engineering prior to working for NASA. All walked into a facility that since its beginning had been predominantly managed and staffed by men.

Born out of the Cold War, NASA was a brand-new institution in 1958, and responsible for the nation's space and aeronautical activities. Formed as a direct response to the Soviet Union's 1957 launch of Sputnik, the world's first artificial satellite, the American agency completed a series of successful spaceflight programs in the 1960s to demonstrate leadership in space. Following the lunar landings, NASA began building the Space Shuttle, which flew for thirty years and was instrumental in the construction of the International Space Station.

NASA, formed out of the three NACA laboratories and two stations, relied heavily on the extensive expertise of the Pilotless Aircraft Research Division (PARD) in Virginia to put a man in space. Robert R. Gilruth, the former head of PARD, accepted the challenge and formed the Space Task Group—a small unit of dedicated individuals that eventually moved to Houston to establish the Manned Spacecraft Center, which was renamed the Lyndon B. Johnson Space Center in 1973.[5] Then and now, JSC recruits and trains the astronauts, designs and tests US spacecraft, and manages missions from its flight control rooms.[6]

The Space Task Group, established at the NASA Langley Research Center in Virginia, mirrored that center's workforce. From the start, women's place in the United States' new and exciting endeavor was primarily as administrative support—secretaries, typists, and stenographers—just as they had been for NACA as well as for the armed forces that had also helped to shape the manned space program. The NACA Langley Memorial Aeronautical Laboratory had few women employees in non-traditional jobs until World War II, when wartime demands opened the doors to expanded opportunities. Women enthusiastically accepted positions previously offered only to men, altering the lab's labor force. Before the bombing of Pearl Harbor in 1941, fewer than five women had worked there as engineers or scientists and were not paid well. By 1954, still only 133 women scientists and engineers worked for NACA, with almost all of them classified as mathematicians.[7] They conceded that opportunities were limited, but they believed NACA treated them better than other employers treated women elsewhere.[8] Mathematicians who were hired to work as computers tended to receive higher salaries than women in other fields, and, besides, working on aeronautical research was new and exciting.[9]

Roles for women in the early days of the space program reflected workforce trends documented by the Women's Bureau in 1958, when women filled nearly all clerical positions in the country.[10] Similarly, other federal organizations, the aerospace industry, and corporations hired women to type correspondence and file the paperwork generated by their organizations.

STEM professional women working on the space program in the 1960s were a rare bunch. Sputnik's launch stressed the value of studying engineering and the sciences—fields where only a handful of women earned degrees. Sex segregation, so prominent in the workplace, extended to higher education and ideas about subjects that were considered masculine or feminine areas of study. When NASA's chief astronomer, Nancy Grace Roman, expressed an interest in taking mathematics instead of Latin, her guidance counselor looked down at her and asked, "What lady would want to study mathematics instead of Latin!" Others—including her own mother—tried to dissuade her from studying astronomy, suggesting that it was a man's field, not a woman's. During these years, even the deans of women at prestigious schools discouraged their female students from taking science or math courses.[11]

By 1962, women comprised about one-quarter of NASA employees, but fewer than 250 women agency-wide were STEM professionals. If divided equally among the ten NASA centers, that would be only twenty-five female scientists, engineers, or technicians per location. During the Gemini and Apollo Programs, women engineers and scientists generally fared worse at the NASA contractors: General Electric had a mere thirteen women assigned to work human spaceflight programs, while North American Aviation had hired only 156.[12] Fewer still were in management or decision-making positions. Surprisingly, there were larger numbers of women in the Army and Air Force and they were well represented in the Departments of State and Health, Education, and Welfare.[13]

When compared to select federal agencies in 1963, only the Federal Aviation Administration had fewer women in positions of authority than NASA; NASA's professional women totaled only 1.3 percent of all the agency's employees.[14] Charles F. Bingman from JSC's Personnel Department explained that they had to work harder to locate those women with technical backgrounds and "make the fact known that Houston was a good place for women . . . to work," because, as he opined, "a lot of them didn't believe that you could go into an old-fashioned engineering shop and ever be given any responsibility or become a real partner in the organization."[15]

Of the women who did work at NASA, their opinions about their treatment there vary, but an overwhelming number insisted (publicly at least) that they never experienced any prejudice. Roman asserted unequivocally that while NASA did not discriminate against women, she had witnessed prejudice against women.[16] She had always wanted to be an astronomer. In 1946, Roman earned her baccalaureate at Swarthmore College, and three years later she received her doctorate from the University of Chicago. Upon graduation she accepted a faculty position at the Yerkes Observatory, where she earned less than men with the same education. The chair of the department did not see that as unfair. He said, "We don't discriminate against women. We can just get them for less." Recognizing she would not receive tenure there because of her gender, Roman accepted a position in radio astronomy with the Naval Research Laboratory (NRL) in Washington, DC. Her supervisors there failed to give Roman any direction for the first few months on the job, and she later learned why. They had previously hired another female astronomer, who had proved to be inept. Because of her, the men at the NRL assumed all female astronomers were similarly "useless," and they concluded they did not want "another."[17]

By comparison, NASA was different. Female mathematicians, sometimes called computers or computresses, found more opportunities. Frances "Poppy" Northcutt worked for TRW, an aerospace contractor, defining trajectories for the Apollo spacecraft return from the Moon. Just a few years out of college, she received a great deal of press for being the first woman to work with JSC's Mission Control Center team. Looking nothing like the men in the room, who sported white shirts, crew cuts, and skinny black ties, did not hurt. A cute blonde, Northcutt wore skirts well above the knee, as was the fashion of the time. One of the few women to support the Apollo flight teams, Northcutt denied she encountered any opposition. "The nature of this business," she told *Life*, "doesn't lend itself to discrimination. If you write a computer program, it either works or it doesn't. There's no opportunity for anyone to be subjective about your work."[18]

Others attributed NASA's receptive attitude toward women to the nation's urgent need. One such outspoken female engineer, Eleanor C. Pressly at the Goddard Space Flight Center, developed rockets that probed Earth's atmosphere. She believed that the space agency treated women fairly and with respect: "There's no discrimination against women at NASA, because we need scientists and engineers too badly."[19] Marjorie R. Townsend worked in the same NASA center, but found the environment less friendly. She was the first woman to supervise the launch of an American spacecraft. She thought that the male managers often set up women to fail, based on the belief that only men were suited for engineering. The first woman to graduate from George Washington University's engineering program, Townsend noticed how few women engineers were in the space program and wondered: was it because so few held engineering degrees or was it the NASA culture?[20]

At JSC in Houston, men and women engineers sometimes struggled to work together. Ivy F. Hooks started working at the center in 1963. A mathematician by training with a background in physics and chemistry, she received the title of aerospace technologist, "which really bothered all the men, because that's what they were too." Engineering had been men's territory up to that point, and when NASA threw a woman into the mix, the men who had only worked with women as secretaries did not know how to treat her. Her first group of male coworkers was cruel to her, including her boss; some of them constantly teased her. To continue working for NASA, she requested a transfer.[21]

For years, most men had little or no real work experience with women in positions of authority or as equals. Relationships with women in their lives had been limited to that of girlfriends, wives, mothers, or daughters, and the men saw themselves in the role of protector, even while employed by the space agency. They wanted to hold the hands of the female engineers in their midst, like Jeanne Lee Crews, to help them. The daughter of an Air Force pilot who had encouraged her interest in science, Crews found herself frustrated by circumstance. Her male colleagues were given responsibilities while she was not, although she was just as knowledgeable. Crews had to work hard to be accepted, to gain more responsibility, and to earn the trust of her colleagues. It was challenging because "if you were cute," she remembered, "they really didn't want you to have to do anything." The men seemed more interested in the clothes she wore to work than her capabilities.[22] Northcutt also hinted about such attitudes, saying that sometimes the engineers would "treat me like a girl, and sometimes like an engineer, but," she added, "always with friendliness and consideration."[23]

Being one of a handful of technical women at the Johnson Space Center was challenging. Hooks found it "difficult being the only" one in her group because she "was too visible." While the men in her section probably never paid much attention to the other men, they observed everything she did because women engineers were such a rarity. They noticed the simple things, like if she drank coffee at someone's desk other than her own. Consequently, she recalled, "I had to be more professional and look busier than my male colleagues because everyone was aware of my presence." With so few women in similar positions onsite, Hooks admitted to behaving "differently" than the male engineers. She realized that men saw women as a monolithic group. "To many males, I represented all women," Hooks said, knowing she had to prove herself. Had she shirked her responsibilities or submitted poor work, other women with technical degrees might have found it more difficult to find jobs with NASA.[24] This was a burden voiced by many of the pioneering professional and technical women in STEM fields.

To cope, the women used different strategies. Crews began swearing so that her engineering colleagues would stop patronizing her. Never a shrinking violet, she fit right into the environment at JSC, where engineers passionately and almost constantly debated ideas and concepts. She enjoyed sitting around a table with her colleagues, including her supervisors, discussing how best to achieve the goal of landing a man on the Moon by the end of the decade. Looking back, though, Crews remembered challenges she faced because of female stereotypes, such as the belief that men are technically superior to women. For instance, management asked her to participate in the preparations for America's first spacewalk on Gemini IV. She successfully completed weightless simulation exercises on a newly installed air-bearing floor at JSC so that the astronaut could practice his planned maneuvers; male coworkers undermined her accomplishment by saying the feat must have been too easy because a "girl" was able to do it.[25]

In the 1960s the focus at Houston was on *manned*, not *human*, spaceflight, and the heroes of the national space program—those with the most-recognized and publicized faces, the astronauts and flight controllers—were all men. Astronauts regularly graced the covers of *Life* magazine, and Flight Director Christopher C. Kraft was featured on the cover of *Time* in the summer of 1965. News broadcasts captured the efforts of the Mission Control teams as they directed historic and complex space missions, including the first lunar landing, which Americans watched live on television. While NASA presented a glamorous and impressive image to the American public, Carolyn S. Griner, an engineer at the Marshall

Space Flight Center in Alabama, found the "totally male-dominated environment" in Mission Control uncomfortable.[26]

Flight controllers had their own unique language, flight rules, heroes, stories, and traditions.[27] Many spent their off hours at local bars drinking beer.[28] Following a successful event, debriefings were held at places that permitted hijinks similar to parties on military bases or at officer's clubs, and, at some point during the night, chugging contests or arm wrestling would begin.[29]

Women were generally excluded from this large and prominent network of men, who made mission critical decisions in the room where they closely monitored spaceflight from launch until splashdown. Leaders of this group tended to be conservative—their wives stayed home to raise their children—and their attitudes toward females reflected their cultural understanding of a woman's place. Crews, who worked there in the early 1970s, remembered overhearing some of the senior members jokingly comment on how "it's certainly good we keep women out of the Mission Control."[30]

By contrast, of the secretaries interviewed, none felt any animosity from the men or heard any comments that their efforts were less valued despite their limited job opportunities. Two secretaries who supported the Astronaut Office described the organization as a family, sharing that the men respected their contributions but—at the same time—were protective. Jamye Flowers Coplin remembered that the chief of the Astronaut Office, Alan Shepard, prohibited her from traveling to Cape Canaveral in Florida as a crew secretary until she turned twenty-one because of all the heavy drinking and partying that went on there. Martha Caballero Speller said the Apollo 13 crew had taken her and her husband's wedding bands (a unique honor) on that historic Moon mission. While working there she felt "loved, cared for, and protected."[31] Still, opportunities were limited for moving into positions with higher pay and rank.

Many of the women interviewed chose not to focus on the challenges or prejudice they faced, preferring instead to concentrate on their roles in a truly once-in-a-lifetime history-making opportunity. Some laughed thinking about incidents, like Shirley Hunt Hinson, who said her boss called her into his office in Houston and told her: "We're going to take you to the Cape to do the post-flight reports. But you have to behave, because if you don't behave, they won't ever let another woman go."[32]

Anne L. Accola, by contrast, talked about the obstacles facing women who entered flight control in the 1970s. Although she held an advanced degree in computer science, she was initially assigned to the Simulation and Training Branch—not only was this area akin to teaching, a career considered appropriate for women, but working at a flight console in Mission Control was definitely still restricted to technically minded men in the early 1970s. Accola found limited opportunities for mobility and eventually transferred to a different office, having sustained "two concussions on Gene Kranz's glass ceiling."[33]

Carolyn Huntoon, a physiologist, started at JSC in 1968, just a few months before Apollo 8, when the first American crew circled the Moon. As a graduate student, she studied the impact of spaceflight stress on astronauts before and during their mission. Given the choice of working at the Veterans Affairs (VA) Hospital Lab or NASA, she chose to continue her exciting research in the space program. If she had accepted a job with the VA, she reasoned her experience would not have been that different because at the time there were just so few women PhDs in science.[34] NASA attracted her not only because the program was thrilling, but because everything was challenging and new, and at JSC there was plenty of innovative work to be done. As long as you worked hard and did your job, being a woman did not really matter, she noted. But she learned there were still barriers to overcome. For instance, Huntoon had to hire male technicians to gather samples from the flight crew after their space capsule had splashed down and then been recovered and brought onboard a US Navy ship. As a woman, she was not allowed on deck.

After a few years at NASA, Huntoon found that there were so few women that they were left alone to deal with any problems or challenges they faced; most of these women had faced similar challenges in college, where very few females were in their classes or graduate programs.[35] With no woman's movement to shine a light on the issues they faced, no one voiced concerns. Problems failed to surface until the introduction of equal opportunity programs, whose officers highlighted patterns.[36]

From the perspective of the twenty-first century and a modern understanding of gender bias, one might scoff at the idea that sexual politics were not at play in the NASA workforce in the 1960s. Generally, women employees assert that that was the case. Charged with the goal of landing a man on the Moon by the end of the decade, NASA had a mission to complete. Personnel issues related to sex or race did not matter, nor did personal concerns.[37] As Jay F. Honeycutt, an engineer in Flight Operations, explained to me, NASA did not "have time to get involved in social justice." He elaborated, "We didn't much care what your gender or race or anything else was. Can you do this work? If so, are you doing it? If so, then you're on our team and we're going to accept you and treat you like one of us."[38] This sense of mission united NASA employees and helped to foster a can-do attitude within the agency that remains to this day.[39]

NASA also believed it hired the best and brightest of its generation and valued hard work. The women—secretaries and STEM professionals—all clocked long hours, and on numerous occasions took work home. Others came in on Saturdays and spent time away from their families. Working sixty- to seventy-hour workweeks was normal for participants in the space program. The average age of those working at NASA at the height of the Apollo Program? Thirty-six.[40]

Outside of NASA, the excitement of space exploration faded over the course of the 1960s—it should be noted that public support for a lunar expedition was never overwhelming—and the American people questioned the costs of Apollo. The United States had won the space race in 1969, but spending millions on space was no longer a national priority during the 1970s. Space had lost its appeal. Americans identified other, more pressing problems in a 1973 poll: health care, the environment, and education. Welfare ranked higher than space, which landed near the bottom of the list. With shrinking budgets, NASA's workforce shrank.[41]

Only a year earlier, Congress had passed the Equal Employment Opportunity Amendment to the Civil Rights Act of 1964. Federal agencies were no longer exempt from the equal opportunity (EO) provisions of the Civil Rights Act; they had to ensure that hiring and personnel matters were not prejudiced against specific groups, including women.

Prior to the amendment's passage in 1972, the agency had given lip service to the concept of EO. Spurred by the passage of the Civil Rights Act of 1964 and an executive order signed by President John F. Kennedy, NASA Administrator James E. Webb issued a directive calling for equal opportunity at NASA regardless of sex. His 1966 policy went beyond any federal mandates of the era.[42] Still, this policy did not result in hiring large numbers of female employees in areas where they were underrepresented.

By 1973 women made up a small percentage—about 18 percent—of NASA's entire workforce, versus 34 percent government-wide. Of those at

NASA, only four women were in the top tier of the civil service ranks, and 310 were classified as scientists or engineers. Finding women engineers and scientists to hire was a challenge. However, NASA's poor record, when compared to other federal agencies, was questionable. Even more troubling was the fact that the agency had failed to hire a significant number of women in nontechnical professional positions—female attorneys, for instance.[43]

Change came only as a result of the negative publicity NASA received when Administrator James C. Fletcher fired Ruth Bates Harris, his deputy assistant administrator of Equal Opportunity, in 1973. Harris, an African American woman, had submitted a report noting that NASA's affirmative action program was "a near total failure." When asked, Fletcher denied that the firing related in any way to her policy suggestions. Instead, he claimed Harris was a "seriously disruptive force." Upon hearing of the story, major newspapers and local Washington, DC, radio and TV stations reported on the "purge." Outraged by the news, women's groups and other organizations wrote to their representatives. Senate committees held hearings on the issue, and eventually Wisconsin Senator William Proxmire required NASA's hiring programs to be monitored by Congress for routinely failing to hire qualified women and minorities. Soon thereafter NASA hired another African American woman, Harriett G. Jenkins, to become assistant administrator for Equal Opportunity. Jenkins had helped integrate the Berkeley, California, school district; her gradualist, nonconfrontational policy helped to bring greater numbers of women and minorities into the agency between 1974 and 1992.[44]

Just as issues about affirmative action heated up at NASA Headquarters, attitudes about women's place in the workforce shifted, more women gained seniority, and larger numbers of women moved into management positions. Hooks, for instance, became head of the Aerodynamics Systems Analysis Section at JSC in 1973. When that happened, men in other areas within Engineering teased the men in her group for having a female boss. Instead of being embarrassed or angry about the decision, the men in her section decided to make her shine.[45] Around the same time, the daughters of many center employees began contemplating college majors. Men who worked with these women began to realize that opportunities for their daughters in the fields of science and engineering were wide open and actually possible, as illustrated by the females they worked with day in and day out.

Key programs instituted at JSC in the 1970s helped women (and some men) in the lower civil service ranks advance into occupations with higher pay and more responsibility. The Upward Mobility Program, begun in 1974, offered clerical employees an opportunity to move into professional positions, providing participants with career counseling and training opportunities. S. Jean Alexander, featured in this volume, benefitted from this program when she left behind her secretarial position to become an equipment specialist for the Crew and Thermal Systems Division, placing her in a highly responsible position that connected her directly with the astronauts preparing for spaceflight.[46]

Gradually, the number of women employed in STEM professional positions at JSC increased, but the overall numbers remained small. In 1974 a mere 105 women served as professional employees, which included both technical and nontechnical fields such as librarians and personnel specialists, compared to the more than two thousand five hundred men who filled those slots.[47] In spite of the center's efforts to hire more women in the fields of science and engineering, through the 1970s nearly all of the women employed by the space agency were secretaries.[48] There were so few female scientists and engineers that Sally Ride recalled the "arrival of the female astronauts" in 1978 "suddenly doubled the number of technical women at JSC!"[49]

Changing a male-dominated environment was more complicated than simply opening doors to women. NASA made a commitment very early on as it designed the Shuttle, and in 1976 the agency announced that the first class of Shuttle astronauts would include females. Women had previously applied to become astronauts, but none had ever been selected; because there had never been a single female astronaut, women had to be convinced NASA was "serious."[50] As some of the agency's most visible employees, women astronauts would demonstrate that NASA had indeed integrated its workforce. Minorities and women were encouraged to apply, and in January 1978 NASA released the names of the first class of Shuttle astronauts. Six white women, three African American men, and one Asian American man had been selected as astronaut candidates, along with twenty-five white men.

The women were mission specialists, a new term to describe scientist-astronauts who would work in space. In the public's mind, astronauts were military test pilots, but the military had not yet opened the doors of its test pilot schools to women.[51] The inclusion of female engineers, scientists, and physicians directly challenged the image of the *"right stuff"* space pilots. An older popular stereotype that linked men's piloting skills to space exploration was replaced by a new vision, one where there was broad participation by a wide variety of people, not just test pilots. Female and male scientists, once considered incapable of exploring space by popular culture and by the astronauts themselves, gradually challenged and reshaped the prevalent image of the hyper-masculine pilot-astronaut.[52]

Although laws prohibited NASA from sexually discriminating against women, cultural attitudes had to evolve at the center. Opinions about what was considered appropriate work for men and women had to change for women to become fully integrated into the Astronaut Office and workforce. Some men, like Alan L. Bean, a naval aviator and astronaut who had walked on the Moon, thought that being an astronaut was a man's job, not a woman's. "A lot of things that you do as an astronaut aren't that genteel," and, having only worked in an all-male squadron and the Astronaut Office, he "thought men are really supposed to be astronauts." But "maybe," he conceded, "we can teach women to do it."[53] Like Bean, few men at JSC had worked side-by-side with women as their peers. Even fewer had female managers; their experience with women at work had been limited to seeing them as subordinates.[54]

All the women astronauts selected by NASA believed they were treated equally by the agency and their colleagues when they arrived in 1978. Kathryn D. Sullivan, one of the first six, believed that her male classmates did not feel as if the women had "been let in free or had been let in easy" and gave them the opportunity to demonstrate what they could do. As for the opinions of others at the center and within NASA, she believed that they received respect because they held the title of astronaut. Men may have never seen a female spacefarer, but at the Johnson Space Center astronauts symbolized the space program itself. If men had dismissed them outright or treated them poorly, Sullivan reasoned they would have had their heads handed to them on a platter.[55]

Nevertheless, Huntoon, a member of the 1978 selection board and mentor to the female astronauts, had worked at JSC since 1968 and worried that a double standard would emerge for the women astronauts. She recalled that changes in attitude had to be instituted through the center. In her opinion, all space flyers had to be treated equally.

Soon after the astronauts' arrival, Huntoon heard complaints from men about the female astronauts, suggesting that the women came on too strong. Reportedly, they were too quick to criticize, something that hard-working women in other roles may not have had the courage to do in the past. The new arrivals also corrected men in meetings, which was embarrassing for them. "You know what she told me?" one scoffed, "she told me that that wasn't right." Or, "She just acts like she knows every-thing." These confident and competent professionals made these men uncomfortable, but Huntoon defended the women and asked the men: "Now, which male astronaut hasn't spoken up in a meeting and told you something?" Privately, she thought, "What were other people expecting of women astronauts?"[56]

Clearly some of the older men, used to women being silent in meetings, were unaccustomed to working with women who had knowledge of space hardware or who had a background in science and engineering. Some may have been resentful or fearful of the competitiveness and assertiveness they were witnessing from these women. The first six female astronauts had no problem working with men, or other women for that matter. They shared what they were thinking. After all, they held advanced degrees in science, medicine, and engineering and were used to being one of the few women in their programs and professions. So this small group of six became a woman's "vanguard" of sorts in the late 1970s, when "there weren't very many women engineers" to weigh in on issues.[57]

As JSC hired more women, the center leadership recognized that the men in management needed training on how to supervise this growing group of women filling professional roles. A two-day pilot course intro-duced in 1980 called "Working with Women" explored issues regarding female employment. Discussions, lectures, and case studies examined supervisor's perceptions about female workers, women's perception of their supervisors, reality versus myth, and how to narrow the gap between perception and reality.[58]

As women came to represent greater numbers of the center workforce, updates had to be made to NASA's facilities. During the 1960s boom, the buildings had been designed without considering the needs of women. Some of the restrooms that had always been designated for men in the Control Center were reassigned one day without warning, leaving a con-fused flight controller wondering how he had walked into the women's bathroom![59] Separate space for a women's locker room also had to be made in the gym.

Women eventually became flight controllers. Flight Director Eugene F. "Gene" Kranz recalled the challenge of introducing women into previously male teams. In the Mission Control Center, "we did many dumb things," he concluded, like one supervisor who introduced a dress code for the women in his organization. "That went over . . . not very well," he admitted.[60] The first women who chose to enter the ranks of flight control tended to choose "the most difficult, most time-critical jobs," which Kranz attributed to the "cockiness" a woman had to have to break into the ranks of Mission Control. One of those women was engineer Jenny M. Howard.

Unlike other team members who might have had more time to evaluate malfunctions and glitches, as a booster officer Howard had only seconds to solve problems during the eight-minute launch of the Space Shuttle. During the liftoff of STS-51F in the summer of 1985, the Orbiter's number one engine shut down. Then a sensor on the number three engine failed. Recognizing the immediacy of the situation, Howard decided to inhibit the engine sensor. Howard received a commendation from NASA for her quick thinking that saved the mission and the crew.[61]

The inclusion of women in the Mission Control teams brought about a shift in the antics and practical jokes, parties, smoking, and cursing. The change was due in part to generational attitudes about women, believed longtime flight controller Jack Knight. "Most of us in my age group were brought up to be nice with the girls. You never hit the girls. You don't use curse words in front of girls. You don't curse in front of your mother." With more women present, he believed that the rowdiness of the early days diminished and the good-old-boys network slowly disappeared.[62]

Although the "brotherhood" of flight control has changed greatly since the 1960s, cultural attitudes about the type of people who fill those positions remain strong to this day. Flight controllers value confidence and toughness; even today, being soft-hearted or overly emotional can be a career handicap. As Flight Director Ginger Kerrick explained to me, "You don't cry in the hallways," of Mission Control.[63]

Change may have come slowly to JSC, but it continued. Throughout the 1980s, the number of NASA-employed women increased, although they generally remained stuck in the clerical workforce and concentrated in the lower ranks of the civil service, where they were underpaid and passed over for promotions.[64] At the same time, however, more women became visible at consoles in Mission Control and on the flight line at Ellington Field.[65]

Unfortunately, obstacles continued to hinder women's career advancement at JSC. In a 1984 meeting with twenty-two women at the center, the most significant barriers were identified. More than half agreed that females were not provided the same opportunity to advance as their male counterparts. Customs and attitudes denied access to key careers, they believed. Another concern was a lack of female supervisors to serve as role models. This led to a smaller pool of women to promote into positions of authority. Nonetheless, the female participants agreed that historical trends proved women had managed to succeed at JSC since 1974.[66]

The most visible indication that the center's hiring practices and work culture had evolved was the selection of female astronauts and their inclusion on missions. In 1983, after more than twenty years of sending only men to space, America watched Sally Ride soar into the heavens and become the nation's first female astronaut. Within the next two years, the rest of the five women candidates selected in 1978 set new records and earned their places in history.

The remaining doors, once barred to women, opened. In 1985, payload officer Michele Brekke began a twelve-month training program to become the first female flight director. Eileen Collins completed training at the Test Pilot School at Edwards Air Force Base in 1989, making her eligible for a pilot slot in the astronaut corps. In 1994, Carolyn Huntoon became JSC's first female center director. She played an especially significant role for women across the center and in a variety of fields, not just science. Mentor and counselor to the first six women astronauts, Huntoon helped to foster a network among that group and provided similar support for other women in diverse fields across the center. Several of the women in *Making Space for Women* spoke admirably of her and the influence she had on their careers.

With an increase in numbers, more issues relating to women began to be addressed. In the late 1980s, JSC began exploring the idea of opening a daycare center onsite. Estella Gillette, a mother of two sons, chaired the committee. Older male engineers resisted the idea, asking why the center should provide childcare—their wives had stayed home with their kids. Recognizing that engineers had a "rocket mentality," Gillette decided to avoid philosophy and focused on the practical aspects.[67] Management came to realize a certified on-site facility for children was a great recruitment tool and a way to retain talented women. The idea was approved, and the daycare center opened in 1990.

Throughout the 1990s NASA continued to hire more women engineers and scientists, but few females served as engineering supervisors, even after two decades of an active recruiting and hiring effort at the Houston center. "All my mentors were all these old, crusty Apollo guys," not women, recalled Julie Kramer White, director for Engineering at Johnson Space Center.[68] By the start of the twenty-first century, however, women had made visible inroads into JSC's professional workforce. The numbers of female employees had grown, as they had across all federal agencies, from 21 percent in 1980 to 34.7 percent in 2000.[69]

The women's movement, affirmative action policies, and social change explain the growth in numbers. Another major reason was the increased educational opportunities for women, especially in engineering. In 1970, women rarely earned bachelor's degrees in engineering, but as of 2000 they represented about 20 percent of graduates in the United States.[70] NASA's entire workforce mirrors these changes, with women accounting for 20 percent of the agency's engineers; their numbers have nearly doubled since the early 1990s.[71] Certain positions remained stubbornly segregated by gender—men fill about 90 percent of the technician jobs and women 99 percent of the clerical jobs.[72]

As the Space Shuttle Program wound down in 2011 and NASA transitioned from its longest-running program, changes in personnel became more dramatic. Women took the helm of several high-profile offices, resulting in the frequent use of the phrase, "the first woman to serve as," to describe the new director of many significant decision-making positions. Peggy Whitson led the astronaut corps. Lauri N. Hansen began directing Engineering in 2013, and that same year former astronaut Ellen Ochoa became the director of the Johnson Space Center.[73] (Interestingly, the only two women to be named as center director both had earned doctorates, one in science and the other in engineering; by contrast, all of the male JSC directors had received only honorary doctoral degrees.) And that stronghold of macho culture—the Flight Director's Office—has a female chief, Holly Ridings.

Further demonstrating just how much the face of JSC has changed, women lead major areas such as human resources and information technology. They have headed the Astronaut Office, Knowledge Management, and Strategic Partnerships. Women now comprise around 30 percent of the center's elite Senior Executive Service positions, and they fill more than 30 percent of the highest management ranks.[74] While the basic mission of JSC has not changed since it was established, the center is

Four women onboard the International Space Station pose for a photo in the Zvezda Service Module. Clockwise from bottom left: Tracy Caldwell Dyson, Dorothy Metcalf-Lindenburger, Naoko Yamazaki, and Stephanie Wilson.

a very different place today. Strong opinions about what women can and cannot do or should and should not do have disappeared. Today, Eileen Collins believes, a woman has "the same [opportunity] as a guy," to become an astronaut.[75]

While attitudes may have changed, one of the emergent themes this book highlights is how gender continues to influence men's and women's communication skills and leadership in spite of women's gains. Women, some of the interviewees recognized, talk differently than their male colleagues, which impacts their conversations with colleagues and customers. "Women speak with a different voice than men," Debra L. Johnson noted, so "women cannot take the tone or approach that men do."[76] She explained how she learned that women could not necessarily imitate a man's authoritarian style of speaking. They had to find their own voice.

Astronaut Pamela A. Melroy explained how ideas about what is acceptable for women and men differed, depending on societal and cultural attitudes about leadership. The Air Force, for instance, encouraged its leaders to imitate the qualities of General Frank Savage of the 1949 film *Twelve O'Clock High*, but Melroy scoffed at the idea. Savage's behavior may have been perfectly acceptable for male squadron commanders, she stated, but also observed that when women mimicked the general's tough style, they were considered "irrational." To be successful, female leaders had to develop their own unique ways of working with their squads. When commanding the crew of STS-120, Melroy thought of herself as "mom" — the person looking out for the crew's welfare, and the individual who put her foot down to keep the astronauts safe.[77]

Many of the women I spoke with provided examples of their compassion and willingness to help their colleagues for the benefit of the organization as a whole. Two of the spacefarers in this volume mentioned leaving the astronaut corps so others who had not flown on the Space Shuttle could have the opportunity they had. NASA benefitted from their choice, too; with more flyers, they could share a wider range of expertise with program managers, mission planners, and trainers.

Well-known stories of America's space program highlight the gutsiness of the men who made important and key decisions for the safety of a crew. Women rarely receive or take credit for their courageous choices because they are generally discouraged from boasting by their peers or society.[78] Even for this project, when asked to explain their specific involvement in prominent management decisions, several of the interviewed women hesitated. Others downplayed their role while crediting others for their endeavors, a self-effacing tactic that women commonly employ.[79] Or they playfully responded like Debra Johnson, who answered with, "You just want me to brag," when asked about the role JSC played in the recent trends in NASA's procurement culture.[80]

Society expects women to be nice to others, not gruff. Several of the women interviewed explained how they strove to be pleasant with their colleagues, expressing appreciation for others and stressing above all else the importance of mission success. So I was not surprised when Melroy told me how she emphasized that her crew not be "high-maintenance," but insisted they be "cooperative" with flight controllers and trainers.[81] She believed it was important to be especially nice to others to demonstrate she was not throwing her weight around. Melroy wanted her crew to be friendly and pleasant because their behavior was a reflection of her female leadership.

Similarly, other women in this publication described how they were reluctant to seek special favors, treatment, or accommodations, even though their requests did not seem overly demanding. Natalie Saiz struggled to make it to a daily seven thirty a.m. meeting held by Center Director Jefferson Davis Howell due to family logistics. "After about six months" she believed she "had earned enough credibility" and asked Howell if he would start the meeting thirty minutes later, because—as she explained—she had to drop her kids off at daycare "and it's just a little easier."[82] Without hesitation, he agreed.

Based on the interviews conducted with these twenty-one women, cultural attitudes about work have changed at NASA over the years, especially the idea of striking a work-life balance. In the early years of the space program as employees strove to achieve President Kennedy's goal, some of the younger secretaries worked all week and then babysat the astronauts' children at nights and on weekends. Decades later, employees began to recognize the benefits of flex time and the importance of having options to be both a good employee and a responsible parent. Saiz—the first female head of Human Resources and a mother of three children—introduced family-friendly policies center-wide when she became director in 2004.

In an era when most center employees don NASA polo shirts and khakis or jeans on a daily basis, clothing might seem insignificant today, but dress exemplifies those societal shifts JSC employees witnessed. Most women wore skirts, hose, and high heels to the office until the pantsuit came into vogue in the early 1970s. The pantsuit symbolized the freedom women experienced in the workplace. As they moved into positions outside of the secretarial pool, less restrictive clothing freed women like Anne Accola to be able to easily climb in and out of spacecraft trainers.

While no woman interviewed for this book talked about her role in the women's movement, the impact of feminism touched the lives of these women in different ways. In the early days of NASA, one testified, men generally did not want women to be their equals, but that has diminished, she noted. The belief that a woman should stay at home and raise her children has changed, and today many of JSC's senior female managers successfully combine work and family. Finally, no one automatically assumes that a woman behind a desk is a secretary.

Throughout this book women share invaluable lessons they learned over the years. One woman emphasized the importance of listening to others to encourage the development of new ideas; intimidating people, she learned, makes it more difficult for people to speak up. Silence might be golden, but if you have an idea or thought—speak up, said another. Make sure you know the whole story before jumping to conclusions, was another woman's advice.

I hope readers will enjoy these narratives as much as I enjoyed interviewing the women who shared their experiences with me. I believe their stories will inspire young women to learn more about the NASA women not included in this volume—there are many more! And the agency and space industry are actively looking for enthusiastic researchers, scientists, engineers, and technicians for the follow-on programs to Space Shuttle and Space Station. For those with a passion for exploration, NASA's future employees will find the work exciting, inspiring, and challenging. Why can't that employee be a young woman?

1

Secretaries and Administrative Support

Secretaries were vital to the space program in the 1960s, even though many were young and at the bottom of the pay scale. Despite their subordinate rank within the organization and lack of status, NASA's Robert B. Voas likened the agency's office workers to the navy chief: they were the ones to see "if you really needed something done," not the ship's captain. As an example, he sheepishly admitted, "I often suspect that my office runs much better when I am absent than when my secretary is absent."[1]

Voas's appreciation of his administrative assistant did not change the fact that "secretarial work was grueling." Although NASA developed cutting-edge technology to send a man to the Moon, their typists and stenographers, by contrast, were sometimes equipped with antiquated

Billie Barmore of the Lunar Excursion Module Office files some important documents.

technology, like manual typewriters. All forms and travel orders rolled into typewriters had multiple carbon copies and "everything had to be letter perfect." If she made a mistake, the secretary started the whole process over.[2]

Estella Hernández Gillette began working at JSC as a secretary in 1964, and after more than forty years she retired as the deputy director of the center's External Relations Office. She is a prime example of how female clerks moved up the career ladder and out of the secretarial pool at NASA. After one year of employment, Gillette left NASA to work for the Federal Bureau of Narcotics but returned after getting married. Over the years she worked in a variety of offices and held a number of posts, including director of the Office of Equal Opportunity and Diversity Management. Gillette retired from NASA in 2006, and in 2012 received her doctorate of education from George Washington University.

Jamye Flowers Coplin worked as a secretary in the Astronaut Office during NASA's heyday in the 1960s. Secretaries were responsible for managing the routine duties of their offices: typing, taking dictation, correspondence, and travel orders. Those assigned to work with the astronauts juggled many unique assignments. As Coplin explains, the secretaries wore many hats—trading Corvettes for the space flyers, arranging private tours for dignitaries and politicians at Cape Canaveral in Florida, and even babysitting for the families, among other things. Coplin, who had no previous work experience, shouldered a great deal of responsibility for the astronauts and their families as the crews flew Moon missions. The secretaries of the Astronaut Office played an integral, yet unsung, role in the success of many Apollo missions.

Estella Hernández Gillette [3]

Portrait of Estella Gillette

I decided to take the civil service exam when I was eighteen. That was an interesting experience because I failed. So I waited, and I took it again. This time I scored in the nineties. I was asked if I wanted to interview at NASA, so I thought, "Sure!" At that time I understood that you had to score pretty high before you could interview at NASA. I came out to Ellington Air Force Base, and I had three interviews.[4] My favorite one was Engineering, and I went into the Structures and Mechanics Division.

I came in as a GS-4.[5] I worked in Building 13 on the southwest corner, second floor. It was a two-secretary office. The first secretary had been out for six weeks with mono. The boss came out, greeted me, and said, "Here's your desk." That was it. A few minutes later he came out, and he handed me a memo. He said, "Would you mind typing this?"

I said, "Of course not." I knew I could type it, but the format was different.

Then he said, "Oh, and don't forget the concurrences." I had no idea what a concurrence was. I was too embarrassed to ask him. I saw the memo. It said Maxime A. Faget.[6] So I thought I'll just call Maxime and ask about that concurrence.

I picked up the phone book, which became my friend. It didn't say Ms. or Doctor. Betty Ensley answered, "Dr. Faget's office."[7]

I said, "Hi. Would she be in or he be in?" (I can't remember if I said "he" or "she.")

"Well, this is Betty Ensley, his secretary. May I help you?"

I explained what I needed. She explained about the concurrence.

I was happy and said, "Thank you." A few minutes later, I heard footsteps in the hallway; it was Marion Loveless, the division secretary.[8]

She came in and said, "I understand you called Dr. Faget's office."

I said, "Yes, and I talked to Betty Ensley. She was really nice."

She said, "In the future you are not to call that office. That is the director of Engineering."

That was my introduction to NASA. I tell that story because it stayed with me forever. It also affected my perspective as I went along in my career at NASA.

I was in Building 13 for just a few months, then Building 49—the Vibration and Acoustic Facility—opened.[9] I was the first civil service secretary in 49, by the way. October 1964 was when I started. The first week of January this young co-op student came in. He asked me for pencils. I was typing. I said, "They're in the cabinet."

He said, "Where?" I got up, showed him. He came back. "Do you have any tablets?" I married that guy. Pete and I got married in July of 1967. I'm sure that happened a lot at JSC.

So our group moved over to 49, a brand-new facility. I had about twenty-one guys, all guys. There were no women engineers that I worked with at that time. It was a great group. They were very caring, very considerate. I had my job, they had theirs, and there was no crossover like we see today, where they're doing a lot of the administrative stuff that we used to do. I had a Selectric typewriter, one of the old kind, not the fancier ones that came out later, and a phone with four lines.

I made the coffee every morning. That was just standard operating procedure. They would stand around and drink their coffee in front of me and talk, sometimes curse. I'd just pretend I didn't hear the f-word. To me it didn't mean anything and I just kept working.

At the time, I was doing all of their travel orders, all of their memos, and everything was multiple copies with all the carbons. They would draft things and give them to me. A couple of the bosses dictated a little bit, but not a whole lot. I would take dictation when they would tell me about travel. After a year I left NASA for a position closer to home.

Pete and I moved to the Clear Lake area when we got married.[10] I got a job back at NASA, and they gave me a GS-4 again. This time I went to work in Reliability and Quality Assurance. Marty Raines was the director.[11] I was the second girl in the program office, so it was your typical "make coffee, take care of the guys" kind of thing. Again, hardly any women other than administrative types in the office. Maybe there were some women in finance, but no technical women that I can remember.

It was a super place to work, but I was still just working in the office. I wasn't involved with other center activities. I got my GS-5 back and my 6. Then I got pregnant. I had my first son in June of 1970, when things were downsizing tremendously.[12] I only took a leave of absence because I was concerned about my husband getting laid off since he was a "last-in" kind of guy, and he actually got a letter from the union that he was on the list to be laid off. We waited until, I think, September before I actually left the rolls. I resigned because I was going to stay home and be a mom and live happily ever after taking care of my baby.

That lasted for about two years, and then I got antsy. It was silly to stay home. The kid really didn't need me that much. He was about twenty-one months by that time. I came back to NASA, and they said, "We're in a hiring freeze." This would have been around May of 1972. I went to work with Kelly Services.[13] I got a job offer every place I went, but I was holding out, hoping that NASA would call me back. Finally, I went to an employment agency. They sent me to Tenneco in Deer Park and offered me five hundred dollars a month when I had been earning seven hundred at NASA. I told the guy, who said, "You're not going to get that much."

"Well, I was getting it at NASA."

"Yes, but that's the government, they pay better. Well, how about if we give you five hundred twenty-five dollars?"

"No, I have to drive all the way over, and my kid is right here." He called my husband that night to tell him that I had refused this wonderful offer.

My husband said, "I really don't want her to go back to work anyway." That was the way it was. Thank God not so much anymore.

So I waited, and NASA called me back. I took a job back in Engineering. It was an organization called the Urban Systems Project Office. Again, it was about twenty guys, no women professional engineers. It was a temporary GS-2 Step 1 position even though I was eligible for a permanent position, but there were no permanent billets then. The pay was a little less than five hundred twenty-five dollars, but it was closer to home, so I took it and had a great time.

The other administrative staff was not very nice. They would make little catty comments about where I'd been before. The guys would tell me. They would snitch on them. They liked me a lot because I always brought cakes for their birthdays. Then in turn they would take me to lunch. My whole birthday month I was taken to lunch, so I became very comfortable with the guys. I guess women just weren't good team players yet; they were still very divided. I really enjoyed working with the guys more than I did with the few administrative women that I had to work with. I didn't have any real friends yet in that world.

A year later they made me a permanent employee, and they gave me a GS-3 Step 1. I made progress. Then I got word of a job in the Earth Resources Program Office, and I applied. Technically, I was eligible for a GS-7 job. This job was a GS-6. I accepted and I went over and I worked for John Zarcaro, who was the deputy.[14] The director was Cliff Charlesworth of Apollo 11 fame.[15]

I was working with John, and the senior secretary was Lois Bradshaw, who was in my opinion another one of those top-notch, don't-make-them-like-that-anymore kind of secretaries.[16] Lois was super efficient. I think she wore white gloves. She was from that kind of school, the Katharine Gibbs Secretarial School.[17]

Mr. Charlesworth—I wasn't his direct secretary but I was still in the office. He never called me by my name. It was "Young Lady." One day Lois stepped out for a few minutes and Mr. Charlesworth—who I always called "Mr. Charlesworth" because he never said I could call him "Cliff"—was in a meeting and he came out. He handed me this piece of paper. "I need this chart right away." He had marked it all up. "I need ten copies, and I need it in five minutes." Lois was gone. Back in the old days, we went to Graphics. We didn't have our own computers that did all this, so I took my Selectric, and I typed it up. It was not a lot of typing. I went down the hall, made copies, took it in to him, and came back to my desk. Lois came back in. I didn't even tell her. After a while Mr. Charlesworth came by. He stopped at my desk. He said, "Young Lady, that was a fine job. I want to thank you for that." That kind of stuff registered with him.

Five months later Lois applied for a promotion. She had previously been affected by the reduction in force (RIF).[18] When you have a RIF, people are laid off, but you can back into one of those jobs that are vacated so that you don't get laid off.[19] She had been affected that way. She was eligible for an admin-type job. This probably was April of 1973. She was leaving, and the job was open. I asked my husband, who's always been extremely supportive with my career and my education, "Should I apply? He scares the hell out of me."

"Well, if you don't, you'll never know if you would have gotten it."

"Well, I don't even think he knows my name. He always calls me Young Lady."

So I applied. There were ten people that were interviewed. One day he came out and he said, "Young Lady, do you have a few minutes?" I said yes. "Just come in here." I got up and went into his office. "Close the door. You know me. I know you. I think that's a good enough interview." I said okay. I was dismissed. It was such a short time that Lois was on the other side of the door thinking the door popped open when I opened it.

I said, "He's done. I tried, what the heck."

A couple of days later he was in the doorway. He said, "Young Lady, come in here." He said to Lois—he called her "Sunshine," he only called me "Young Lady," but he called her "Sunshine"—"Sunshine, get up." She stood up. He told me, "Sit down." I sat down. He said, "The chair fits. You got the job."

That was it. I got the job. That was a GS-7, finally—top of the rank for secretaries at the time.

It was a very interesting assignment, because John transitioned out, and Gene Rice came in.[20] He became the deputy. That's about the point where the job became a career. It was no longer just a job. I was really engrossed in what I did. I loved the information flow. That's about the time I realized that I didn't have to ask too much. Information would just fall in my lap. I began to learn a lot about the organization. I became very involved with what was next, the strategy, and who we worked with. It was a very interesting evolution for me because I think that's when I made the turn into, "I'm really part of the Johnson Space Center, and I'm here to stay."

When Glynn Lunney was head of the Shuttle Payload Integration and Development Program Office he was asked to go to NASA Headquarters in Washington, DC.[21] Dr. Chris Kraft, who was the center director at the time, asked Mr. Charlesworth to go fill in for Lunney.[22] I thought I was going with Mr. Charlesworth to his new job.

He wouldn't tell me, but I suspected something was going on. He said, "If there were anything, I would tell you." I was in class, and I got the word that he moved. I came back after class and just wanted to see him. He said, "When I'm told I can't tell, I don't tell. By the end of the week you tell me if you want to come with me."

I said, "I don't have to wait till the end of the week. I'll come with you."

He said, "No, I want you to think about it."

I said okay. So, all week I moved a few of his little things, because secretaries were responsible for packing up the boss and making sure he had everything.

Finally, the last day I went in and I said, "You're not going to ask me to come with you, are you?" I started to cry.

He said, "Shut the door." So I did. He said, "Young Lady, you know, sometimes you can't do what you want to do. I'm not going to disrupt somebody else's life just because of what I want. As much as I want you to come with me, I'm not going to bring you with me. You have the corporate memory of what goes on in the office. Gene needs that for transition, so you need to stay with him." (Gene was going to be the acting director.) I was so brokenhearted. I was so loyal, and here he just left me.

I had a great opportunity with Gene, though, because Gene was the kind of boss that let me do anything that would make my office run better. Charlesworth wasn't that interested in the details. Charlesworth was just like, "Just do it." Gene was interested. I could talk to him about the divisions and the "girls" and everything they were doing together. So, there was a big difference there.

A few months passed, and Charlesworth came in one day. He said to Gene and me, "I want to talk to the two of you." We went into his old office. He said, "Chris has just told me that Lunney is coming back, and I will be his deputy." He looked at me and said, "See, Young Lady? Now you would have had to make a decision to take a downgrade. That's why I couldn't bring you with me."

I said, "Well, you could have just told me that, saved me all this suffering." That was a shift for me. Charlesworth became confident and started to share what I considered big management kind of things with me, talking to me about what Dr. Kraft was saying.

I would venture to say that many of the secretaries of my era were in the same boat. We were mature enough, professional enough that we were part of that management team, and management really relied on us to be part of that team and not let them worry about those kinds of things. I took care of who was going to cover what when vacation time came around. Building a team with my coworker secretaries in Earth Resources was key to me as far as developing my leadership skills, because I saw myself as one of them. If I wasn't very busy, I would ask them, "Do you have anything you want me to do?" Now we had a deal that we did it their way. The formats were such, but there were still specifics that we would do it our own way.

This was a very respectful transition for me from the other women that I had worked with who were very catty and not very nice, to this team of Earth Resources women. There were only four of us. We teamed up very well. For me that was the beginning of developing real leadership skills, where I was also mimicking some of the skills that I saw in my male bosses. Well, we all had male bosses.

The other important person in this era was Marilyn Bockting.[23] Marilyn had come with Dr. George Low and was an executive secretary.[24] She was a very professional woman. She was the supervisor admin for five program offices, including Earth Resources. Maybe it's different now, but program office directorate-level secretaries didn't have a lot to do then. A lot of those manual tasks weren't there. There wasn't a lot of typing. Correspondence was taken care of by the mailroom. They would open it. I went to Marilyn and I said, "Is it okay if I do some of your manual labor, like HR (Human Resources) requests for action, and if I could coordinate the moves for my folks and all that."

She said, "Oh, absolutely." She was like a GS-12, and I was a 7.

I said to her, "It's not like I want your job or anything. It's just that I would like to have the skills so someday I can claim them." She was really super as a mentor. She was probably my first real female mentor, and we both acknowledged that I was her protégé. The others I had just observed from the sideline.

I brought her up because she was one of the few mentors that we females had, and the fact that she was important to me as far as learning skills that I would need eventually. She and I had a very good relationship. She went to California after the Apollo era was over. She had cancer, and when she died I asked if we could change the name of the Outstanding Secretary Award to the Marilyn Bockting Secretarial Excellence Award. For many years it had her name because she was instrumental as a role model for most of the secretaries in that era.

I stayed with Gene. Gene was super. Then there was talk about reorganization, as always. There's always reorganization. I thought I would leave the secretarial field because I was so tired of my career being decided by others through reorganizations. Well, guess what? It happens to everybody.

I don't care what position you have. Reorganization comes and you're all affected, but I wasn't experienced enough to realize that at the time.

At that point, Space and Life Sciences, which was being run by Dick Johnston at the time, merged with us.[25] They swallowed Earth Resources up, so we were no longer Earth Resources. I think there was a remaining little branch. At the time, a couple of senior directors talked to me about moving when that merger took place. They thought I was going to be left without a boss because Gene went over as a deputy, but Dick Johnston asked me if I would be his secretary. So I became the directorate secretary for Space and Life Sciences.

One of the things I had done throughout my time at JSC was talk to my HR rep, Greg Hayes.[26] One day in 1979, Greg came to see me and he said, "I've just been asked to be head of the HR side of astronaut selection. How would you like to come be my assistant?"

I said, "Oh, absolutely."

"No promotion."

"That's fine." I was a 7.

"But I think you'd do real well."

"I'd love to do it." So I went over.

Greg was another tough boss. I told him he was a little Mr. Charlesworth. He was very tough. Maybe it was because we had been—not peers—but just buddies in a way. Now all of a sudden he was my boss. We had an operation out in Building 266. He didn't ask us to do anything more than he did. He was very dedicated to the effort, so we were constantly working. It was a humongous effort. I think we had thirty-six hundred astronaut applicants, so the HR guys came over and reviewed the applications, but primarily Greg and I went through all those.[27]

Let me back up a little bit. While I was working in Earth Resources, HR asked me if I would be in the secretarial rating panels. The secretarial rating panel provided me with some background on reviewing applications, looking for relevant things.

When I got to astronaut selections I had to make that switch to engineers and what to look for, instead of secretaries. Greg was an excellent teacher in that respect. Going through applications wasn't really my cup of tea, because as a secretary you're all over the place. I had to sit for hours reading and reading, so he taught me how to review the applications. I was very grateful, although it got to be a little stressful sometimes. The

Secretary Helen B. Statz answers the phone in the Apollo Managers Office.

other thing that happened was my introduction to George Abbey and the Astronaut Selection Board.[28] I'd heard about Mr. Abbey, so I was scared to death of him. He was introverted.

Of course, Greg's job and my job were to serve this board. They would come and go, and we had to make sure that all the files were in order, and there were drinks, that comfort kind of stuff. One day Greg came in. I think we had five hundred highly qualified applicants. He said, "Mr. Abbey would like copies of every application made for the members of the committee." It was four thirty in the afternoon. "By eight in the morning. Do you think you can do it?"

I said, "Yes, we can do it." So I called in the team of women that were working with us. I said, "Okay, here's the deal. We have to go through five hundred applicants. What should we do?" That's one of the things that I've always enjoyed, asking others to give me their ideas. I don't have to know everything. Others have good ideas, too.

We came up with a strategy that we would divide it A through C, and so forth. There were enough of us. One would go to Building 1 second floor, another on the first floor, and use the copiers there. Then bring it all back. I think we worked to probably midnight that day, but when they came in in the morning, they were all in boxes. Did anybody say thank you? No. But we knew they noticed, because when they left in the evening our files were not there. It was those kinds of challenges that George put on us.

I became the morale captain. They were working so many hours that we didn't do dinner and stuff like that, so I started keeping a drawer of cookies and licorice. The guys knew where the drawer was so they would come and help themselves.

After the 1980 selection, I was a staffing specialist. I was about thirty-five years old, no degree. After going through the astronaut applications with all these young people who had masters' and PhDs, I thought, "It's time." I had about thirty credits of night school.

Then I saw an announcement for the Federal Women's Program manager position in Equal Opportunity (EO). I'd gotten my GS-9 with Greg, but the job was an 11. I applied for it, and I got it. I worked with Dr. Joseph Atkinson, Jr., who was a true inspiration for me in many, many ways.[29] In fact, he probably planted the doctorate seed because he was working on one of those NASA-sponsored programs when he got his doctorate. He and I would talk a lot, philosophically. He loved to philosophize, but because of him I understood a lot about the racial issues, the perceptions of inequality, the background. He was a Morehouse College man, and he had known Martin Luther King, Jr., so he provided a history for me that I would not have had. Plus, he told the story in a very nice way—why things were the way they were. There wasn't any controversy or hate or anything like that, so I loved talking to him. I went to work for Joe, and he had been on the selection board when I had been on the HR side.

One day I get a call from George Abbey. He said, "Estella, I understand you just made a move, but I wanted to see if you'd like to make another move. Would you like to come be the admin in the Astronaut Office?" Cy Baker had just died.[30] He was a longtime admin there.

I said, "Oh, Mr. Abbey, I'm so flattered you called me. But I just made this move, and I just couldn't leave Dr. Atkinson. It wouldn't be very nice."

He said, "Well, in CB it's an all-women administrative group.[31] They've had some major issues. They just need somebody that would put it all together." I was very resistant about leaving. He still scared me, quite frankly, so I wasn't too sure I wanted to go. He said, "Well, think about it."

The next day the chief of the Astronaut Office, John Young, called me. "Ma'am, I hear tell you might be coming over to the Astronaut Office."[32]

I knew him from the selection process, too. I said, "Mr. Abbey called me and I'm very flattered. I'm so flattered you're calling me now. But like I said to him, I just came over to EO. I just got a promotion, and it just wouldn't be right."

He said, "Well, ma'am, we got a lot of female problems over here. It'd sure be good if a woman could come over and handle them." That's just the way it was. It was about thirty administrative women, about half contractor, half civil servant. One of the HR guys had told me, a good friend of mine, said, "Estella, if you go over there, it's tough. Think about it."

I said, "I'm really not thinking about it. I'm happy here with Dr. A."

George called me the next day. "So, did John Young call you?"

I said, "Absolutely. I cannot tell you how flattered I am. This is incredible. Jeez, I'm so sorry, but I really just can't leave Dr. Atkinson."

He said, "What if I make it a supervisory admin?"

I said, "So when did you want me to come?" Because former secretary, no degree yet—where else was I going to get that kind of experience? I transitioned over about a three-month period where I was doing both jobs. This was around October, November. In January 1983 I went over there full-time.

He also gave me free authority to do whatever I thought needed to be done. Talk about a challenge. Some of the people I was going to supervise were very resentful of my being there. There were all kinds of rumors about how I got the job. There was all kinds of speculation. They were concerned that I was a George spy.

John Young was super to work with. I absolutely adored working with him, because as technical as he was, he was also very people-oriented. I had to laugh at him sometimes, because he was so caring. For example, the award money that HR gives to civil servants (HR gives so much to each organization)—John and George said, "Astronauts don't need any of that. They get perks through the year. Give it to your admin ladies." I would take the full amount and split it among my ladies, because that's how George and John felt: they were the net that kept the astronauts going, so they deserved more.

But when I got there, it was absolute chaos. Part of the problem was that when they formed a crew, they took astronauts from all kinds of places, and then they had to put them together. Building 4 North, the old building, wasn't configured as it is today.[33] It was just great big offices. And the corps was growing. We had hired the "Thirty-Five New Guys" in 1978, so offices were filling up faster.[34] There was also the growth of the contractor support that had to be onsite. Payload specialists were coming in, and there was actually a Payload Specialist Office in that area.[35]

One of the main issues that I had was the space. It's still a big issue, I think. Building 4 is pretty much a square. It had all the outer offices, and then we had what the guys called the snake pit because it was in the middle. It was the core. At the time, I probably had about eighty astronauts and about nine secretaries. I had to somehow balance out the support. I don't know how I came up with this, because I didn't have any prior experience and certainly didn't have the formal education yet, but it made sense to me to move all the astronauts out from the core area. They all wanted windows, so I put them in the window areas. They also didn't want modular desks. They wanted regular desks. They didn't have any computers. Neither did the secretaries. For me, it was easier to make the workflow for the admins a little bit more defined.

Astronaut Appearances had four thousand requests a year at that point. This was 1983, 1984. We had a staff of about three civil servants and another three or four contractors.[36] The travel was all done by one lady. That one lady civil servant would call the Travel Office, and then they'd put her on hold. She'd get the information, eventually get to the astronaut, and the astronaut would say, "No, I don't want that travel arrangement, go back." She was on the phone constantly. One of the things I decided was, why can't she have one of those SABRE computers so she can look up the information and just queue it up to a point?[37] She doesn't have to be on the phone with the office. I went to Mr. Abbey. He said, "How much?"

I said, "Five thousand dollars a year."

"Get it."

Well, then I ran into, "It's not CB's area." Everybody was always fighting with the Astronaut Office because the Astronaut Office was always trying to do stuff without going through the right channels, so I made a deal with Center Operations that we would not cut any tickets. We only wanted to queue the ticket up. Then we'd release it to the Travel Office, to the real travel people, so that they could take care of it. They agreed. We get this computer. The lady who was in charge in CB didn't want to use the computer. We gave it to one of the contractors, and then the civil servant overlooked all of that, because she knew what to look for. She just didn't know how to use the computer.

Then we had the issue of no computers. Mission Operations Directorate was going to the Xerox Star computer. The few computers that were in the office were for the astronauts, who had probably gotten them through their contractors. Mr. Abbey, when I told him that the astronauts wanted computers, said, "If I wanted typists I would have hired typists."

I asked Mr. Abbey if I could have computers.

He said, "Well, how many would you need?"

I said, "Well, one per person, and I have thirty."

"Oh. Well, you're going to have to show me that they really need them."

So, I put together a presentation. He gave me two computers. There were nine secretaries, there were nine people in Astronaut Appearances, and then there were other people. I worried about those two groups first, and I gave one computer to the secretaries. I had an office. I cut one-third of my office off, made an entrance into this little room, put all kinds of acoustic stuff in for soundproofing, and set up the computer there. We had a sign-in board. I said to every secretary, "You get five hours a week. There are nine of you. That's forty-five hours. If you want two hours in a row, you guys negotiate. I'm not going to get into that." So they did. Then the astronauts wanted to know when it was their turn. I said, "When they're done at six o'clock, then it's your turn."

The other computer, I gave it to Astronaut Appearances because they had a lot of work. I called in that team and they said to me, "We don't need any computers; we need more people."

I said, "Well, we're not going to get more people, so we have to figure out how to do this with other things."

"We don't want any computers. We want more people."

There was a lady who was probably my age now. She said, "I'll take it. Put it at my desk."

I said, "Oh, thank you! Thank you." We put it at her desk. Now we've taken care of that for a little bit. The computer with the secretaries seemed to be working. They were cooperating with one another. The one in Astronaut Appearances—after a few days I went over and I said, "So how are you doing with the computer?"

She said, "Well, girl, it's working out so good. The only thing is I haven't figured out where I roll in the paper." For every file she had a little icon across the top of the screen because she hadn't figured out how to print! But we muddled through all of that. A few months later, the ladies from Astronaut Appearances who had said we needed more people came in irate. If they all didn't have computers they were quitting. That's how I felt. Eventually Mr. Abbey gave me one for everybody, but that was a struggle just to set up some efficiency.

The other thing was the middle area in Building 4. I made it entirely administrative support. In the corners of that middle area I put two secretaries per corner. "You're going to work seven thirty to five. You guys decide which shift you want. I do not want to get into that. You guys decide who's going to go to lunch if you have an appointment. You guys decide who's going to cover, because it constantly was, 'Well, I was going at eleven thirty, and then she beat me and went at eleven fifteen.'" That kind of stuff. Remember, I said I worked much better with the guys. Well then God punished me and I got thirty women!

I gave them decisions to make on their own, and that seemed to help. With the middle section I gutted all the walls and put cubbyholes in. The workflow was such that everybody walked out into the hallway from their offices before, so pieces of paper—of course it was all paper then—would not get to the next place. They'd stay in a basket, and an appearance request would not be finished because somebody didn't pick up that piece of paper. Of course, computers started to make it a lot more efficient. At that point, the new layout started to help. The secretaries' work areas were flowing better.

The other thing was coffee. Everybody wanted coffee. A couple of the ladies would go to the store across the street, cart the coffee back to the office, get a basket, and carry it upstairs. I'm like, we're not going to be doing that, so I paid for a coffee service and they delivered the coffee. I charged the astronauts twenty dollars every six months for coffee. If they didn't drink coffee, they still had an assessment of fifteen dollars because their guests drank coffee. You should have heard some of the resistance I had for that one! But the bosses were behind me, because it was just more efficient to do that and not worry about if we had enough money to buy coffee. From that fund I also had money to buy cards and flowers, and I opened an account at the bank so we had interest that it was earning. At one point I had twenty-five thousand dollars. The fund was making money.

Plus, I started to collect money up-front for the silver and gold medallions, which only astronauts could buy.[38] I wanted the money up-front so that when the bill came for the finished ones I had it. I didn't want to be in debt with the medallion company. You just couldn't do that. Somebody had to be responsible. There was some resistance. "Why should I give it to you when I could be earning money on it?"

I said, "Yes, but I could be earning money on it too, because it goes to the astronaut fund for flowers and all that." The one thing that I always had was management support, but I would always go and explain my reasoning. John Young was extremely supportive. George Abbey's response was, "Just tell me what you're doing and how much you need." They really trusted me, and I just loved that there was this trust.

It took me about two years. There was this real tight clique there. Some of the women that I supervised had been there for a long time, so I had to break through that. I remember the STS-51C flight was the first crew that was nice to me.[39] The crew included Loren Shriver and Ellison Onizuka.[40] They would come in and tease me and joke with me, so they started to make me feel comfortable. I had all these issues with the women going on, but that's why I had gone there, because there were some significant issues, mostly caused by lack of defined roles and responsibilities.

Then came the Space Shuttle *Challenger* accident.[41] By this time I was well into the organization. We ran the astronaut gym, the patches, and the logos—everything that we supported in my organization. I had a super team of women. I was in my last semester of school for the bachelor's. In January of 1986 I had signed up for my last nine hours for that semester. I was very ambitious. I was going to graduate. Well, then we had the accident. I said to my husband, "I don't even want to go home." I just wanted to stay at JSC.

I remember the day of the accident I went to Ellington, truthfully, because I just didn't want to go home.[42] Mr. Abbey and Carolyn Huntoon came in on a plane.[43] I'll never forget. They got off the plane, and it was like the big brother and little sister getting off the plane. It was so sad. For Mr. Abbey, I know it had to be devastating because he was so tied to those crew members. Plus, they had all been part of his selection groups. They were all special. I was around for the Apollo 1 fire, but it wasn't the same impact.[44]

I'll never forget Jack Garman was sitting next to me in Building 1 in their conference room.[45] We were watching the launch in a meeting. He said, "It exploded."

I said, "No it didn't."

He said, "Yes, it did."

So, I got up and I went into Ron Berry's office because I wasn't sure what I was going to do.[46] He said, "Yes, it exploded."

Then I went to the elevators and Chris Critzos was standing there.[47] He said, "We lost it."

It was just one of those things that couldn't have happened. Then I walked over to Building 4 and everybody was sitting in John Young's office. We didn't have multiple TVs all over the place. Everybody was sitting in John's office, and some in the conference room. It was very quiet; everybody was fixated on this.

Brewster Shaw came up the hall and he gave me a hug.[48] He said, "Are you okay?"

I said, "Yes."

He said, "Well, I'm going to go talk to my wife at school and then when I come back, if you need anything, you let me know." How thoughtful. He was going to talk to his wife so she would know firsthand what happened. Mary Lee Meider—Thompson now, she was my lead for the mailroom— we were sitting in my office and we were both just boo-hooing away because we didn't know what else to do.[49] Nobody had told us, "Do this."

I said, "People are going to get hungry." It happened around ten thirty a.m.

She said, "You want me to get some food?"

I said, "Yes."

She said, "I'll just go to the cafeteria."

"Yes, see what they can do for you. Just tell them you want something and they'll bring it into the conference room." Within minutes they had cold cuts, bread, chips, and punch. They put it in the conference room. For a while nobody came. After a while, everybody started coming and took a plate.

We started getting telegrams and then flowers. They just started trickling in. In the afternoon, they really started coming. So, there we were again. Mary Lee and I were sitting there. I said, "We're going to have to acknowledge this somehow. We should come up with something. How about just a card?"

She said, "I just had the astronaut symbol done in graphics," because, again, we couldn't do that ourselves. "I have it in all sizes."

"Bring it in, let's work on that." We came up with a card that fit in a legal-size envelope. We put the logo on it, and "With grateful appreciation" as an acknowledgment. We had John Young sign it. By the next morning we had one thousand printed and cut by the print shop. We gave it to the secretaries, who were all sitting around like we had been, wondering what to do next.

We had cards. We had origami birds. One card came in that somebody had watercolored. It was a couple looking up at the sky with pretty words. I really, really liked that card. I think it was one of the first that came in. Immediately we had a collage done for the families with that card and the last picture of the crew leaving out of Ellington Field. It's only the partial crew, because the payload specialists had gone ahead, Christa McAuliffe and Greg Jarvis, but it was the NASA crew and then the logo.

That picture was from the same photographer at Ellington who always took each crew's picture as they left Ellington for the Kennedy Space Center. He had sent it to us. The next day we were getting individual cards for the Scobee family, for the Onizuka family.[50] We had the same card made, the one Young had signed, and took the card over to the surviving spouses to sign and did the same thing with the cards. We gave them to the family so that they could have friends fill them out. It kept us busy, and it kept us from floundering.

P. J. Weitz eventually told the rest of the astronauts to deploy.[51] They started naming the casualty assistance officers (CAO).[52] Then the memorials started to jell, and the ninth floor directed the memorial at JSC. It came together very, very beautifully. We were told that every astronaut could get a ticket for himself or herself and then a spouse. The night before the memorial Mary Lee came into my office crying, "It's not fair. We were not included in the memorial."

I said, "That's because only the astronauts get to go."

She said, "Yes, but it's not fair because we're part of this. It's hurting us just as much."

I went in to P. J. and I said, "It's not fair, P. J. We're part of this just as much as anybody else. Our ladies are in just as much pain. They did all this stuff, the patches, the crew stuff."

He said, "What do you want me to do?"

I said, "Is it okay with you if I talk to John?"

He said, "Sure. I'm not sure what he can do."

I went in to Captain Young. I told him.

He said, "You can have my tickets."

I said, "No, that's not what I want. But I would like to ask you, because I know a lot of the spouses are not going to come, if I have Mary Lee issue those as the astronauts come in, can I have any extra tickets that are left over?"

We had exactly the number we needed.

Then I told my ladies, "You do not sit next to President Ronald Reagan!" Because some of them would have! I said, "You sit to the back." So, we all went over and we all got to sit in that memorial group.

I learned it's not good to just stay quiet. If you have an idea, tell it, because maybe they'll accept it. I could have just told Mary Lee, "Sorry, we can't do anything about it." But I did ask permission and it all worked out very nicely.

Then the memorials out of town started happening. Because I had met Lorna Onizuka just shortly before the 51L crew took off, I felt responsible for her somehow. Loren Shriver had volunteered to be the CAO.[53] I asked Loren Shriver, "Are you going to need help taking her back to Hawaii?"

He said, "Well, I think so. It's her parents," who were elderly, and then her two little girls. "I could use some help."

I said, "Will you ask Mr. Abbey if I can go?"

He said, "Okay."

We went up to see Mr. Abbey. I waited outside. Loren comes out and I said, "Did you ask him?"

He said, "No, I couldn't. I just couldn't get around to it."

I said, "Okay, I'll go in." So I said, "Mr. Abbey, Loren is going to go to Hawaii with the Onizukas. He's leaving tomorrow."

He said, "So he needs your help."

I said, "Yes, he thinks he does. I'm not sure what I would do, but I can help."

He said, "Go."

So I got to go. I escorted Lorna and the girls, and Loren took care of other things that were going on. We flew together. We arrived in Hawaii. This community of Ellison's was there, and people were thanking me, just hugging me and loving me because I had volunteered to do this. I felt like

I was on a boondoggle, but I really did want to go. Going to Hawaii was just—oh, wow! I would see what Ellison had talked about!

The whole Hawaiian community was just so special. The governor's wife, I think, was there. They all were so grateful that I had made this effort to go to Hawaii. We flew into Oahu and then flew to Hawaii Island. That was home for Onizuka, so the community was just very sad, but they were really honoring him. Lorna's girls were little. Lorna was really trying to stay brave. I became her assistant. I would vet her calls. I noticed there were no newspaper reporters. I said to her, "It's funny that you have no newspaper reporters."

She said, "Well, that's because Mr. Fujimoto had told them not to call." Mr. Fujimoto was like the George Abbey of Hawaii. He was passing out the word: "You will not bother her."

I got to know Lorna very, very well and the Hawaii family, his parents. We went to his childhood home. That was the first trip. This was in February.

We went again in May, I think, when we did the actual burial. That time we went in one of the planes that they use for the president and vice president. Not the big one, but a KC-135. They had the casket there, and Loren wore his Air Force blues. My job then became to look after him, because it was hot and he was sweating. I was getting him water. It was just whatever I could do, again helping Lorna with the girls. It was such a sad time that I didn't know how to act or what to do. That time there were several other NASA people who went on that plane. They were treated royally. The Hawaiians all knew who NASA was. That was 1986.

By that time I was almost finished with my degree. I had started with nine credits and I dropped one. My husband, when I told him I was going to drop the whole semester, said, "No, talk to your teachers, they'll understand. Tell them what's going on." Sure enough, I think I missed five classes that semester but the teachers understood and I made up my work and I aced the two courses. Then in the summer I took my last course. I finished and graduated in August of 1986. I was two weeks from being forty-one. I was still forty, the goal I had set for my graduation.

Then John Young got moved, and politics set in. I was so brokenhearted because he really just wanted to be an astronaut. He was moved to the ninth floor of Building 1. There were three offices there; one was a middle office and then two side offices. Rumor was that the other office was for George. I went ahead and made up three offices. A few months later

I think George was moved to that spot, so the prediction was right. He didn't stay long because one day I got a call out at Ellington, and it was Abbey calling, but he missed me, or I missed him. He was on his way to Headquarters, and he didn't come back for a while.

Then Don Puddy came in as the director of Flight Crew Operations.[54] I didn't know Don very well. I knew who he was, but by that time George had made me chief of all administrative activities for all of Flight Crew, not just the Astronaut Office. When Puddy came in it was a night and day difference. I had said that I was trusted, and I was left alone to make decisions because I knew what was best in their opinion. You get a new boss, and he doesn't really yet know the ropes. So there were constantly questions about why don't we do it this way. Well, we already tried to do it and it didn't work. Well, why didn't you do it this other way? So I was not happy. With George, when I had an admin status with him, I would spend hours with him. Business would continue and I'd be on item two, but he'd take another phone call. I was privileged to be able to listen to what was going on, so that was another way of my knowing what was going on in the big picture and think that I was a part of it.

With Don, when I would finally get in to see him, he'd say stuff like "What *administrivia* do you have today?" Maybe he didn't mean anything by that, but to me it was just a big night and day difference between him and Abbey. It was a difference in management style that I just couldn't get used to. I made myself a promise that I would stay until Return to Flight, STS-26, then I would seek another position.[55] By that time I also was thinking if things don't work out maybe I'll just leave, get my master's now that I know all about school, and then go teach.

It wasn't too long. I muddled through. On Return to Flight day I get a phone call—I hadn't told anybody this—it was Harvey Hartman.[56] Now that I had my degree, how would I like to go back to HR and lead up the secretarial development and communication skills and human resource development? I said I would love to. I went over to training in Human Resource Development, and I had a really good time with that.

Jamye Flowers Coplin[57]

I was a senior in high school and my sister was working at NASA at the time. She told me that NASA was hiring and that it would be a good opportunity for me if I wanted to apply. So I applied in the spring of 1966, was hired, and started work in June.

I knew very little, to be quite honest, about NASA at the time. I remember when the announcement was made in the early 1960s that the Manned Spacecraft Center was going to be built in the Clear Lake area. I was a teenager at the time and very naively thought they were going to launch from Clear Lake. It wasn't until two or three years later that I finally realized that Mission Control would be in Houston. The launch was actually going to be from the Cape. I remember thinking that at the time. In March of 1965 the first Gemini flight flew; John Young and Gus Grissom flew that one.[58] They made the flight on my birthday, the 23rd of March, so I always remember that mission.

As I said, my sister was working at NASA. In the spring of 1966 she married another NASA employee. He was working in Procurement when I went out to NASA and interviewed for a position. There were, I think, several openings at the time. One of them was in the Procurement Office, and the other one was in the Astronaut Office. My new brother-in-law worked in Procurement, and I thought, "It may not be a good idea for me to go in and work in the same area that he's working." So by default I picked the Astronaut Office. At that time, I really wasn't completely aware that the astronauts' home base was in Houston. They always had been associated with the Cape and Florida. I really wasn't aware that the Astronaut Office was *the* Astronaut Office. I thought maybe it was some type of support unit. It was quite a surprise to find that out!

When I started, it was the first week in June 1966. I had not interviewed in the Astronaut Office. I was interviewed at Personnel out at Ellington Air Force Base. I remember going out there and processing and finger-printing in one of the old barracks buildings. My first day after processing, I reported to the Astronaut Office. Sarah Lopez was the admin person.[59]

I remember she was a wonderful lady. She took me down the hall and introduced me to John Peterson, the chief of staff.[60] He was in the front office, and his secretary was Maxine Henderson.[61] Sarah introduced me to him, Maxine, and Gay Alford, and then took me in and introduced me to Alan Shepard, chief of the office.[62] That was the first time I realized exactly where I was, when I went in the office and was introduced to Shepard. I still remember my knees shaking, and it was quite a surprise. Quite a surprise. This was my very first job. I had just graduated from high school, and I was standing in Alan Shepard's office about seven days later.

I just adored Shepard. He had a reputation back then for his cool looks and steely eyes—he could shut you down in a minute just by a look. He ran a tight ship. Most of the guys at the time—and I think they will all freely admit it—were a little bit afraid of the man. He had a very casual elegance about him and a presence. He promoted that. He worked at keeping himself just slightly away from the group. He wanted everyone in the office to know that he was in charge. He was friendly and would stop and visit in the office with you, but there was no mistaking that he ran the boat.

The same week that I started, Lynn Cross started in the Astronaut Office and was assigned to the Appearances section.[63] I was hired as a stenographer for the steno pool. There were two of us, Georgie Huepers and I, when I arrived. Our immediate supervisor was Sarah Lopez, and we were the floaters. We took dictation and also backed up the flight secretaries. They had just hired Charlotte Maltese, now Ober, to be the secretary to the new nineteen, the "Original Nineteen" as they called themselves.[64] That was the astronaut group that had been named in April 1966. They had also just started arriving in the office when I came. I always considered myself to be part of that group because we all arrived at the same time. Shortly after I arrived we moved upstairs because those nineteen men practically doubled the astronaut corps. I think we had around forty-nine or fifty astronauts at the time. We had taken over the third floor in Building 4, and I remember the first few days and weeks were spent mainly packing and getting the offices ready for the move.

The flight secretaries were secretary stenographers. If they were busy and the guys needed someone for dictation, then Georgie and I were called. We backed up the secretaries if they were out on leave, vacation, at lunch, or when they were on a temporary duty (TDY) at the Cape. We were all down the hall from each other, so we could easily back up the secretaries. When they brought the nineteen in, they had grouped them all in offices wherever they could. When we moved upstairs, they completely reorganized everything. Charlotte was no longer secretary to that nineteen group. They integrated the nineteen in with all the other astronauts.

After that move, everyone in the astronaut group was reassigned alphabetically into five flights. Each secretary had eight to ten astronauts. When we moved upstairs, Flight A secretary was Toni Zahn, and the chief of Flight A was Scott Carpenter.[65] Flight B secretary was Charlene Stroman, and the chief of Flight B at that time was Gordon Cooper.[66] Flight C chief was Gus Grissom. That position was vacant when we moved upstairs. They needed another secretary, and that was the flight that they chose to have the vacancy. Charlotte Maltese was the secretary to Flight D. Jim McDivitt, I think, was chief of that flight. The last flight, Flight E, included Penny Study as secretary and Wally Schirra as the chief of that flight.[67] For the most part, the secretaries backed one other flight—Flight A and Flight B backed each other. Flight C backed Shepard and his office. Flight D and Flight E backed each other.[68]

When we moved upstairs, it was going to be several weeks before the Flight C secretarial position was announced. They put me in that position temporarily, which was a big surprise and a big adjustment because I really had not become familiar enough with the office. For a few days it was stressful being put in that position, but it ended up being great because I got to know Gus Grissom, and would not have otherwise had the opportunity to work directly with some of the guys in Flight C. Martha Caballero was eventually hired into that position, and I went back to the steno pool. I did not have enough time in my pay grade—nor the experience—to qualify for it.[69] So that was the reason I got sent packing again. I stayed in the steno pool and floated doing that type of work for several months.

Then I was put in the position of handling the technical correspondence and mail that came into the office. If I remember correctly, Shepard decided to create that position. (We had the mailroom that handled all of the requests for the astronaut photographs and letters, and they also sorted all the mail, technical and personal, that came in.) That was also another neat opportunity for me, because about once a week I would get to go in and sit down with Shepard. He would tell me what the changes in the assignments were and who was supposed to get which particular pieces of mail. At the time probably only Al Shepard, Gay Alford, and I knew which guys were getting which assignments.

At first I spent quite a bit of time in those weekly meetings with Shepard, and he was extremely gracious and would explain a particular subject to me. I learned a lot. It wasn't too long before I got to where I could read technical memos and knew who they went to. I found myself picking up more of the technical information and learning about how everything worked. I got to the point that I didn't need to check with Shepard with regard to who needed a particular document. It didn't take me long to pick up on the jargon or the acronyms. That was a big adjustment, learning all those acronyms and going through the mail. A lot of times, especially before the official announcements were made as to which astronauts were officially assigned to specific programs or missions, we would get cc'd on everything. The memos would be sent to the Astronaut Office. A large part of my job at that time was to read these memos, figure out what it was about, and who was working on it.

I filtered that information out, like the accident reports. We had secret clearances. I remember I had to hand-carry those things to the astronauts. They had to read them and then return them directly to me. We had to sign back and forth, back and forth. The other secretaries did not officially see those reports. I was the only admin person besides Gay Alford that saw the accident reports.

We generated, of course, a tremendous amount of correspondence within the office. The individual secretaries kept their files, but it was also required that CB—that was our mail code—master files be kept. I was in an office by myself to begin with, and it was lined with cabinets to maintain the CB files and also the incoming mail distribution. I continued to fill in for the secretaries when they were out.

I think it was a few months into that assignment that Goldie Newell—she was Don Gregory's and Tom McElmurry's secretary—was out for about six weeks because of surgery.[70] I was sent over to Deke Slayton's office for six weeks to work with Mr. Slayton and Sue Symms, Mr. Slayton's secretary, which was another fortunate break for me early on because I got to know Mr. Slayton very well, and he got to know me.[71] It was a lucky assignment timewise, as I got to know a lot of people—and a lot of people in Building 2—that I would not have had the opportunity to have that close of a working relationship with otherwise.[72] By the time I got back over to the Astronaut Office again, I had been around the block a few times. The opportunity to fill in served me well. I was grateful for the opportunity.

Sometime in 1968, I believe—I had been there less than two years, probably a year and a half—Charlene Stroman, the secretary to Flight B and Gordon Cooper, left and took a position with the State Department in Washington, DC, and that position became available. At that time, I had met my year in my pay grade and was eligible for a flight secretary position. I applied for it and did get it. My first secretarial position of my very own was secretary of Flight B with Gordon Cooper.

We had the responsibility of answering the telephones, technical and personal correspondence, and arranging travel. Working with ten astronauts who traveled at least one trip a week, it was hundreds of trips a year. We had to do the orders, including overseas travel, especially after the flights. The travel vouchers and travel arrangements took up a lot of our time. They would do the round-the-world trips, and that meant working with the State Department to get all of the visas and passports.

We were also responsible for maintaining the astronauts' manuals and checklists, especially for the crew-assigned astronauts. As their missions neared, the changes came and were updates to the actual checklists that they used for flight. Updates would come in the office in the mail. We would go get their checklists, pull the old sheets out, put the new sheets in, or make handwritten changes. They were little binders that didn't have covers on them, but they had the single rings. We had to do those for each one of the guys. I remember it was time-consuming because you had to be very careful to make sure you pulled the right pages. I remember stacking that stuff and taking it home with me. I just didn't have time to do it during the day. That was something that I could do at night while I was sitting at home watching television, updating checklists and manuals. The current crew was the priority on updating the checklists. The checklists and manuals were strictly technical, and most of the time it was a procedural change as to how this switch was going to get flipped instead of the other one.

Technology was so different then compared to what we're using now. We typed on IBM Selectric typewriters and had to type in at least triplicate—which meant erasing carbon copies when you had any type of error on them—for anything that went out of the office. They were fairly lenient with regard to using Bic Wite-Out and that type of thing. Most people now don't even know what Wite-Out is.[73] If anything actually left the office, then we did it until it was perfect. Anything that went to Mr. Slayton's office—we used to laugh, because having been over there and worked with Sue Symms, I knew she really did get a ruler out and measure the margins to make sure anything that was going to the director's office was exact. We called her the ruler lady. Sue was wonderful. We knew what was required, and we produced good documents.

We also worked with the astronaut families. We actually were a family within the Astronaut Office, and the guys traveled—I still call them "guys"—the astronauts traveled so much and were out of town. When they were in town, they were in meetings and the simulators for hours. We were the contacts the wives had as far as being able to get to their husbands. We developed very close relationships with the wives and the families, especially during launches. We babysat. I babysat for the Roosa family and for the Evanses and for the Engles and for the Anders family.[74] There were two or three of us in the office that would babysit, so we really did get to know the kids and the families. Of course, the astronauts' sisters and brothers and mothers and dads would call the office, and we all knew them very well too.

I remember I babysat one time for the Evanses because Jan's mother was very ill and Ron was at the Cape training, so I went over at least a week, maybe longer than that, every night after work.[75] I stayed with their two kids, got them up and off to school the next day. I then went to work, came home, and fixed dinner for them, like I knew what I was doing. I was twenty-three years old. But I was like everyone else back then, I didn't know that I couldn't do it.

Part of my job was protecting the astronauts from the public. We were a bunch of mother hens in that regard. We screened calls, but we didn't screen calls to the extent that the guys didn't get messages. We never said, "Throw this one away." They always got the messages, and we took very detailed information with regard to who was calling. We did protect them from the public, but back then Building 4 and the Astronaut Office was an open area. There were no security guards. Anyone could walk up to the third floor and walk down the halls. The five flights occupied the outside offices. Two astronauts per office, except for the chief of the flights. They had their own offices. The flight secretary generally sat outside of the chief's office, so you had to enter through the secretary's office to get into the chief's. But otherwise you could walk up the stairs or come up the elevator and walk down the hall and go in and visit with any of them. Now you can't even get in the building without having a cipher code.

Our work hours were terrible. For seven years, I had terrible hours and very full days. You worked until you could get to a stopping point, then you went home. I think my hours were supposed to be seven thirty a.m. to four p.m., and I rarely got out of there before five thirty or six p.m. We put in a lot of hours, worked a lot of weekends. The only time I ever remember getting paid overtime was when we were involved in an accident board. Then they would ask the secretaries to volunteer to come in on the weekends in support of typing up the accident reports and assisting with that. Money was always allocated for that. Otherwise it was part of the team effort. Everyone worked those hours. We weren't the only ones. Everyone did. When the astronauts would be in the simulator from eight in the morning till five in the afternoon, the time that they could get their work done as far as what they needed to do at their desk was after that. We all wanted to be available.

I was very involved in the Apollo 10 mission.[76] Let me explain. The flight secretaries took turns accompanying the crews to the Cape. It was generally the secretary to the commander that went. We rotated, and I had been assigned as the secretary to Flight B and Gordon Cooper. Gordon Cooper was the backup commander of Apollo 10. The backup crew was Donn Eisele and Ed Mitchell.[77] I also worked for two of the support crew. Penny Study was secretary to Tom Stafford, commander of Apollo 10, at the time. She was unable to go to the Cape, and I had been asking Shepard if I could go. He kept kidding me, saying, "Not until you're twenty-one, not until you're twenty-one." I turned twenty-one in March 1969, and when the opportunity came for a secretary to go to the Cape with Apollo 10, Shepard let me go. That was my first time to accompany a crew to the Cape. It was not the first time I'd seen a launch, but it was the first time I got to go and work as crew secretary. I worked with Tom Stafford, Gene Cernan, and John Young.[78]

When I went to the Cape the first time on Apollo 10, I was still living at home. Believe it or not, my mother and daddy drove me down there. The Cape had gotten an apartment for me to stay at while I was down there. That was the first time I had ever been on my own. My parents drove me to Florida and got me settled, then returned to Dickinson. I had a gray goose government car with no air conditioning in Florida; it was hotter than Hades, and I could go only to work and back. If they caught you anywhere else in that thing you were history. Shepard was adamant about us following all the rules while we were on TDY at the Cape. You didn't want to get into any trouble whatsoever. Those that did never went back. After having been at the Cape for the first time away from home, I got my own apartment when I came home.

The secretaries generally went down about six weeks in advance of the launch. When we went with the crews, we worked in the offices that were right outside the crew quarters where the guys lived when they were at the Cape. It was different than working in Houston in that everything was mission oriented. In Houston, you might have three guys out of your ten that were on the current mission, on current crew, and the rest of them were on one coming up. When you went to the Cape with the crew, you were solely working with that crew for that mission and other astronauts at the Cape training for upcoming missions. It was very, very focused.

A large part of our assignment down there was the guest list and working with the astronauts supporting their personal matters, which needed to be handled at the last minute before the missions. They each had their own family guest lists, their official guest lists, and their personal guest lists. We had NASA Headquarters guest lists, politicians, and celebrities. We had to find them rooms, cars, and transportation. That was a large part of what we did while we were down there. It was challenging because back then there weren't many hotels. We were renting apartments and blocking rooms. This all would start months in advance of when we had to get these things set up. By the time we went to the Cape, we had the major portion of it done. It was just all the changes that would come up.

The guys, for the most part, just spent hours in the simulators while down at the Cape. It was really crunch time, those last six weeks. But the good thing about that crew was there was a little bit of levity to it.

At the time, in the late 1960s, there was a lot of unrest in the country, and I felt like I was very insulated and not really a part of all that. It was going on and I was aware of it, but we were so focused on NASA's mission. I was so involved that, quite honestly, I didn't pay a whole lot of attention to things that were going on in other parts of the country.

But the one thing that we all enjoyed was Charles Schulz and the

Peanuts cartoon. Snoopy had climbed atop his wonderful little doghouse and had become the astronaut going to the Moon. At the time, I was cutting out the *Peanuts* comic strips and saving them. It was absolutely a delight for me, and I think for the country, when the crew selected Charlie Brown and Snoopy as the call signs for Apollo 10. Apollo 9 was the first mission that had call signs, Gumdrop and Spider, because the command module (CM) looked like a gumdrop and the lunar module (LM) looked like a spider.[79] Apollo 8 did not because they did not carry a LM.[80] Charlie Brown was to be the command module and Snoopy was the lunar module that was going to fly close to the surface of the Moon.[81]

I asked Gene Cernan about it recently. We were all fans of the comic strip. We talked about it. Charles Schulz was a presence. We had the Silver Snoopy Manned Space Flight Awareness Awards.[82] He said that because of the popularity of the comic strip he always thought John Young was Charlie Brown, so that's what planted the seed to see if they could get approval to use Charlie Brown and Snoopy as the call signs. I think that the use of *Peanuts* in the space program captured a lot of attention within the country.

Let me tell you about a prank that became an historic Snoopy moment. I think the Apollo 10 guys had already left the crew quarters on launch day. Dave McBride, the team leader, apparently had told Captain Cernan that I had something I wanted him to take to the Moon for me. Apparently, Captain Cernan had told Dave, "Sure, I'll be glad to, but at this point the only thing I can do is stick it into the pocket of my spacesuit. Get it down to me and I'll take it for her."

That was going to be the gotcha on Gene Cernan, because Dave McBride had this very, very large, wonderful stuffed Snoopy. The plan was for me to stand at the door of the crew quarters as the crew was coming by with this Snoopy, and the "gotcha" would be on Gene Cernan. He was going to have to get this very large Snoopy in this very small pocket on the side of his spacesuit. It was going to be a fun time. The crew suited up and was coming back down the hallway as I was standing just inside the door of the crew quarters with the Snoopy in my arms. Just right at the last minute, Dave McBride gave me a push, and I ended up out in the hallway.

Stafford was right there, and he stopped briefly and patted Snoopy's nose. That was a picture that defined that mission. I think Captain Young patted him when he went by. Out of the corner of my eye I could see

Captain Cernan heading in my direction, and he was a man on a mission at that point, so I knew I was in trouble. He came and instead of patting Snoopy on the nose, he turned the prank into a gotcha on me, because he grabbed me and Snoopy and tried to get us in the elevator. He was going

Apollo 10 astronaut Gene Cernan tries to take Jamye Flowers and Snoopy to the Moon.

to take Snoopy and me to the Moon if he could. When you see Apollo 10, that's the picture everyone thinks about. Tom Stafford still says that photo is one of his favorites, if not his favorite.

We went out and saw the launch while Snoopy guarded the crew quarters. It was exciting. Of course there was concern, but the success of the previous missions gave you confidence. The guys were excited. They were well trained. They were ready to go. There was a great deal of apprehension, but excitement, very much so. Seeing a *Saturn V* launch was just—I can still hear it. You don't forget that sound. The *Saturn V* launched beautifully, flawlessly.

Once the guys left the crew quarters, we left shortly thereafter, too, and went out to the private area where we could view the launch. We, the secretaries, were spectators at launch time. As soon as the launch was over, we went back into the crew quarters, and at that point I think they considered the secretaries' duty at the Cape complete. Once the crew launched, we returned to Houston. Sometimes they gave us a day before we needed to be back. We would return to Houston and work the mission from our office.

Most of the time we were like everyone else. We watched and followed the missions as much as we could. During Apollo 10 I also worked for Mike Collins, and he was getting ready to fly Apollo 11.[83] I also worked for Pete Conrad, and he was commander on Apollo 12.[84] So a lot was happening in the office as 10 was flying; 11 was getting ready to go, and 12 was shortly behind. We were turning around those missions pretty quickly. We were going from a May 18 launch for Apollo 10, and shooting for a July 16 launch for Apollo 11, so it was quick.

I returned to the Cape for Apollo 11. It should have been Toni Zahn, because she was working for Flight A at the time. Neil Armstrong and Buzz Aldrin were both in Flight A.[85] As I said, I was working for Mike Collins at the time. There was an issue with Colonel Frank Borman, who had taken over as the chief of the flight when Scott Carpenter left.[86] There was something that he was assigned to do, or needed to do, that Toni had to support, and she was unable to go. The people at the Cape, it's my understanding, requested that I return to work that mission. I was back in Houston only a couple of weeks and then I turned around and went back to the Cape to work with the 11 crew for the mission. At the time I was thinking, "Oh, boy, here we go again," but also, "Oh, my gosh, am I lucky to be here on this mission!"

Apollo 10 was wonderful and had very special memories because it was the first one I had gone down to the Cape to work. But from the historical standpoint, to be involved in the Apollo 11 mission and to be the crew secretary was quite an opportunity. I had an opportunity to meet the guests and dignitaries that I had arranged to be there. They would come into the crew quarters, and we would give them private tours. There were also private parties that were held the night before launch that became quite famous and popular. The invitations were coveted. The guys had some very popular friends that were well known. We always took very special and good care of them.

Compared to the 10 crew, the Apollo 11 crew was so quiet. Armstrong was very soft-spoken, a very quiet man. Buzz also was at the time. Mike Collins was just the sweetest man; he's just adorable. But they were a very quiet crew. They had a tremendous amount of responsibility. There was so much anticipation with them being potentially the "first." The atmosphere was considerably different than it was with 10 down there. I don't know how many times the White House called. It seemed like every time we turned around the White House was calling to check the phone lines, because President Richard Nixon was going to call the guys, and we needed to make sure that all the lines were okay. I think there were a million people out there on the causeway to view that launch.

I stayed in Florida after the launch. I was very, very tired. We worked very long, long hours leading up to Apollo 11. I had my car down there, and I was going to have to drive it back to Houston. I didn't want to be on the road if something didn't go right. I wanted to make sure I was where I could get home quickly if I needed to, so I stayed and was in Florida when they actually landed, and when Neil took his first step on the Moon. Then I returned home shortly thereafter—I think the day after. I had to get Apollo 12 ready.[87]

There was quite a difference between commanders on Apollo 11 and 12, Neil Armstrong and Pete Conrad. Pete was Pete: he was one of a kind. He was delightful. They were all just wonderful. They were all overachievers. All had their own unique individual personalities. Most of them had a great sense of humor and appreciated that a good laugh and a lot of levity helped to ease the stress in the situation. We had some fun times. You had to have a sense of humor because they would play jokes on us just for the sport of it, so you had to learn to give as good as you got.

Apollo 16 was the last time I went to the Cape.[88] I was working for Gene Cernan at the time Apollo 17 flew. I did not go because I'd just been for 16. After Apollo 12 flew, Pete Conrad decided that he was interested in flying Skylab. He was named commander of the first Skylab crew, with Joe Kerwin, and Paul Weitz, "P. J."[89] I quickly got into the Skylab mode also, which was really good for me because I got to learn all about that program.

Skylab was going to be the follow-up program to Apollo.[90] I worked for Conrad, Kerwin, and Weitz, so I got to go to the Cape with that crew. After Apollo 13, and having to switch out T. K. Mattingly with Jack Swigert because of his exposure to measles, they put the crews in quarantine not only afterward, but also before.[91] When I was down there on Apollo 16, we had to stop before we could come up to crew quarters. It was very limited access. You were on the personal contact list; you had to stop in the lobby and get your temperature taken every morning before you were allowed to go up into the crew quarters. If there were any problems whatsoever, then you weren't allowed in direct contact with the crew.

Most of the Skylab training was done in Houston, and so they quarantined the Skylab crew in Houston. For about a month before Skylab flew and we all went to the Cape, we were quarantined over in Building 5 where the simulators were. I got to go home at night, but it was very limited access while we were actually locked up in that building for about three weeks. Those guys were just crazy. They delighted in making sure I brought them word from the outside world because they couldn't get out and go anywhere. That was a different aspect of Skylab: we were quarantined tightly before we ever left to go to the Cape.

Another thing, too: when we were finally at the Cape, it was brief. We went down just a few days before launch. We watched the launch of the Skylab itself and the crew was supposed to launch the next day. Then they had the problems with the launch.[92] That was a unique launch because we were standing on the roof of the Operations and Checkout building with Conrad, Kerwin, and Weitz, the crewmembers, watching the Skylab workshop launch. I had never stood with the crew watching something launch. That was a unique experience.

I mentioned driving to the Cape. Today it's nothing to think of a woman driving that far. Oh, but back then! Jim Rathmann, the Chevrolet dealership in Melbourne, Florida, leased cars to the guys. Most of the guys would get Corvettes or Cadillacs. The crews of Apollo 12 and 16 had their Corvettes all painted alike. One of our little perks—jobs—we would switch the Corvettes out. One of the girls from the Cape would drive the new ones, and we would drive the old ones, a year old, from Houston and meet halfway. We'd spend the night and have a good time in Biloxi, Mississippi, or Mobile, Alabama, or somewhere else, and switch cars. Then we'd drive the new 'Vettes back to Houston and they'd take the old ones back to Florida. Women making that type of road trip was unheard-of at the time, and yet we had the opportunity to drive it in Corvettes. The guys trusted us with them, which was probably a huge mistake. Driving to the Cape, back and forth—women did not do that. I guess I had a lot of confidence.

I was young, working in a very male-dominated office with high achievers, military backgrounds, and a lot of egos. Most of those guys were test pilots. We had a couple of them that were called in from duty in Vietnam.[93] They found out when they were in Vietnam that they had been selected.

They weren't used to desk jobs. These guys were pilots. They were flyers. We had to help them understand what it meant to have an office. It was an adjustment for a lot of them to be in offices and have a routine and go to the meetings that were required.

I mentioned the office was like a family. There were not really any rules about socializing with the astronauts on a personal basis, because Shepard, his philosophy was—and he promoted it, too—that we were a family. I don't know if anything was said to the astronauts about socializing. There was nothing official or otherwise that I was ever aware of.

Astronaut Office people went out together. It wasn't that we were try-ing to keep separate, it was a convenience factor, and we were working the same hours. We had opportunities to mingle. Back then happy hour was really popular—especially around here at the Nassau Bay Hotel and the Kings Inn. There were also some darn good splashdown parties. The launch parties at the Cape, that's where everybody wanted to be, and then the splashdown parties here in Houston. We partied hearty, but families and the wives also participated and partied. We had special times where their kids had an opportunity to go to the circus. I remember riding in a bus with a busload of the astronauts' children to the circus at the old Sam Houston Coliseum.[94] Oh, it was the rowdiest bunch! I don't think I volunteered for that duty again.

We did have a lot of interaction. It made us all feel a little bit more con-nected. Gay Alford was very good about keeping the secretaries working closely together because we needed to cover each other. We called her Mother Superior—and still do. If the astronauts were coming up the back hallway from the simulator, and you were closer than going to ask their own secretary to do something, the guys didn't care—they'd pop in your office and ask you to do something, and you did it. That's why we all had to be pretty close and work as a unit, as a tight unit, because you couldn't say, "No, I don't work for you. Get around to the other side."

There were territorial attitudes among the women, but you got over it. The secretarial positions were highly coveted jobs, and we were reminded that they were highly coveted jobs. There wasn't competition per se, but I think there was, because I know that, personally, I wanted to be known as the best of the group. I think each one of us did.

I decided to leave NASA in 1973 because I had been in the same pay grade for a number of years, and I had gotten all my promotions on time.

But there was no potential for a higher pay grade in the Astronaut Office unless Gay Alford left. I didn't think that, short of hauling Gay Alford out by her heels, as long as Alan Shepard was there that was going to happen. I had looked at several other positions within NASA. Affirmative action was just starting, but it was not in full force. I had applied for and been offered several other positions at JSC, but, ultimately, I didn't feel I would be happy working at NASA in any other place beside the Astronaut Office. That may not have been right, but that's the way I felt at the time. I had an opportunity to work with M. David Lowe Personnel Services in Houston. Then when Pete Conrad left just months later, I went to Denver with him when he joined American Television and Communications Corporation as chief operating officer, and spent a year and a half up there with him.

The most challenging aspect of my job during those seven years at NASA—probably that we were required to wear so many hats. That was really the most challenging part of it. The work was so diversified, and it varied so much depending on which astronaut you were working for, too. A lot of them had outside interests that you tried to help them with. You were public relations, you were mother hen, you were a screener; it was a difficult job. I think that was probably the most challenging part of it in addition to having to be able to recognize the priorities when you're working with ten astronauts with varying responsibilities on different missions and programs. Each one of them thought his work was the most important and his mission was the most important. Having to prioritize, that was really tough, really tough.

In retrospect, if I could have picked any seven-year period to work at NASA and in the Astronaut Office, that would have been the exact seven years, the end of the Gemini Program through the first Skylab mission. How lucky can a girl right out of high school get? That seven years was, to date, NASA's greatest adventure: the end of Gemini, getting ready to do Apollo, and the Apollo Program. Until we return to the Moon and go on to Mars, I think Apollo is going to be the crown jewel in NASA's history. It certainly captured the public at the time. It was a very, very positive thing for the country when there was so much that was going on that was not. I think the Apollo Program helped the country as a whole.

I think everyone associated with NASA at the time felt the same way, that we were all very lucky. As a young woman, I was extremely lucky; I did feel like it was a golden opportunity for me. It was probably years later that I fully recognized how lucky I had been, because it was my first job and I had nothing to compare it to.

Mathematicians and Engineers

Mathematics is one of the building blocks of a STEM career and it is one of the fields where women were better represented during the early days of human spaceflight. Women with engineering talent often came onboard as "computers," whose work had them reducing data from flight tests and then plotting the results. All of the women in this chapter majored in mathematics and became engineers working on the historic spaceflights of Mercury, Gemini, Apollo, and the Space Shuttle.

Over the years Dorothy B. Lee made significant contributions to the advancement of human spaceflight, beginning in the late 1940s when she went to work for the Pilotless Aircraft Research Division (PARD) at the NACA Langley Memorial Aeronautical Laboratory in Virginia as a human computer. After World War II, PARD conducted flight tests on rockets at their research station on Wallops Island, Virginia, to learn how high Mach

Cathy Osgood reviews rendezvous plans for an Apollo mission.

numbers and temperatures impacted rockets. Lee went on to become an engineer and an authority on aerodynamics and aerothermodynamics.[1] In 1969 she was one of a handful of experts designing the country's new spacecraft, the Space Shuttle Orbiter. She was responsible for the design of the nose, and to this day it is known affectionately as "Dottie's Nose." Her design, unlike previous iterations, allowed hot gases to travel smoothly across the vehicle as it reentered Earth's atmosphere, thereby resulting in lower temperatures over the entire spacecraft surface. Once the spacecraft design had been finalized, Lee became the subsystem manager for Space Shuttle Aerothermodynamics, coordinating the nation's most extensive wind tunnel program to validate the design requirements for the spacecraft's protective tile.

Ivy Hooks was one of two female engineers to serve on the original Space Shuttle design team in 1969. As the reusable spacecraft was being designed and tested, she became the Shuttle Separation Systems integration manager and then headed the Aerodynamic Systems Analysis Section.[2] Hooks began working at the center in 1963 after earning a degree in mathematics at the University of Houston. An expert in aerodynamics and flight dynamics, she received the coveted Arthur S. Flemming Award for her engineering contributions to the space program.[3] After working more than twenty years for the agency, Hooks left NASA for private industry, becoming an expert in requirements development and management.

Dorothy B. "Dottie" Lee [4]

As a child, I was interested in astrophysics; I read George Gamow and other astrophysicists.[5] That word, "astrophysics," didn't exist when I was a child. When I was ten years old, I knew that we were going to go to the Moon. That was in 1937. No one outside of *Buck Rogers* had stories of those tales in those days. My childhood dreams became a reality by virtue of being at the right place at the right time, which is the story of my life. My entire life has been lucky.

I was recruited while in college at Randolph-Macon Woman's College to go to work for NACA. I went to work at the Langley Memorial Aeronautical Lab, and I was put in the best division there. It was called PARD. We had exciting opportunities to launch vehicles, to test different configurations, and determine how those configurations would ultimately design spacecraft by virtue of the shape of the nose. We went from cones to blunt bodies.[6]

We launched spacecraft from Wallops Island, which is off the coast of Virginia in the Atlantic Ocean. I had to do everything in the design, the trajectory, and the stability analysis of the spacecraft, plan the instrumentation, and, when the data returned, analyze and document the data.[7] As our Mach numbers increased, our knowledge increased, and that was how we developed how we could go to the Moon and Mars. It was just by virtue of that experience at PARD.

Back in 1948, when I graduated from college, there were not many gals at NACA. We were hired as "computers." Computers didn't exist.[8] We had calculators. Later, we were classified as mathematicians. When NACA recruited us and we got to Langley Field, I had never heard the words "Mach number" or "Reynolds number" or any of the other terms.[9] My job was to do mathematical support for the engineers.

One day my project was to solve a triple integral for an engineer; it didn't require using my calculator.[10] I could just do it at my desk. At that time, Max Faget's secretary was going to get married, so I was asked to be his secretary for two weeks while she was on her honeymoon.[11] (I don't know how to type, still.) I answered the phone, distributed the mail, and worked my triple integral. I did this for two weeks.

At the end of my two weeks Max asked, "Dottie, how would you like to work for me all the time?"

I thought he was being funny, because I don't type and I knew that this was the last day. His secretary was returning Monday. I said, "Sure," in a very flip way.

He got up, went downstairs to talk to the division chief, returned, and said, "Dottie, you start working for me Monday." They found me a desk, and I was put with some engineers who taught me how to be an engineer. I learned on the job.

Brilliant men have touched my life throughout my career, and I couldn't help but win and do well because these people guided me and continued to guide me even when I got to Texas. When I went to work for Max Faget it was just a marvelous experience. That also led me to becoming a project engineer, which is rare today—male or female—anywhere. Projects are so large that you can't have just one project manager.

During that time I was the only female engineer in that group. We did get another gal from our section out of the computer group. I can't remember when her classification became engineer. I don't remember seeing her until we moved to Houston. There was one, to my memory, one female college graduate engineer in another division, whereas I gained my classification on the job.

From the time I worked for Max until I became a project engineer, the projects became more complex. I started out with what we called a single-stage vehicle. Before I left Langley and came to Texas, I had a five-stage vehicle in which the payload, if you will, or the nose, was going to be the shape of the blunt body of the Mercury. I progressed in the complexity of the firings or the different flights that we had.

I remember one flight in particular. It was a five-stage vehicle. We were waiting for another rocket to be launched before we could launch. Well, we waited all day long! The wind picked up, and we wanted to make sure that our vehicle's trajectory would not be jeopardized by the wind. We wanted to orient it properly.

We'd sent up balloons to see which way the wind was blowing and to get atmospheric data. Finally, by dark we were able to launch. A night launch was exciting because you could see it as each stage separated. Of course, after a while they're too far from you, but you could see the fire from their rockets. We never saw the fifth stage ignite. We would have to wait to get the data telemetered back to us to see what we could do. It was midnight. We were on this ferry, and we were going through these charts trying to determine why that fifth stage did not ignite. This is what we would call a failure. It turned out that it did give us data, but the rush and the thrill of trying to solve a problem was the fun part of our job. I don't think it will go down in history as anything important, but I learned with each experience, of course, and that's what I did every day of my life.

The first rocket I launched was a Thiokol rocket called the *Cajun*. It was a single-stage, and it had never been flight tested. They usually do what they call static firings. They're tied down and they just fire them to see how long they'll burn. I was given the project. It meant designing the fins so it would be stable and the nose so it would survive. We thought it would go to some Mach number, maybe 3.

I was the last passenger ever flown on the *Goose*, which was a Langley amphibious plane that flew to Wallops Island. I had the rocket's shroud with the fins on the floor beside me and the nose cap in my lap. We got to Wallops Island, and then I stayed as they screwed on the fins and the nose and put it up on the launcher, but the weather came in. They said, "Dottie, we've got to fly back. You've got to leave."

I didn't get to see my flight, but we had photographers. They were very talented. They could follow a lift-off, and we had multiple photographers, fortunately, because this rocket took off faster than they expected. It went out of sight with one photographer. We actually got a Mach number higher than Thiokol had advertised.

I wrote up the data. Because they wanted to know the results as fast as possible, instead of having a review committee (five or six people from around the center)—which is standard procedure when a paper is being put out—they review it for its technical acceptance as well as its English and everything else. I had one reviewer. Guy Thibodaux was my reviewer, and it was so appropriate because he's from Louisiana.[12] I think that was the first time anybody had ever written a report that had a single reviewer so we could get it out. That was a fun thing.

Things have progressed over the years from my time at Wallops. One year at Langley, true, they were simple vehicles, one-stage, maybe two-stage vehicles, but I actually launched three vehicles, analyzed the data, and wrote them up. That's moving out. That's really going fast, doing it with a Friden calculator. A Friden is a very noisy contraption the size of a small television used to calculate the trajectory. Oh, goodness! You can't begin to appreciate what it takes to calculate the trajectory of a vehicle, and you do it for every second of time in flight. Today you just punch a number in, and there's the answer. Back then you punched lots of numbers. To do that, and to do three of them in one year, was quite an achievement.

I mentioned Wallops Island. I worked there when women couldn't wear pants to work, and as I said there were not many of us. In the telemeter room, where we watched the launch or received the data, there was one bathroom for all these men and me. The time would come when I had to go to the bathroom. It's right there. One day I thought, can I walk to the barracks? I was too embarrassed—I was only twenty-something—to open that door and go to the bathroom with all those men there. So there were little inconveniences.

In the early days when I worked for NACA, we didn't work past eight hours a day. No one did. You might stay a few minutes late, but no one worked until eight and nine o'clock at night, as they do now, or on the weekends. When it was my project and I had the responsibility for the entire vehicle, I couldn't wait until Monday morning to get back to work. It was that exciting. I was married and ultimately had two daughters. I was fortunate enough to have full-time help because that was how I could work outside the house. I had someone at my house at all times. I just loved my work.

Sputnik took Americans by surprise in 1957. When it launched, my husband, our children, and I were driving to Florida to his twin brother's graduation; he was getting his master's. We were excited because we knew it would help NACA. Quite frankly I was thrilled to death. NACA would get funding to work on those ideas that Max and Bob Gilruth had.[13] NACA was an advisory committee for industry before NASA was formed. We would advise and help companies in airplane design: the P-51s, for instance, and, I think, the P-47. I immediately saw a change. In fact, I remember the director of Langley, Tommy Thompson, getting up and saying, "We have a tiger by the tail, and we've got to swing it so it won't bite us."[14] We had to really get in there and work. We were challenged. Men who had ideas were able to bring them to fruition because of Sputnik.

When Max designed Mercury in 1958, he made paper models of the capsule as you now see it, and he would put little tabs out—we called them trim tabs—to make the models stable. I remember him getting up on his desk, and he would drop them to see if they were stable. We were doing a lot of the work in-house. I think, before we had MacDac on board, Max Faget and Bob Piland stood behind me as I calculated the trajectory for the Mercury capsule and how many Gs the pilot would experience.[15] They stood there because a man sat in a centrifuge in Pennsylvania, waiting to find out. They were going to test him to that capability. That was exciting. You don't make a mistake because there's a human involved. Mercury was fun.

I became part of NASA in 1958. Several years later we moved to Houston. I wasn't looking forward to moving to Houston, where it's hot, even though it was going to be a new adventure, which I enjoy. I had mixed emotions about it. They flew a group of us down to find a place to live. My husband and I were in the motel and I thought, "I wonder if there's a way I can talk John into not moving."[16] He had already joined the Space Task Group.[17] I was queen bee at Langley. I had my projects and was happy as a lark. Then there was a knock on the door, and it's Max. He said, "I've found where we're going to live. Come on."

The three of us got in the car and drove to a community south of the center called Dickinson. He had found two lots on the bayou. He said, "Now, Johnny, I have to go back to Virginia tomorrow. I want you to negotiate for my lot, and if you can't get it, then you negotiate this lot for me. Otherwise, that'll be yours."

Johnny and I went back to Dickinson the next day. He successfully got the lot for Max. Our lot had a sign, so we pulled up the sign and went to the agent, who also had a savings and loan. This man's office was glassed in, so he saw us coming. He had someone in there he dismissed very quickly, because, as he said, "I've had a lot of people ask me about property, but I've never had them bring me the sign before." The three of us went to lunch, and we became very good friends and bought that lot.

From then on, it was fun. I designed our house. I returned to Virginia, and every day at noon I would do the floor plan on a piece of graph paper. Ultimately that graph paper was given to our architect, who built the house from that after they put it on blueprints.

We were still flying Mercury when I came down here. I'd been given the project to design a calorimeter to put in the last Mercury capsule that

was going to be launched, to make measurements in this ablative body.[18]
Mercury had an ablative heat shield.[19] We wanted some measurements
because you could not get good measurements in ablative material. This
calorimeter was to recede, if you will, with the ablator. We were going to
the Moon, but we knew that the only material that could let us go to the
Moon was an ablator, which we had tested at Langley. This material was
a honeycomb structure, which little old ladies literally gunned.[20]

We then attempted to do it on Apollo, and it was a disaster. The Apollo
design had what we call "graphite wafers." In fact, we did put sensors in a
flight vehicle to get measurements at this high velocity. Coming back from
the Moon, the spacecraft entered at thirty-six thousand feet a second. We'd
get temperatures, but the moment that it would be getting interesting for
our analysis, the wafer would be gone. I'm sorry to say we didn't get any
meaningful data from one Apollo capsule. All we could do was gauge how
the material charred, and with our theoretical predictions try to rationalize
our predictions with the char depth and slope of the curve.

Instrumentation has not been able to achieve the degree of technological advancement that the spacecraft have. We need sensors that are
capable of handling higher temperatures than they are presently designed
for. Sensors haven't kept up with us, and it's been a challenge, one that
we haven't quite achieved. I'm thinking of the lee side of the vehicle
away from the blunt face on the Shuttle—the pressure measurements
were just so low that we couldn't get them to the level of accuracy that
we wanted. Hopefully someday it'll all come together and we'll have the
instruments. The materials are advancing rapidly that can withstand high
heating. They'll get the instrumentation. I won't; I hope I'm not designing any sensors anytime soon! Each project has been a natural transition
depending on what's thought up or whatever the goals are of the nation.

It was a natural transition to move to Apollo. One of the first people on
a design is an aerothermodynamicist, and you have to work hand-in-hand
with the aerodynamicist. It has to be an iterative process because you might
want it one way, but you'd interfere with the accomplishment of another
discipline.

I knew we'd land on the Moon in spite of the tragic fire. The Apollo fire
was an awful experience, but one that apparently was necessary.[21] Johnny
and I and another couple were driving to New Orleans to go to a Mardi
Gras ball. When we got to the motel, the radio, for some reason, was on
in our room, and they were talking about the fire. We went to dinner and

then Johnny had to catch a plane to come back to help the investigation, so I stayed with this couple, because we were in our car. We went to the ball. It's a shame that some lessons have to be learned through tragedy, but I had no doubt America would pick up its program and begin again. Everyone was motivated to succeed, to do it right.

Over the years I coauthored numerous papers with my colleagues. Charles B. Rumsey took me under his wing.[22] He showed me how to do the very basics, and I credit him for that.

Max and I coauthored one paper together. I remember that paper. The vehicle had gone to angle of attack and had rolled, which we didn't want, and when you have instrumentation, which consisted of thermocouples — thermocouples will measure the temperature of the outer skin and from that temperature we translate that into heating rates, and that's how you design spacecraft, in very simple terms.[23] With the instrumentation on a certain line and the vehicle rolled at angle of attack, it challenges the interpretation of those data. That was a very difficult paper, I remember. It took quite a while to analyze the data. I found somebody else to type it.

Interestingly enough, I made the figures. That meant taking these data points and fairing a curve and then putting them on charts. Sometimes the charts already had the values of the parameters that you were plotting. I typed the numbers, spaced an inch apart, and sometimes it worked. I did everything. In fact, that's the beauty of how we all started out. We had hands-on engineering. Every engineer faired data, and they could feel it and understand it and understand the meaning. Today, projects are large, you have contractors who do all the work, and the engineers are just monitoring the contract.[24] Now it sounds derogatory, the way I said that, but it doesn't give them the opportunity to really get the feel of it.

I'm really skipping ahead — just before I retired, I was the local manager for a project called AFE, Aeroassist Flight Experiment. We initiated it with the idea that it would give the engineers a hands-on engineering opportunity. We had twenty-six people here at JSC design and manufacture the brake part, the blunt entry face. I'm proud to say that. It was a four-center project. The propulsion system, which was behind this brake, was to be done by Marshall Space Flight Center in Huntsville, Alabama, and then we had primary investigators from Ames Research Center and Langley.[25] Unfortunately, the paperwork associated with this project, which another center imposed on us, became so costly that it cost more than the hardware, and the project was canceled.

This project, to me, was going to help justify the Space Station, which meant you could have a vehicle at the Space Station. Now, this was my idea. This wasn't part of the Space Station or the AFE's definition. If a satellite up in geosynchronous orbit went out, we could launch the vehicle from the Space Station and retrieve that satellite.

Now let me turn to the Space Shuttle. I remember my division chief called me to his office on a Friday morning in 1969, and said, "Monday you're to report to Building 36. Don't tell anybody where you're going or what you're going to do. Read this document," which had other thoughts and concepts about a vehicle that could go up to orbit and return to land.

I had to call my Apollo contractors and say, "I'm going to be out of pocket for a little while. I cannot say where I am." It turned out that when I would need their support, I couldn't identify the project. It was just "job number ten," and I would get in a taxi and go elsewhere, maybe to my old building, so they wouldn't know where I was.

Bob Gilruth had asked Max to design a vehicle, and Max did this in his garage, unknown to anyone. At one meeting Max had a garment bag, and he took a model out of it. That first model of the Shuttle was a straight wing. He stood up on the desk. He declared it stable in two attitudes, zero degrees angle of attack, and then he flew it across the room. For sixty degrees angle of attack, he angled it by eye and threw it again. I thought, what goes around comes around. I remembered him standing with that Mercury capsule, and here we were again on the Shuttle.

In this room in Building 36 we had a representative from each of the disciplines—a trajectory man, a heating gal, a structures man, a thermal protection system man, different people.[26] Max, meanwhile, had three men on a drawing board doing changes that we would come up with. A design would pass around the room in a matter of hours. Sometimes the design would change three times in a day because I might bump someone's design and then he, in turn, might do the same as we went around. That was exciting. That's how you design a vehicle—a small group of people pass the design around the room. Of course, we had to go from the straight wing to the swept wing that we have.[27]

Jim Chamberlin was in charge. He and a group of Avro men from Canada had come to us at Langley.[28] He was an amazing man.

We were locked up, literally. There was a guard outside the door. This building had no windows. It's a high bay next to an office building. We landed on the Moon for the first time when we were locked up in that room. I can remember saying before I knew I was going to be in the Shuttle Project that I was going to really celebrate our landing on the Moon. We were going to celebrate after the return of the crew. That day I worked past hours, then got in my car to drive home. I was living in Dickinson. I didn't know where John was because he was driving his car. As I drove down NASA Road 1, cars were parked everywhere in different bars and places. I wasn't celebrating, I was just going to go home. I had to relieve the babysitter. I didn't get to celebrate. Interestingly enough, I didn't think about the time when I was ten and sitting out in New Orleans looking at the stars, knowing we were going to the Moon. That didn't enter my head then. I think I was tired; I felt sorry for myself that everybody else was partying and I wasn't.

I talked about design, so let me talk about testing. Testing the Shuttle, when compared with our previous spacecraft, was different. When we finished Apollo, which was a marvelous experience, we had several facets to the heating world, such as radiative heating. When you have a strong shock in front of the vehicle, coming back at thirty-six thousand feet a second, you didn't know all of the contributors to the heating. You have convective and radiative heating.[29] Behind that shock you have a high spike of radiative heating, and then it will reach equilibrium and get less intense as the flow approaches the spacecraft.

That was quite a challenge, trying to determine how high that heating would be for Shuttle. When we started, many of us were asked to determine how complex is the Shuttle going to be relative to Apollo. I put a factor of ten times more complex, because you have the Orbiter, which is the airplane-like structure, the external tank (ET) that houses the liquid hydrogen and oxygen, the two solid rocket boosters (SRB), plus we had nozzles (engines) at the back end of the Orbiter. Those are the four elements of the Shuttle system. The air flow around and between each of those elements is complex. As a designer of a spacecraft, you want your flow to remain laminar as long as possible, because once you get turbulent flow, that increases the heating significantly.[30] To describe that environment analytically or theoretically is very complex, so the Shuttle was much more significant in its challenge than Apollo.

With a manned vehicle, we had to make sure that the sensors would not interfere with the safety of the vehicle. Putting pressure sensors and

Dottie Lee points to
an illustration of the
Orbiter during reentry.

thermocouples in tiles was relatively easy. (We had tiles on the Orbiter.) Trying to instrument the solid rocket boosters and the external tank was a challenge because you had to avoid penetrating the skin so that you would not jeopardize the safety of the crew and vehicle.

The fact that the crews were in the vehicle did interfere once with a couple of the orbital test flights.[31] The aerodynamicists wanted to get various effects of angle of attack on the aerodynamics of the vehicle and came up with the idea of what we call pushover-pullup.[32] I didn't like that because that was going to disturb this nice flow that we had so carefully maintained with smooth tiles so it would not prematurely trip the boundary layer from laminar to turbulent.[33]

Joe Engle, who was one of the astronauts, had me go over to the trainer and get into the simulator with him, and, believe it or not, he said, "Now, Dottie, we're going to go five seconds down," and he did, counting "one thousand one, one thousand two, one thousand three," and then four, five, and then up.[34]

I said, "Well, that's all very good, but you could still trip the boundary layer." He wanted to show that he wasn't going to do anything to my vehicle. Well, I was outnumbered. It must have been STS-4 when they did the pushover-pullup, and it tripped the boundary layer.[35] Fortunately the flow re-laminarized, i.e., went back to laminar, but we had that trouble for those few seconds. It wasn't enough to cause so much heating that we exceeded the capability of the thermal protection system, but we did upset the flow and I knew we were going to do that.

I made a contribution to the Shuttle's design: the nose. I have two stories about that. In the heating world we want everything to be smooth. I made a trip to Los Angeles once a month to meet with the people that worked to support me and to see their progress.[36] As the design was developing, I went to the drawing board to see how they were progressing, and the designer that day had put a sharp corner on the Orbiter. Well, we couldn't take that. The aerodynamicists might want it for better lift, but aeroheating people didn't want that. I got with my counterpart and I said, "They're going to have to change this design." Fortunately, the man in charge of the design recognized that if I said so, his response was, "Yes, ma'am." We called a meeting, and in forty-five minutes that design was changed so that it would be a smooth configuration.

Soon after that, I was at JSC and Hans Mark, who was the director of Ames at the time, and Glen Goodwin were in Houston, and they had with them Dr. Edward Teller.[37] Why he was there, I don't know. I'm trying to describe this gradual radius of curvature change as you look at the side of the nose so that it maintains its smoothness, and Dr. Teller said, "You mean like a French curve."[38]

I said, "Yes! Like a French curve! I could kiss you for saying that." I didn't. Later, as I started out, Dr. Teller rose from his chair and bowed. I thought, how charming. Charm still exists in this world. That's how it became "Dottie's Nose," because I stopped the design on the drawing board. It's smooth and maintained its laminar flow. It's worked out well.

The second story took place at the contractor's again. The heating people tell the thermal protection people how hot something's going to get, plus the pressure loads, and they design it. They had carbon-carbon from the nose to aft of the nose landing gear door. That's very expensive and very heavy to make this carbon-carbon leading edge. It's laid up in layers, literally. I met a little old man in Dallas at Vought Corporation who was working in one section of the wing.[39] He was so proud of what he was doing. He laid up a layer, and he had to get inside to do it. He explained exactly what he did.

Well, I knew, based on wind tunnel data, that the heating drops off rapidly from the nose as you progress aft, and negotiated with my counterparts to move that carbon-carbon forward. That took all day, because you don't just tell them to change a design. They're the designers. They just have to meet our specification. You have to be diplomatic and negotiate.

"Here's what the theory says." They agreed, at the end of the day, to take that carbon-carbon off and move it forward as it stands today.

Of course, it turned out to be a headache. The tiles immediately behind that carbon-carbon nose, what with trying to apply them on this curvature and not having any steps, required a lot of retiling. They would have to take it off and put it on, take it off, so I bet they hated me for doing that. That saved us, in those days, 7.3 million dollars. That was significant savings in those days.

Before the Shuttle first launched it seemed like I lived in Florida for a while. I was there for nine days before we ever lifted off. Our task was to see that those tiles were put on to the nth degree. It's a terrible burden we put on the people, but we did not want to have a rough surface because that can cause tripping of the boundary layer. I memorized practically all thirty-five thousand tiles and crawled up on scaffolding and looked at things once they were up in the Vehicle Assembly Building, which was fun. Some of the scaffolding would move. Fortunately, I'm not afraid of heights!

There was never a day I wasn't learning. Everyone was learning as we progressed. I remembered talking with Max Faget in his office, and he asked, "Dottie, how do you know such and such?" It had to do with boundary layer transition.

I said, "Well, Max, it's a judgment call."

He said, "Yes, that's a good word."

Judgment has to come from experience. There are no theories that can predict boundary layer transition. It's been the nemesis of all aerothermodynamicists. Today, not everyone has the opportunity to look at data because there are other people doing it for them. I don't mean to talk against contractors. I'm now a contractor. The engineer at NASA, regardless of the center, needs support, because projects are so large and you have to have computer and other support.

When I became a project subsystem manager for aerothermodynamics on the Shuttle, there were fifty-six men at the contractor who had to respond to me because I was responsible for the heating to the vehicle both during ascent and during entry. Quite a few of the men and I have remained friends even to this day. We still correspond, or they'll come in town and give me a call. I, in my endeavor to do my job, would have each man come to me and inform me how he was progressing, because we worked as a team. I would just ask questions, sometimes knowing the answers. I think it's because of Bob Gilruth's influence.

Bob Gilruth had been my division chief at one time at NACA. I worked in an office with perhaps seven men, and I was in my little corner. Gilruth stopped at the door, where I could see him, as the men were discussing something, trying to solve a problem, and he listened. Then he asked a question that turned their thinking around and headed them down the right path. He turned around with a smile on his face and walked out. I thought, there goes a big man. He didn't tell them how to do it; he just asked a question. Well, that became my modus operandi when I became a manager over men. I knew never to tell anyone how to do his job. I would ask questions instead.

Just before we launched the Shuttle for the first time, a group of us went to Ames to present to the Aeronautical Space Advisory Panel, composed of directors of NASA centers or leaders of industry, a very august group of men. Each of us was to present on the status of our discipline in regard to preparation for the launch of the Shuttle. I was last on the program and I was the only woman. I had forty-five minutes. I started putting charts together for this group of people who did not necessarily know all the nuances of heat transfer, and I ended up with forty-five charts. Well, you don't give a chart a minute, but I had to tell this story. I looked at my watch, and then I started talking and turning my viewgraphs. They all listened attentively. When I finished that last chart, I looked at my watch and I had done it in forty-five minutes.

As I put the charts back together, people were milling around. After a while one of the men from the panel came over and said, "Based on what you have told us today, we're going to recommend to the president that we launch the Shuttle." I still get chills thinking of that. I was quite proud of that.

I mentioned work hours at Langley. The hours changed when we moved to Houston. We worked more than eight hours or an occasional weekend at Langley, but it wasn't the routine. When President John F. Kennedy said, "before this decade is out," we worked because we wanted to. John and I drove in separate cars, fortunately, so he could stay late or I could stay late, and, as I said, I had full-time help until our girls were through high school. We worked many hours. When you like your work, it's not work, you just do it. I never saw anyone gripe or complain. You just knew it had to be done.

For instance, I remember the first flight of the Shuttle when they opened up the payload bay doors, and we saw the tiles were off of the orbital maneuvering system pod.[40] I actually cried. I'm sitting in a room with eight or ten men, and there's silence. I was sick to my stomach. Immediately the question was, can they come back? I got on the phone with my counterparts at the contractor's office. I had to determine the heating as the contractors were in California. Their bosses were at JSC with us.

At about eleven o'clock at night that first day, after only drinking water all day and working steadily, we (the contractor bosses and me) got on the telecon and declared that the Orbiter could reenter safely. "You will not burn up the spacecraft." We had been told, "Be prepared to return to the office at four thirty tomorrow morning," because at five o'clock we were going to tell NASA Headquarters in Washington, DC.[41] Well, fortunately I lived right across the street from the center by then.

I remember getting in the car at midnight and then walking into my house. I had not eaten any food all day. I opened the refrigerator door and, like a child, stood there. "What can I eat?" There was a sausage patty. I fixed a Tanqueray on the rocks, ate that sausage, drank my drink, went to my bedroom, took a shower, and got in bed. Three hours later I got up, got dressed, and was back at JSC at the appointed time to say, "Yes, they can come back." I was tired, but I had to do that. Nobody complained, nobody griped. It was something you had to do.

I was responsible for determining Orbiter entry heating because I was the subsystem manager for Aerothermodynamics.[42] I worked for the Structures and Mechanics Division in the Aerothermodynamics Section. The head of that section, Bob Ried, felt that I could be the person to work with the contractor, to see that they met our requirements.[43] The contractor was responsible for the design. In my particular area I had to see that they did it correctly. Milton Silveira, the deputy to Aaron Cohen, who was in charge of the Orbiter Project, had been assigned over the subsystem managers.[44] Milt approved my appointment; he felt that I was capable of being the contract monitor for aeroheating. Again, I was there at the right place at the right time. The men responded to me well. Previous to that, I'd had a small group of contractors support me on the Apollo Project, so that gave me managerial experience.

We had panel meetings frequently for the Space Shuttle in which we would call upon experts from other centers to help solve a problem. I'd have a panel meeting every two or three months. I was in Virginia having a panel meeting when *Challenger* happened.[45] On my panel I had two men from Marshall who were experts in plume heating from the exhaust of the SRBs and the main engines. One of the men went out to get coffee, and when he came back he said, "Dottie, the Shuttle has exploded." We fortunately had a TV in our meeting room for in-house review and turned it on. We sat there in horror as we watched. We played it over and over again. Meanwhile, my two men from Marshall knew they had to get back. We didn't know what had caused this explosion, where, why, or what. Of course, they left immediately.

It was quite a traumatic experience. When I returned to JSC, they called all of the people together, managers of each discipline, because we were going to have to investigate each of our inputs to this flight. It meant traveling to different contractors all over the country. We'd be changing planes, and we worked constantly, even while waiting for the next plane. I'd forget what city I was in, we traveled so much.

It turned out that our world was not the culprit. The segments of the SRB have seals between them and the gas that escaped from that seal where the plume impinged on the attachment between the SRB and the ET, melting it, caused the nose to go into the liquid oxygen tank, and that caused the fireball. The thing that was so amazing about it was had that opening been at any other segment or any other degree at that segment, we would not have had that explosion.

We used to play "what ifs." Suppose such and such happened. What if this? What if that? We would have never come up with that scenario— never. I sat on a plane next to pilot Michael Smith's widow on my way to an AIAA presentation.[46] We had never met. I knew who she was. She, of course, didn't know who I was. I told her that story, because I believe what's to be will be. I don't know whether it was any comfort to her at all, but I said, "It happened for a reason, because this is the worst on worst on worst case. You would never have come up with that ever happening." We've had other leaks and we didn't lose the vehicle. That was a terrible, terrible time, and a lot of people did a lot of work. You had to work from the bottom up and the top down, trying to determine what did happen.

After nearly forty years of government service I decided to leave NASA.[47] The paperwork involved with the AFE project inundated everybody, and suddenly it wasn't fun anymore. That's the reason I quit NASA. I was still doing Shuttle work, doing an aspect of the ascent heating, as well as being the program manager for the AFE simultaneously. I was sixty. My mother was getting older, and I thought, "It's time to travel and play." So I retired.

The night of my retirement party at JSC, one of the speakers said, "Dottie's retirement is a line item in our budget." I did not know what he meant, or why he said that. I found out later I was replaced by ten men—three NASA guys and seven contractors. I'm so proud of that.

I thought it was time to retire, but I was wrong. About a month after I retired, I got a phone call. "Dottie, there's a company up in Detroit that would like some aeroheating support. Could you do that?"

"Yes, I'd be glad to do it." I had to make my reservations and get my car. I'd been spoiled because at JSC gals did that for me, and I really appreciated it.

As soon as I got back, Bass Redd of Eagle Engineering called me.[48] He said, "Dottie, I thought you were going to play for a year. I hear you worked. Come talk to me."

After having made my own reservations, I said, "I'll be right there." So I went to work for Eagle. Three months later Eagle got a contract with a company in Italy and sent nine of us to help them design an entry spacecraft. They had put up satellites, but they had never retrieved one, so they didn't know how to do entry heating, per se.

There was only one woman engineer in this company. She was charming. They could all speak English except one. Of the nine of us, we each had a separate discipline, mine, of course, being heating. We had a thermal protection man and a retrieval man and an aeronautics man, various disciplines. One of the young men in my group couldn't speak English, but he sat and watched me the whole time. We were there for nine days. I thought, "I wonder if he's trying to learn English and with my Southern accent?"

We sold Eagle Engineering, and about two years later I received a phone call. "We need an aerothermodynamicist. Would you come talk to us?" I thought, "yes, because I want to keep busy. I'm going to work to my grave. I'm just not going to stop." I was hired part time. I can work eight hours, two hours, whatever. It's a nice way to be retired and work. I don't ever want to quit. I don't ever want to retire. I like the way I'm retired, because I can work when I want to.

It would be hard for me to pick one area of my career that I found more challenging than another. Each had its own challenge. I can't say that I was happier any one time than another. I enjoyed it all, absolutely loved every day. I had a boss tell me—and he was assistant division chief—"Dottie, if you were a man, you would have been a boss." I don't remember what I said, but I would not have wanted it. I had the best job at JSC. Being a manager toward the end did involve some personnel activity and some necessary paperwork, but I liked doing what I did.

When I started my career, American women were not encouraged to go into engineering. I did not receive—in the beginning—raises as the others did because I wasn't a graduate engineer. Ultimately, I did. It just took a little longer. And the men: oh, they liked me—they just didn't really want me to be their equal. Chauvinism existed more then than now. They'd want to help me. I don't object to being helped into the car. I want to be a woman first, but I don't want to be recognized because I'm a woman; I want to be recognized because I can do the job. That would be the only thing that would make me resent being rewarded—if it was only because I'm a woman.

There were some inconveniences, although the men were not ugly. I was very fortunate. Never in my entire career was I ever embarrassed by a man. If they embarrassed me, I didn't know it. I was too naïve, and I brushed it off and dismissed it as if it was a joke for the day. Of course, I worked with professional men, and NASA really provided a very sheltered world. I never had a problem.

There are a few stories, however. When we moved here, our offices were at Gulfgate in apartments, and I think my office was in the living room.[49] A man came in with telephone equipment, and he started doing something with my phone. I was working, doing my thing. Finally, he said, "Now you will be able to answer so many phones."

I looked at him like, you have lost your mind. He thought I was the secretary. He quickly found out that I was not. I was too busy to be diplomatic. I said, "No, you're not going to do that." I went and found the secretary, and she set things straight. Well, he wondered what happened. It was ugly.

In this man's world I had to take ten giant steps. I didn't advance as quickly, but I wasn't a graduate engineer. Here in Texas every one of my coworkers in my section was either getting his PhD or his master's. There were days I was the only person in the office. But even with all those PhDs, I had the same General Schedule grade they had.[50] They finally gave me the same recognition that they gave those PhDs. The men respected me. They loved me, but I think that those my age had great difficulty letting me be an equal, particularly since I was a math major and not an engineer. Today that isn't true. The young people give women a little more credit, not that our men didn't. By "our men," I mean men my age. Women still were the wives and the homemakers then.

I find today, and this is a criticism—I went to JSC to a meeting and there was a gal chairing the meeting, and when I left I was with one of the men, as I worked for a contractor at this point—I said, "You know, that gal was tough." What she had left at home was charm. Both men and women still have to have charm to deal with other people. Now, that might sound old-fashioned to these young gals, but it'll work every time. My family said, "Dottie, sugar will get you farther than vinegar any day." And that stuck.

The first media interview I had at Langley, I hadn't been there long. Why I was selected, I do not know to this day. A woman came in, and as she walked in, she said, "Do you believe that women working with men have to think like a man, work like a dog, and act like a lady?"

I said, "Yes, I agree with that." The article came out in the Sunday paper, and in quotes, "Dottie Lee says," and she repeated that. I thought, "I didn't say that. She said that." But that is true.

If there was one contribution I'd like to be remembered for, I would want to be remembered that I did my work professionally, with a sense of humor, and graciously. I am appreciative of the men who touched my life, and I'd want to be remembered by those whose lives I touched. I'm extremely proud of what I did.

Ivy F. Hooks [51]

When I graduated from college I had no idea what I was going to do. I was sure I was not going to teach school.[52] That was an absolute; I was not going in the classroom. I didn't like that idea one bit. I looked around for jobs and wasn't offered anything that looked like anything I wanted to do, so I just kept going to graduate school. Then NASA, the Manned Spacecraft Center, moved to Houston, and a friend's mom cut out an article from the newspaper that said NASA's looking for women scientists and engineers. I assumed they were looking for me. I appeared on their doorstep and said, "I'd like to apply for a job."

I think I first filled out an application at the University of Houston, and then I got called to come in for interviews. I went for one interview that was in a building off of Telephone Road, and previously they had manufactured boxes in this building—you can imagine how elegant the building was—no windows, no nothing. It was really terrible. I interviewed with a man there, and there was a woman who was in one of my math classes in graduate school in the room and another guy. Quite frankly, I thought they were a little strange. I thought, "I don't want to be in this box," and the work they were doing didn't particularly appeal to me.

I'd met a woman at the beauty shop whose husband worked at NASA, and she knew he was looking for people. If you want information you go to the beauty shop, right? I went and interviewed with him, too, and I said, "Well, that looks pretty good. I think I'll take that engineering job." The money looked wonderful after making no money as a student.

That's how I got to NASA. It was probably one of the best times I could have gotten there. The only thing that would have been better was to have been hired a little bit earlier, with the guys building the Mercury and the Gemini, but I wasn't old enough to do it then. I'm just lucky to have gotten to work with them and to learn from the people that started the space program.

I hadn't a clue what I would be doing when I first started. The odd thing was that I was working in a group that was working on advanced stuff. It wasn't structures or mechanical systems or propulsion; it was a systems-type group that looked at advanced things. The organization was in disarray. A number of people there had job experience from Chance Vought Aircraft or other aircraft manufacturers, so whatever they were working on, they knew what to do. I got very boring tasks like, "Go plot this." It didn't seem to have structure to me, and it really bothered me.

You have to remember there weren't very many professional women at the time. There was one professional woman in my immediate group and another one in a building across the way; we were working in a set of apartment buildings at the time, up in town. In fact, at the time I started, I think the whole population of NASA MSC, not its contractors, was like three thousand five hundred. There were thirty-five professional women. There weren't very many.

Now, I was really lucky, because there were professional secretaries. They were people who had experience, had worked up at Langley Research Center, and had come down with the NASA group to Houston. They were very professional people, and they were really a huge help to me all through my career. I had a real cheering section from the women that were secretaries, administrative assistants, and all. I think it was a mutual thing, because I felt what they did was important, and I respected what they did. A lot of the guys didn't. If you were not an engineer, you didn't count.

The guys told me I couldn't take more than thirty minutes for lunch. That's all we were allotted. You couldn't leave that building and go to a restaurant and come back in thirty minutes. It didn't matter. I was poor; I brought my lunch. They would go off and stay gone for hours, some of them anyway. One day they came back, and I was sitting at the drawing board doing this graph for them. I'd gone upstairs where the drafting guys had their table because I was just dying to try that electrical eraser they had that I'd never seen before.

The guys came in, and I didn't realize they'd come in. I think there must have been fifteen or twenty guys in there. They'd gathered up half the employees from the building. All of a sudden somebody behind me cleared his throat. I looked back, and then I looked to the other side of me, and all these guys standing there were snickering. "What are they going to pick on me about now?" I have four younger brothers, which I found was a big benefit in going into this career because of all the dumb things that people did to try to annoy me, frighten me, or scare me. I'd probably been through it before, so it wasn't new. They bugged me a little bit, and I said, "What?" They pointed to the top of the drafting board, which was far away from me. They had a computer printout or something laying over something. They pulled it away, and there was this snake. It was just a garter snake, for gosh sakes. I grew up in the country with four brothers.

Earlier I had seen people around the railing at the apartment looking at something, and I'd gone and looked. I saw there was a snake in the pool. I already knew they'd found the snake that had fallen in the pool. I just reached over, picked up the snake, turned around, handed it to the guy nearest me, and said, "Go away."

That probably was a very good start to my career in NASA, because an awful lot of the guys in that room would have fainted if I'd tossed that snake at them. I've often wished I had, but that wasn't nearly as effective as just cutting it off and saying I didn't care. There were those kinds of pranks off and on for quite a while. I'd go in a new group, and I'd get some more of that again. But the people that were professional were so professional.

Lunar lighting was the first project I remember working on. They came in and said, "We need to know what the lighting is like on the Moon so we can build simulators to train the astronauts." Heat, Light, and Physics was one of my worst classes in college because the professor never taught it. He always taught the mechanical stuff that had been the class before, which I understood, but I didn't get the heat or light. So I did what any good student would do—I went to the library.

Our library at the time was in the same apartment complex we were in, and it was in an office approximately twenty feet by twenty feet. That was it; that was our library. I went through all the reference cards—we had cards—I don't even think we had microfiche at the time. I found something on lunar lighting and asked them if they would get it for me. They said, "Is it classified?" That was a big thing. That was probably one of my most un-fun things working with the government, having to deal with classified material and worry all the time about whether your safe was locked and not saying anything to anybody.

I said, "I don't think so, because it was written by the Russians." It was a report written in 1924 by Russians who were studying the albedo for the Moon, how it reflected light. It was totally mathematical. That was wonderful for me. I could totally understand that; I could convert that into what they needed to do their modeling for the Moon.

Of course, I also found out about a jillion other people had dumped that project and didn't want to work on it because it wasn't what they'd studied in school. I thought it was so neat when I came to work at NASA that I never did anything I'd studied in school. I already knew how to do that. I was afraid when I went to work I would keep doing what I learned in school. That didn't sound very interesting. But it wasn't; it was always something new. That was my first something new.

I left that group. They were just a little too trivial, and I got really sick of it. I went to work on the cost model. The cost model was for Hum Mandell, the guy whose wife I had met at the beauty shop.[53] He came out of the aircraft world, so he was building a cost model for NASA to use for spacecraft. We ended up building the first cost model that NASA had. He continued doing cost models forever for NASA. He was really into cost models. With all of his cost models, the heavier a spacecraft got, the more it cost. I said, "Well, that may be true on airplanes, but it looks to me like it's costing us money to make things weigh less." I didn't know anything, right? We had a guidance and navigation system, and I knew we had structures. I generally knew what those things did, but I didn't know how any of them worked. I certainly didn't know what they cost.

I would just get the phone book out, find that organization, and call up somebody and say, "Can I come talk to you about this?" for things like guidance and navigation or reaction control engines.[54] Of course, here I was, a twenty-two-year-old in a miniskirt, so I didn't get too much turndown. There were some real advantages at the time to that. I found people love to talk about what they know, and if you shut up and listen, you can get a wealth of information. I started learning how this worked or how that worked. None of them knew much about why things cost what they did, to be frank, because that wasn't what NASA engineers dealt with. They knew technically what happened, what was hard, what was simple, and what was the same as something else. People were so generous with their time explaining things, and I often went back and asked more questions.

I worked in that area for a while. That was probably the starting point of when I really started finding my niche of loving to work on things that involved lots of systems, loving to work on things at the front end, when you're just starting something. I've never been a really great detail person. Who wants to find out the last little decimal point or something? But I appreciate those people because things don't get built without them.

Somebody has to do the front end, think upfront, think big picture, and be willing to think about things nobody thought about before. I loved it, and others don't.

Then I was told by Caldwell Johnson, the deputy division chief, "You need to get back in the technical world," so I ended up going to work in a branch that had been in the same building and in the same group with us.[55] We were all in the same division, but they'd gotten moved out to Ellington Field.

So I went out to Ellington Field to work with them. As I walked up to the building, somebody was dropping teeny-tiny models of parachutes off the second-floor balcony. That was where I got introduced to the guys that did flight dynamics and flight mechanics. At that time, our branch chief was Milt Silveira.[56] I got to work with some of the brightest engineers I think NASA ever had. When I would be told to do something, here were all these people I could talk to. They would come to me for the real heavy math problems because my degrees were in mathematics and physics, so we worked together very well. Most of them were engineers. That was a really wonderful environment to work in. It was a group of people who loved to solve problems, who would go find problems that people didn't even know they had.

Probably the best dynamics engineer I ever knew was a man named Joe Gamble, and his degree was in civil engineering.[57] He was very, very smart. He could learn anything, do anything. He was my officemate for a long time.

Our group was trying to build empirical models for parachutes. Could we have steerable parachutes? One of the issues on the Apollo flights—not on the first ones where we were launching the unmanned capsules off the launch pad that was closer to the water, but on the subsequent ones. When we were going to the Moon, we had to launch from a launch pad that was further back. If we had an abort and we couldn't control the parachutes, we were afraid that, depending on winds, we would drift back over land. Control or not, we didn't even know how to model the parachutes and how much they would drift back. There were people doing modeling, and they were doing things with the controllable parachutes, like the parasail.[58] We were also taking every piece of data—I was working with Joe Gamble on this—and modeling three parachutes, different winds, and all the conditions.

They were really fun times. The group that I worked with, they pulled a lot of pranks on each other, teased each other. They didn't pick on me badly. Whatever they did to me was exactly the same as they did to the other guys. I was just one of the guys. That was a very fun environment to work in. We all helped each other and worked together.

Another thing that I worked on early in that period was something that they call plume effects, the exhaust that comes out of an engine. If it hits something, what does it do to the thing that it's hitting? For Ed White's walk in space, there was a real fear of what was going to come out of the plumes if the reaction control jets fired on him.[59] They gave me the job of running this computer program that could make some of these predictions. It was another technical thing that I got involved with that nobody else knew anything about, and I didn't either. What difference did it make if I didn't know anything about it? Neither did anybody else. I certainly could not have created that program, but I could set it up and run it. I could make estimates from it and do things with it. That was the only thing I remember doing in the Gemini Program.

I was working mostly aborts during Apollo. Worrying about coming back and hitting the land was big. For the Apollo 8 flight, when the astronauts circled the Moon, I was in the Mission Control Center in a back room looking at the winds to see if there was any problem at the Cape.[60] We came up with a way to estimate if the crew was in danger of landing on land instead of in the ocean. There were no winds that morning, so there was no problem.

The Engineering Directorate came up with the concept of determining the wind conditions under which it was safe to launch. I worked mostly with John DeFife on that.[61] We came up with that, and then we taught the Ops guys.[62] We didn't ever go back over and do it again because it was a very straightforward thing to do. You just plotted a plot, and if winds were on one side of the line, it was okay to launch; if winds were on the other side, please don't do it, it might be a risk. Your hope is that you can make something that you can use for the whole mission, and that was used for all of the Apollo missions.

The other job they gave me was the Apollo launch escape system. You had to set the motors, the little thrusters, so that when the system came off the pad, it would pull it off right, depending on the center of gravity of the spacecraft itself, which you didn't know until they got it all together down at the Cape and ready to go, almost. For every flight I had to do this set of calculations. I came up with my numbers every time, and I said, "I hope I did this right," but we never knew. We never had to do one of those aborts. I worked on a lot of things that never happened. If they'd happened, it would have meant we didn't do the mission, so you didn't want them to happen, but I didn't really know if I was ever right.

When we started on the Shuttle it was April Fools' Day, 1969. The phone rang, and I heard, "This is Betty Ensley."—that was Dr. Faget's secretary—"You need to go to Building 36, the third floor at ten a.m."[63] We were having fun because it was April Fools' Day. Two or three others near me got the same call, so we looked up Building 36 on the map on the back of the phone book because we didn't know where 36 was. We drove down there, and it's only a two-story building, but next door is the high bay that clearly had three stories. There were people walking in that I knew, and we went in there. We went up this high bay. All they had used it for since they built it was to store furniture. It was filthy, and I was in white. We were just standing around. "What are we going to do? What are we here for?" I would have thought it was an April Fools' joke, but there were people there that you would not pull an April Fools' joke on.

There were people I knew from thermal and aerothermal. The group I worked in had the aero guys, I was doing flight mechanics stuff at the time, and there were people from propulsion. I knew a lot of these people going back to the cost model days; I'd met and known these different people.

We didn't know what we were there for, and then Dr. Max Faget walked into the room carrying an airline garment bag, unzipped it, pulled out this funny-looking little balsa wood plane, flew it across the room, and said, "We're going to build America's spacecraft. It's going to launch like a rocket and land like an airplane. It's going to be reusable. It's going to go about this high, do this," the whole thing. We spent six months locked up, trying to figure out if we could really do that. It was really fun. The only thing wrong was no windows. We weren't allowed to tell anybody what we were doing and what we were working on. We just stayed in there with one telephone in the room, one secretary, and us. I spent a lot of time going to the library.

Jim Chamberlin, who Faget had put in charge, gave me every odd job there was. One was to look at—since it had to land like an airplane—putting jet engines on it for takeoff and landing. I whispered, "The propulsion guys are over there."

He said, "They don't know anything about jet engines. They just know about launch vehicles and reaction control jets. They don't know about engines." I had a pilot's license; I think I was the only person in the room who had a pilot's license—not that I'd flown anything but a Cessna 150—but he knew that I knew at least some of the Federal Aviation Administration regulations. Then he said, "Go get *Jane's All the World's Aircraft.*"[64] He would tell me which books to go get in the library to look things up in; he just kept heaping jobs on me.

One of those jobs was the separation part because I'd done some jet impingement on the lunar module. He said, "Well, you know jet engines, so you need to do this, and then you know about jet exhaust, so you need to look at this. Go take care of this." I had a list a mile long of things I was working on. About once a day he'd come over and ask for one of them, never the one that I'd been thinking was the most important.

Finally, when he would give me a job, I would ask, "Where would it go on the list in order? Do you want a ten-minute, a half-an-hour, or a two-day answer?" I tried to teach people who worked for me later, "If you don't know that from your boss, ask." I think a lot of people mess up because somebody asks them to do something, and they go off, and they're going to do it to the *n*th degree, and all somebody wanted was just a summary and they wanted it fast. What's more important, accuracy or fast? You've got to know that to do a job for somebody.

Of course, this was all very front-end stuff, and I had been doing analysis of all kinds of spacecraft for the last year or two before that of what kind of spacecraft we might have, what we might do, and the performance of different things. I had knowledge to work from, but I then had a complete lack of a lot of things he was asking me to do. So I'd just go back to that library and start looking things up.

We stayed locked up for six months. It was the secretary, Dot Lee (who was aero/thermo; she came from Langley), and me. We were the only three women in the room. It didn't matter at all in that group; it was just get it done. It was really fun.

Because I'd worked the separation system, that's where I ended up going, working on the separation systems for the Space Shuttle. Because I had worked with jet plumes—exhaust plumes before in Gemini and Apollo—all the exhaust stuff fell in there. I was in the group that did aero, and the Shuttle got broken up into the major things like the entry aerodynamics, ascent and the aerodynamics for that, and then the separation systems. So it really wasn't, I don't think, a surprise to me that I became a section head, but I do know that at least one of the guys I worked for kept thinking I should have been on staff.

Years later I found out people in the group, managers up above me, thought I should have been in a staff position. "Put her in a staff position and let somebody else manage the men." Dr. Faget said no, so I was left to manage the work, and I had a wonderful time. I did. And I did a lot of great things, but I always did them with other people. I can't think of anything I ever did that I didn't have some help from somebody. We turned out some really, really good work, and besides that, they trained me how to be a manager.

Years later I found out that they got really harassed. "Oh, you have to work for a girl. Nah, nah, nah, nah, nah." Their whole attitude was, "We're going to do the best job of anybody on this planet. Everything we do is going to be superior to everybody else's. We're going to make Ivy look so good." They did. They never let me know—never, never, never—until just a few years ago, one of them blurted out something about it.

I truly do not think there was anybody sitting in Personnel at NASA or in many organizations saying, "I don't want women." I think there were pockets of people that maybe would never say it but thought it. But that's life, like somebody doesn't like somebody that's tall or fat or has blue eyes, or whatever. People have a lot of prejudices that they don't know they have, and if you ask them, they would say, "No, I don't do that"—but they do. It's the old I-want-somebody-just-like-me syndrome. I think it was recognition of that that made it a little easier for me to get through a lot of things, because when I realized they thought, "I'm a guy, and I want somebody who looks and talks and walks and thinks like I do as my deputy or to work with me," then there were no surprises. I work best with people who love surprises, who like something different, who don't care if we go off on some crazy track that we've never tried before.

People who have a very tenuous comfort zone don't do that, and they don't want to. That's the reason I think people pick employees the way they do. If I picked everybody who was just like me, where are we going to get any new ideas? If you think just like I do, I must not need you, because I can already do that. I need somebody who thinks things I can't think about, that does things I don't do. But for an awful lot of people, that's not in their comfort zone.

You have to understand, there was always a group, I think, who didn't care if you were a woman as long as you got the job done. It didn't matter one bit. You could have been a little alien and they wouldn't have cared just as long as you did your job. They did theirs; you did yours. Then there were other people who were very much trapped in that male/female role-playing model.

About ten years after I came to NASA more women moved into scientist and engineering positions. As we had more co-ops there was much more recruiting of women. There were more women in the engineering schools by then. I remember when I first became a section head I ended up with one woman who was just out of school, and then there were two or three other women in the building who had graduated within the last few years and come to work at NASA. By that time, there were more women coming in, and some of them were coming directly out of school though a few of them had gone to work for contractors and then came over. In the group I was in, everybody was so supportive.

The thing that made it fun working at NASA was going to the Moon; nobody had ever done this, and the challenge of it was so unbelievable that NASA had to have a bunch of people who weren't set in their ways. We had a lot of them in very high positions, like Faget, or Guy Thibodaux, head of propulsion. There were a lot of people who didn't have those hang-ups, so I fit in. If you'd thrown me in an oil company, I don't think I'd have ever gotten out of the back room. Somebody took really good care of me and let me go to a job that was awesome, straight out of school—you couldn't beat that. I looked around, and I thought, "Oh, the rest of you aren't going to get to do that, and that's terrible." It was a great opportunity at a great time.

In 1973, I joined the Society of Women Engineers, and I saw what was happening at NASA was not unique; it was in every industry that women engineers were in. It wasn't a NASA thing, it was a people thing, and it was our society. A lot of ideas about women's place in the workplace had to do with all that brainwashing at the end of the Second World War. Times were changing; eventually it would get better. When I was in college in a sorority, when protest marches were big, they would tell us, "You can go march for any cause. You're free, and you can go do that, but do not wear your sorority pin when you're doing it because you're not representing everybody else." I thought, "Well, I can understand that. You can't do that and say you're representing some group you belong to when you're just doing it because that's what you believed in." But I always found it was rather difficult being a woman because I couldn't take that pin off that said, "I'm a woman."

I think being the first woman in a nontraditional field stressed, "Whatever you do, if you really mess this up, it isn't just you you're messing up. They will use you as an example." They don't say, "Guys gossip." They might say, "Joe gossips," but they'll say, "Women gossip." We just get classified as a group. That became something I was really aware of, and especially as guys would introduce me, "This is Ivy. She's really done so well, blah, blah, blah." I'd think, "Boy, I'm glad they're not saying the opposite of this," but when they say that and they're introducing me, they're giving me a step up with whomever they're introducing me to. You become a little wary. That has to matter. You have to think about how you're not doing it just for you; you're doing it for other people, too.

In the 1970s, I shared an office with four or five male engineers. They had *Playboy* calendars on their desks. They had pinup photos inside the big cabinets—those big cabinets that had big doors, so when you opened the doors, you'd see the pinups. They'd always had that; that was just the way it was. I never paid any attention to it. You know, they're just there; it's okay. I didn't care. I grew up with four brothers. What is the big deal here? I just never made any big deal about it.

One day I came back—I'd been on a trip—and I walked in the office that morning, and the men were in a huddle. That either meant that they were going to pull a trick on me or that they were telling a dirty joke that I was not supposed to hear. They just backed away and snickered, and I sat down at my desk. My desk faced outside. I sat down at my desk, opened things up, and waited for whatever the next thing was.

The next thing I know, they gathered around my desk, and they were *ahem*-ing a lot. "What is going on?" They pointed to the wall beside me, and I looked up, and there was Burt Reynolds's picture torn out of *Cosmopolitan* magazine.[65] I had only heard about it; I hadn't seen it. I just broke up laughing, and I said, "Who did this? Who put this in here?"

"Not us. Not us."

Now, they were adamant it was not them, which meant it wasn't, but they weren't going to tell me who it was, so I waited until everybody had gone to lunch. One of the guys was still there, and he couldn't keep his mouth shut. I just kept bugging him. "Who did it; who did it?"

"I'm not going to tell you. I'm not going to tell you." I just kept nagging. Finally, he told me it was Joyce. Joyce Koplin was the division secretary, and her office was right next door to ours.[66]

So, I went over to Joyce's office and sat down and said, "Do you want your picture back?"

Of course, she's a very good professional secretary. She just looks me straight in the eye and says, "What picture?"

So I start in on her, teasing. I said, "Look, I think it's really funny, but how in the heck did you get that picture? And why did you bring it in there?"

She said, "Who told?" so I tattled. She said, "Bruce Jackson asked me to do it."[67]

I said, "What?"

She said, "Well, I was talking about having seen it at the beauty shop one day, and he said, 'You know, that would be so neat if Ivy had that picture over her desk, because all those guys have all those pinups in that room. Ivy could have her own pinup. You think you can get it?'" So she did. She tore it out of the magazine, brought it in, and put it above my desk. It stayed there.

Then one day I came in, and the guys were back to the same huddle. I went to my desk, and Burt was still there, so I went to work. Then the guys started *ahem*-ing, so I said, "Okay, what is it now?" Well, Bass Redd, Bruce's deputy, had come in, and he had decided it wasn't nice. Now, he never decided this all those years before, but now that I have a picture up, it isn't nice for them to have all those Playmate pictures and share an office with me. He made them take all theirs down, so I had the only picture on the wall.

Later that decade, I did the separations for the Approach and Landing Tests. On all of the vehicles that NASA had separated before, they were stacked. Let me illustrate. On the bottom of a *Saturn* rocket is the booster, the one that's going to go off first. It burns and pushes everything up until it burns out. Now it's not pushing anymore, so everything starts falling. Then you sever the connecter and fire off this second stage, it goes, and this one falls down, and you do that however many times you've got stages.

Now with the Shuttle stack, you've got the Orbiter. When it flies, it has the tank under it. On that are the two boosters, and they're firing and so is the tank. When the boosters quit firing, if we did nothing but let go, this

Using a model of the Space Shuttle, Ivy Hooks explains how the solid rocket boosters separate from the vehicle during ascent.

tank would hit those boosters. The boosters have to be pushed away so that we can't hit them. They had to do it really fast. Rockwell's solution was to use jets that the crew would fire back toward the Orbiter and tank, and that would push the boosters away. They were going to fire these big hummer jets that had a lot of solid particles in them right back at this Orbiter with dainty tiles and a tank. It was apparent we were not going to do it that way.

I went to Marshall Space Flight Center, who was building the boosters, and said, "Don't use those solids that have all those solid particles in them. I need as little stuff in them as possible. I'm going to need you to cant them out and do other things with them so they won't spray," and they said they couldn't.[68] Even Rockwell said I didn't know what I was talking about. I went to Bob Thompson, who was the Space Shuttle program manager, and said, "You're going to destroy the Orbiter with the separation system, so it's got to be changed. I need to run some tests to prove how bad that is."[69] I had people at Marshall who were working on the tank and the boosters—the tank in particular—and people who worked in materials who believed me. I worked with them. We got a few tiles, some carbon leading edge, and some motors. We took them to the Tullahoma Arnold Engineering Development Center.[70] We ran a test and showed it would all be gone if you did what they wanted.

The way I knew how bad this was going to be—I hadn't even started looking at it yet—was when a rep came in and said, "Do you know you can't do that separation?"

I said, "I haven't even looked at it yet." It was right after the contract was awarded.

He said, "You need to see what it looks like when a Titan IIIC sep-arates," because a Titan IIIC has a center and two booster rockets to separate while the Shuttle has an Orbiter/tank and two booster rockets to separate.[71] The boosters' separation motors blow toward the center and push away from the center in both cases, but the Titan IIIC center is hard metal, while the Orbiter is covered with fragile tiles and the tank is aluminum. He showed me films, and it's just nasty, dirty stuff. You go down where they fire off the launch pads with Titans, and it just eats up launch pads. It was obvious you couldn't do that. Other suggestions had been mechanical systems, to make them move out, and I thought, "No, no, no, no, no, no, not going at Mach 4!" That would never work. How do you solve it? This was a really hard problem. On Titan, if you had trouble with the booster, you could blow off the front end—the nose cap. If you don't have a front end, you don't go any further; it just quits.

People wanted to talk about thrust termination, so we looked at thrust termination. If we had tried to do thrust termination, we would have destroyed all the tiles on the Orbiter, and then we couldn't bring the Orbiter back in because it would get too hot. That idea didn't look good, and they could go off when you didn't want them to.

The hardware and software of the Orbiter first came together when it flew on the Shuttle carrier aircraft. They started talking about needing to fly the Orbiter before the first flight. Oh, there were all kinds of drawings on people's walls. Rick Barton had walls just covered with funny drawings and a slingshot-looking thing with the Orbiter sitting on it, a new airplane built just to carry it up and drop it off.[72] Then they had it hung under a C-5A. They had it on top of the 747. They kept doing all these drawings. Then they said, "Ivy, to make this work, we have to separate. What are you sep guys going to do?"

I said, "We'll look at all of your ideas." I figured they weren't going to build another airplane for it, so we looked at the C-5A and the 747. The C-5A has a T-tail. Well, with the Orbiter sitting on top of that T-tail, the wash from the Orbiter was going to do something to that thing—who knew what? We hadn't run any wind tunnel tests, but we thought, "We really don't want to fool with that T-tail if we don't have to, because we'll have to look at a lot more things than we would if they used the 747."

I was at Downey one day when they were having a telecon on the subject.[73] Bill Schlish, who was my counterpart on separation systems out there, left the room, went to a telecon, and came back. He said, "Well, we're going to go with the 747. We're going to buy it from Continental Airlines and have them move their tail for us." The Continental ads then were, "Let us move our tail for you." We all giggled about that a whole lot. A few days later, they came back. They had bought the 747, but they weren't buying it from them; they bought it from American Airlines. You could always barely see "American Airlines" on the side if you looked under the paint.

We needed wind tunnel data; Boeing had a facility where they thought they could get some data. Of course, they had 747 models, which was handy. We took the Orbiter models and the 747 models and did some testing; I didn't go out on that test. It took a while to get some decent wind tunnel data.

Then everybody decided it would be really neat if we could practice this separation. Of course we were running simulations, putting our wind tunnel data in, and seeing how we thought it would work. They actually did some simulation at Boeing, where they had a two-plane simulator. A. J. Roy and Fitzhugh Fulton and other NASA pilots who were going to fly the 747 went up and flew that part, and Joe Engle and other astronauts flew the Orbiter.[74] Out of all the separations, it was absolutely the easiest one to do. We had the tail cone on the first time we were flying it around before we did a sep because we were worried about the effect on the tail, that we really hadn't gotten good wind tunnel data.[75] It turned out to be fine, so thereafter we pulled off the tail cone and flew the Orbiter.

Every time they flew, a couple of guys and I went out to California and monitored the testing. We got loads from the sensors between the two vehicles to verify they were pulling apart at the rate we were trying to pull apart. They went up and flew through the pitch-over maneuver; you could tell that the lift between them was exactly what it was supposed to be. Of course, we had some of the best pilots in the world flying—didn't hurt to have them.

On the first flight we lost a computer. Gordon Fullerton was in the right seat, and he said, "Computer"—whatever number—"big X." I thought, "Ugh, I'm so glad we simulated hundreds of these." As soon as I got back to Houston, one of the safety guys came up and said, "You know you were one failure away from losing your vehicle." I never understood how the software worked. This was before I went to the Flight Software Division.

I thought we had four multiple computers and they were all listening and all talking to each other so they would all get the same data. It turned out that with one particular computer, if it failed, we would not get the signals to say that the Orbiter had released. We had to get a signal that said the vehicle released before the elevons would start moving on the Orbiter. When it failed, nobody was getting that signal, so it was one signal away from failure. If anything else had gone wrong, we wouldn't have gotten it.

Of course, I'd simulated all these things—a computer going out in the Shuttle Avionics and Integration Laboratory and in the Shuttle Mission Simulator.[76] They always made one computer go out, and it wasn't the one that had any of those measurements on it, so I never saw the loss of the measurements. I said, "Wait a minute, that's a bad way to simulate this, guys. Since different computers do different things, we ought to make different computers fail at different times to make sure we're getting it all. Or at least run all the traps for it." But Shuttle's so complex, you don't get it all at once.

There's another story about watching the vehicle doing the tests in California. Out at Edwards Air Force Base we were inside a room to call sep, to say we had the right conditions. We were on a console like the Ops guys. We had to call, "Go for sep." Then we would drop our headsets, run through the building, run up the stairs, and get on the rooftop. I never could find the Orbiter coming down until the dust was coming up. The Orbiter did it, but I didn't see it. Then I'd watch the playbacks.

We'd had a lot of questions from NASA Headquarters, "Was it really going to work?" At that time we had an Adage computer. We could actually take our trajectory data and do a 3D drawing on the screen. This was in the 1970s; this was way out then. You would actually see the Orbiter and the 747 flying along and doing the pitch-over, and then you'd see the Orbiter come off. Because it was three-dimensional you could see it from head-on and all other directions. I got so tired of the questions, and I was tired of cutting out little paper airplanes and pasting them together to show them how it was going to work, so we did the film. We put the camera under a hood and took a picture of the screen and played it in Center Director Chris Kraft's conference room one day for the folks from Headquarters.[77] No more questions.

When it did separate, that evening everybody was all over us. My guys and I were the center of attention. "It looked just like the movies!"

I said, "Oh, I'm so glad it looked just like the movies." It would have not been good if it had not looked just like those movies!

If I had to pick my most significant contribution at JSC, it would be the design of the sep systems, and I didn't do it by myself. The most challenging project that I had was getting that test run on the sep motors to show that there really would be a problem. It was terrifying; it was exhausting. When we fired the engine and finally went down there and saw what happened, we saw all these tiles destroyed and huge pieces of metal melted and everything else. I just sat down on the floor. Then here come the guys that have been telling me I was wrong and that nothing was going to happen. They said, "How are you going to fix that?" How are *you* going to fix that? I felt like the Little Red Hen, because that's what I was.[78] That's what they meant, too. They weren't going to help. The others were, who had participated and gotten involved, but it was mine. It was my baby, and I was going to have to go fix it.

We weren't under pressure because the main engines and the tiles kept everybody else busy, so we never were that pressed.[79] We held quarterly meetings on the separation systems. The various contractors, the various centers—everybody would all come in for a day or two of meetings to go through the status: where we were, and where the problems were.

In 1978, I was on Center Director Chris Kraft's staff when we selected the first women and minorities as astronauts. Until the Shuttle we mostly hired pilots from the military. There weren't any women pilots in the military, so why would we have hired anybody to be an astronaut that was female? There were darn few minorities too, and, again, we were picking from that already-filtered list. I don't think it was a prejudice as much as it was that it just didn't come up on anybody's screen that maybe somebody else could do this job.

In 1981, I was on Dr. Max Faget's staff, and I took over a job that looked after the interns and the co-ops. What I inherited was a bunch of angry interns and co-ops, because the person that had the job before me, that wasn't his interest, so he focused on some other job and just let that go. I was appalled. It was 1981! I'd been working nearly twenty years at that time. I knew things were better. We had women astronauts by that time—still didn't have a woman flight director, but eventually we did.

A group of three or four women came to see me, maybe not all together— they came probably one at a time—that were all interns. They were in one division, and they were miserable. They were being treated very badly. One of them had a boss that had said he wasn't going to give any important jobs to women because they'd just get pregnant and leave. I could have hauled his you-know-what in front of the Equal Opportunity officer, gotten him in all kinds of trouble, and maybe even lost his supervisory position, but that didn't seem like a very good thing to do. There were other guys who said other things. At least one of them we sent to sensitivity training. I didn't think it'd do any good; I just thought we'd use it as punishment, but he came back from there a new person. He said, "They said all the things my wife and daughter and all have been saying." I was surprised that the sensitivity training actually worked, so I would never hesitate to try that again because you like to get somebody out of that mold.

The last ten years I was at NASA men would come to me and say, "My daughter is in high school. She's very good in math, and I'd really like her to consider becoming an engineer. Can she come talk to you?" So "my daughter" became a new way of looking at the world: "I want this for my child, and that fact that she's female doesn't matter; I can see that it doesn't matter because I know Ivy or I know this woman and I know that woman. They made it, and my daughter could make it, and this is such a fun world to work in. I would like her to be here." That was one thing that I saw. People got into that mold that women were going to work, and if they were going to work, let's give them the opportunity to use their talents.

Also by the time I left NASA, which was just a little over twenty years after I'd come to NASA originally, a large percentage of the people coming to work, and even some who had been there a long time, had working spouses.[80] When I first came to NASA, there were almost no working spouses. The wife stayed home, had children, and raised children at home. There were just one or two guys I can think of whose wives worked outside the home. They had no model. If a wife did work outside the home, she was supposed to either teach school or be a nurse. Now I don't want to put down either of those professions. I'm very grateful to everybody in those professions who's helped me in my life. That career would not have been my choice, but I'm really glad somebody else chose it. That was their only model, but it changed over the years, so that made a difference.

Flight Controllers

Flight controllers, the people who monitor and manage on-orbit operations, are known for being "tough and competent." After the tragic loss of the Apollo 1 crew in 1967, Flight Director Gene Kranz directed his flight control teams to write those words on the blackboards in their offices and never erase them.[1] The words emphasize the core values that remain the foundation of flight operations to this day. Tough—controllers are responsible for their actions and, when necessary, must take unpopular or difficult stands. Competent—teams are prepared and 100 percent dedicated to the mission.

These qualities describe Ginger Kerrick, who became NASA's first Hispanic female flight director in 2005. She continued to push forward, even when the path seemed rocky: after the loss of her father when she was eleven; when her grades slipped in college and she was unable to

Trajectory Operations Officer Ansley Collins sits on console in Flight Control Room-1.

join the co-op student program at NASA; and when, for health reasons, she was disqualified from astronaut selection, she refused to give up. Her experience highlights the importance of determination and perseverance.

Sarah L. Murray is equally innovative. She began working in Mission Operations in 1988 when Rockwell Space Operations Corporation offered her a job working in Mission Control. The hustle and bustle of real-time operations in the Mission Control Center felt like home to her. A former emergency room nurse, Murray had spent six years in the United States Army, working in the 547th Emergency Room in Grafenwoehr, Germany. The lessons she learned there could be applied to her work as a flight controller, where she only had eight minutes to make critical decisions during the launch of a Space Shuttle. Since then Murray has worked as a lead in the Shuttle Communications Group and the International Space Station (ISS) Electrical Power Systems. She led the *Columbia* Recovery Office at JSC, and in 2007 she became deputy chief of the Spaceflight Training Management Office.

Ginger Kerrick[2]

When I met my minimum requirements for astronaut candidacy, which is a master's degree and one year of technical experience, I went to talk to Duane Ross.[3] He told me that it might make me more marketable if I had some experience in Mission Operations. He explained the concept of the rotational assignment to me as well. I had no idea you could do that.

I went back to my boss in Safety, Reliability, and Quality Assurance (SR&QA) and explained the situation to him. He said, "I can support that. You go off for six months and see if this helps you." I started that rotational assignment, I think, in September 1995.

I got a call around October 1995 from the Astronaut Selection Office that I had been selected for interviews that year. I was twenty-six years old. I did a dance, I called my mom, and I was really, really excited. Back then they did the interviews a little differently. The process was a week long. The first day you meet with a psychiatrist for about four hours and take a psychological exam. It's basically just to see how comfortable you are talking to folks. I sat there, no problem at all. The exam was fine. After that day they went into a series of physical and medical tests. For physical tests, I was fine because I was an athlete. Had the treadmill test, some strength tests, and did fine on those.

During the medical tests they were doing an ultrasound of my lower abdomen area and they saw something they weren't sure of. I went back for a subsequent X-ray and CT scan. At the CT scan level, they were able to see very clearly white dots on my kidneys, and I was informed that I had kidney stones. I didn't know I had them because I had never passed one before. What I did know is that that year they'd instituted a new medical disqualification, and it was the ability of your body to actually form the kidney stone. It was a lifetime disqualification from the program, regardless if you get them cleaned out, based on the fact that your body forms them.

It was traumatic. I don't remember how I got home that day. I remember crying out of control in the doctor's office, and I had one more day of interviews to complete. I completed that day. Then I just stayed in my apartment for the weekend crying and talking to my mom, talking to my friends, not knowing what I was supposed to do next.

Until that point in my life I had found a way to continue going down the path at each turn. When I didn't have money for school—I went and found money for school. NASA didn't accept co-ops from UTEP (University of Texas at El Paso), so I needed to transfer—I found a way to transfer. I dork up my first semester at Texas Tech—I get myself back on my feet. I get a NASA internship, and I finally figured out how to turn that into a co-op position. Then NASA had a hiring freeze, and I couldn't get in. Eventually they offered me a position. But in this particular scenario there was nothing I could do, absolutely nothing. It was hard for me to accept that.

I thought back. As I watched my dad die when I was eleven, there was nothing I could do. If that didn't cause me to cave in, then nothing would. So I allowed myself to have my pity party and to grieve the death of the astronaut dream. Then I said, "What's my backup plan?" After talking to Duane Ross about how much I was enjoying the opportunity in Mission Operations, I explained that I was concerned that it was only a six-month rotational assignment. "What can I do? I really like it here."

He said, "Maybe you could look for somebody to swap." I learned that if I could find somebody in Mission Operations that would trade places with me in SR&QA then maybe I could swap.

I went door-to-door; I started on the first floor of 4 South, and I walked in the different cubicles.[4] I said, "Hi. My name is Ginger Kerrick. Are you happy here?" I interviewed several different folks and met some neat people. By the time I made it to the second floor I found an individual who was not happy there. When I described what I did in Safety and Mission Assurance she thought that sounded great. I took her over there, introduced her to my boss and the things that I did. I asked my boss if he would sign the paperwork to perform the swap, and he did.

It took probably about a month to find somebody. I knew that my six-month rotation would be up in February, so I was on a clock. Once she went over there she was winning awards. She had a great time. It was really clear that that was where she was meant to be. And it was clear I was supposed to be in MOD (Mission Operations Directorate).

When I had first arrived in MOD for my six-month rotation, I was notified that I would be an instructor for the environmental control and life support systems (ECLSS) for the Space Station.[5] I thought, "Wow! We haven't even designed that system yet. How am I supposed to learn it myself and teach it to the astronauts?" Well, that was the job. Back then we didn't have any of the training materials developed. Our job was to comb through the software documents, these requirement documents, talk to the operators, and figure out how we thought the crew would be operating the Space Station. We had to develop classes based on that and start teaching.

The first crew I taught was the Expedition 1 crew.[6] It was very early on, so the operations concepts were not as mature. As we matured them, we would revise the lessons. That crew was in training for about four years because we kept delaying the start of the Space Station.

After I got my transfer, I was still working on the ECLSS. It was shortly after the transfer became permanent, maybe about a year, that I taught my first lesson to the Expedition 1 crew. That led to an opportunity for me to create a new position. I had taught Captain Bill Shepherd a couple of lessons.[7] Then I'd gone over to Russia to work on some dual-language training manual development for the life support system. While I was there, I observed some things.

When they put up a diagram of a carbon dioxide removal assembly in class, they were using different symbols for valves. They were using different symbols for pumps. All the graphical symbols were different. The theory of removing carbon dioxide from the air in space is the same, yet they would go into a deep level of theory. On the US side, we would only cover a certain amount of theory. I thought the theories are common, perhaps we can integrate our training classes and save the crew some travel time.

When I started looking at the written documentation, again I realized the discrepancy for the training manuals we were making in the US and the training manuals they were making on the Russian side. It was up to the crew to integrate the information; same thing with our displays. The crew was being told, "You're going to control the operation of these pieces of hardware via the laptop displays, but we're going to design the US laptop one way and we're going to make you learn a totally different set of standards for the Russian laptop."

The reason nobody had seen this was because we were siloed. The Russian counterparts were developing their things. We did have some operations folks, but they did not get the level of access that I had from going to classes with the crew. They got whatever their counterparts would feed them.

It was a really great opportunity to see things from the crew's eyes. I think it allowed me to explain the level of frustration that that first crew was seeing as they were going through this training process. I put together a summary, and I talked to my boss. "I think there are opportunities here for us to improve integration, not only in the training products but also in the operational products. I'd like to create a new position." I wanted to keep my instructor title so that when all this was over I could go back and teach, because I loved that aspect of my job. I said, "Let's call it Russian training integration instructor, RTII." That sounded good. I wrote up a position description. I gave it to my boss, who then gave it to the head of Mission Operations—at the time, Randy Stone.[8] He said, "Sure."

It was such a unique opportunity. When I talk to co-ops I explain to them that when we're developing a new program, we don't always get it right the first time. It takes folks that can go in and survey the situation, figure out where the holes are, and then be able to put a comprehensive story together to pass on to management about plans to address those holes. That was a perfect example. It worked very well. That RTII position still exists today. I think we still have four folks that are doing it, and it has proved invaluable in getting the integration with our Russian counterparts.

When I first started this, all the Space Station modules were on the ground. We hadn't flown any yet. I got to go to the technical reviews inside the modules, basically the acceptance testing to declare them ready for launch. The training diagrams that I have verify that these are indeed where the lines run. I got to go inside the Functional Cargo Block (FGB), got to go inside the Zvezda Service Module when they were getting ready for shipping in Kazakhstan.[9] I also got to go inside Node 1 and the Laboratory Module when they were at Kennedy Space Center.[10]

Wherever the crew was, I was. I'd explain to the crew some of the tasks they were going to be performing on orbit. "You're going to be installing this particular system or this valve or this rack." I got a lot of hands-on experience that not a lot of other people did. I spent so much time in there that when something fails in one of those modules onboard ISS, I can still see it in my head. When the crew calls down and says, "I can't open this valve." I think, "Oh it's right there." It was very valuable experience, especially for an ISS flight director. But I had no idea of that at the time.

I spent about 65 percent to 70 percent of my time in Russia during that four-year span between 1997 and 2001. I guess it was about a total of three-and-a-half years. The prime modules that were going to be up for Expedition 1 were mainly the Russian modules. The Laboratory Module didn't come until later in their increment.

I was in Kazakhstan for the launch of Expedition 1. It's an embarrassing story. After spending four years with these guys, they were like my kids. When I was at the launch, I could watch them getting in the capsule on the TV. As soon as they shut the door on the capsule I started hyperventilating. I turned to Scott Kelly, who was one of the management astronauts there.[11] I said, "Oh, I know how my mom felt when I left for school the first time. You raise these kids, you put them in a spacecraft, and there they go." I was very, very nervous.

One of the Russian generals saw me. I think I was shaking at the time. He says, "Come." He pulls me through this crowd so I could see a bigger TV screen. As we're going through the crowd he's like, "*Eto mama eki-pazha, eto mama ekipazha.* It's the mother of the crew, it's the mother of the crew." He plopped me down in front of this TV and gave me some tea. He said, "It's going to be okay."

And it was fine. It was a foggy day. About two minutes before the launch the fog lifted, and we could see liftoff. Their adventure began.

I stayed in Russia until January. I worked in Mission Control Moscow as a crew support engineer. Bottom line, they wanted somebody there, someone that had watched their training and understood how they would respond to certain situations if the crew had specific questions. It came in handy a couple times. I came back in January when the lead control of the ISS transferred from the Russians to the US. That occurred with the launch of the Lab Module. Up until then Moscow was in charge.

I came back, and there was not a slot for me in Houston Mission Control, but in the background, I assisted here and there. Once the Expedition 1 crew returned to Earth in March, I took three months off. I was tired.

I was investigating other jobs that I could pick up in Mission Control. Each one of them was interesting, but I felt like I had been given this opportunity and I had this broad breadth of knowledge across the Space Station, plus a broad breadth of knowledge about the crew and what they're thinking and what training they receive. I wanted to be in a position where I could utilize that. I didn't see any individual console position in Mission Control that would allow me to do that. I met with Randy Stone and talked to him about what I thought I wanted to do. He said, "I think you'd make a good CapCom (capsule communicator), because a CapCom is supposed to take all these technical discussions that are going on in Mission Control and formulate them into words that make sense to the crew.[12] It helps to have a good baseline knowledge of the systems and good understanding of the crew. That'd be perfect."

I said, "Well, Randy, yes, that would be perfect—if I were an astronaut, because only astronauts are CapComs."

He said, "Oh, we'll talk to them. Do you know why only astronauts are CapComs?"

I said, "No."

Randy explained, "Well, historically the folks in space wanted someone on the ground that had seen what they had seen, been where they had been, and walked in their shoes. How many people do you know that have flown on Space Station so far that can say that?"

I said, "Oh, that's a good point, Randy."

He said, "You're the closest thing we have to that, so give it a go. Maybe, if you're not too bad at it, this could be a new career path for other people."

I thought, "Well, there's no pressure there!" He negotiated it with the head of the Astronaut Office. They let me do it. I worked my first shift during Expedition 3.[13] I remember being so nervous.

I walked in for my shift. Who was the flight director on console? Norm Knight—who's my boss now, ironically, in the Flight Director's Office.[14] Norm is very serious. I sat down. He asked, "Who are you again?" I explained to him that it's a new concept. No one had seen me because I'd been going back and forth to Russia, so I was unknown.

The crew makes their first call. Frank Culbertson calls down.[15] I respond, and there's a pause. Frank says, "Ginger, is that you? Today is your first day, isn't it? How are things going?"

I look at Norm. "Can I answer?"

"Sure."

So I answer. Then, again, there's a pause. A little while later we heard, "Gin, is it really you?" It's our cosmonaut Vladimir Dezhurov, who had not spoken to Houston Mission Control yet on his mission.[16] He says, "Oh, Gin, it is like present from MCC (Mission Control Center) to hear your voice." Now Norm is looking at me like I'm really nuts, but it was a great icebreaker for my first day.

I think it was the very next day my technical expertise helped out—so Norm didn't think I was just there to be the chatterbox friend of the crew. There was a Freon leak in the Russian air conditioner. As I'm listening I'm calling up a diagram, because that's part of my life support system. As soon as the crew says, "Oh it's in the large diameter hose," I show Norm.

I said, "This thing is leaking right here. Based on the amount of Freon that's in there, all the Freon could leak out and you'd still be below the SMAC (spacecraft maximum allowable concentrations) limit. So there's no safety hazard to the crew."

He looks at me like, "How do you know this?"

It was great because I could understand the Russian that was going on, and I knew the system. It was that day where I thought, "Yes, I'm here to help." I knew something that I could share with the team.

I did that job for several increments. I think it was Expedition 3 through 10, about four years. I loved it, I loved it, I loved it. Sitting next to the flight director, watching what they do, I thought, "Hmm. I like what I'm doing now, but one day I might want to do that. I might want to lead this team." I chewed on that for about four years. Then I started talking seriously to folks about three years in, and I decided I was going to apply. I had the knowledge of the room, the systems, and the crew. I had a desire to be in a leadership position. I thought that would be the next best thing to go off and try.

No CapCom had ever been a flight director before. That would mean an astronaut had been a flight director, and that certainly hadn't happened. But then no non-astronaut had ever been a CapCom before, so I figured hey, I'm going to give it a go.

There had been a few folks that had grown up in the Training Division that had crossed the lines. I had talked to them before my interviews just to try to understand. Historically you would work a console position in Mission Control for a number of years. You commanded the vehicle. You'd get a management position and then that would qualify you to be a flight director. Because I'd taken a different path, I needed to be sure what areas I needed to emphasize, either on my application or during the interview process, to make sure that my skills were equally weighted with the standard skills.

The process for applying to be a flight director is just like for any other job here, although it's not like any other job here. There's a call for applications, and you fill it out. I was selected in February of 2005, and I certified in September of 2005. It takes seven months of training to become a flight director. I was just certified on Station. I trained on Shuttle after I was in the office about five years. I was the last flight director to go through and certify on the Shuttle.

My boss called me into his office, and he said, "I have an opportunity for you."

I swear I almost cried. I accepted immediately, "Yes, I would love to do that." I certified—I think it was four months of training. I just dove into the systems and spent a lot of time in that Shuttle Flight Control Room, which was very different from the Space Station Flight Control Room, learning the lingo. The communication style was very different; the management reporting very different. I worked my first mission in March 2010.

Let me give an example. Communication: On Shuttle, the crew is there for a twelve-to-sixteen-day mission and they've got a job to do. I don't have time to be having extended conversations. We've got this mission timelined and just have to make it happen. On the Space Station side, my guys are up there for six months. They're looking to us not only to guide them through their workday but to provide some source of psychological support and entertainment, as opposed to the crewmembers on a two-week Shuttle mission. So we talk to them about different things in the Station environment and we have a different pace, unless you're facing a critical situation, in which case you flip right back to Shuttle mode. There's a distinct difference in how you approach the crew, how you talk over things with the crew, even on the loops with the flight control team as well.[17]

Management: Our Space Station program manager and our Shuttle program manager are very different; they're different individuals with different expectations. We usually work with the IMMT (ISS Mission Management Team) chair for most of the standard ISS activities and don't usually work directly with the ISS program manager.[18] Whereas in a Shuttle mission, the Shuttle program manager was up in the viewing room and conducted daily MMT (Mission Management Team) meetings.

We were more autonomous on Station, but we had to be cognizant of some of the decisions. That's why we have our flight rules and our procedures. We're autonomous operating within those constraints. When you go outside of that, you want to make sure that all the right folks, at least, are consulted or have an opportunity to voice their opinion. A Shuttle flight is often compared to a sprint, with the Space Station being a long-distance run.

The first crew that I worked with as a Space Station flight director was Expedition 12, but the first lead assignment that I got was Expedition 14.[19] I got certified in September 2005. One month later I was notified I would be the lead for Expedition 14, which would be flying in a year, so it was pretty quick.

Michael "L-A" Lopez-Alegria was my commander, who was also my first boss out in Russia when I went out to work in Star City.[20] So that was ironic. Then my other crewmember was Suni Williams. She was an astronaut support person for the Expedition 1 crew, so I'd already worked with her for four years. The third crewmember was Misha Tyurin, who was my backup cosmonaut for Expedition 1. Again, I'd already known him for four years. So I didn't have to spend any time at all getting to know my crew. I already knew them quite well, which was an advantage.

You get handed a set of requirements by the Station Program about a year out. You work with the teams to design your training plan and figure out how you're going to get all these objectives accomplished in the time that you're on board. Six months out it'll change. Then it changes again, and it changes again. So you spend most of your time trying to keep up with the vehicle. There would be Shuttle missions that would start off in your increment, but those plans would change, so you would make adjustments to the training plan and the operations to account for those. About four months in is when it started to settle out and we were able to get a handle on what we were going to have in store for us.

As lead flight director, I lived in Mission Control every day. Now we have what's called a flight director suite. About six weeks before launch I was in presentation readiness mode, where I had to do the stage operations readiness review, the flight readiness review, give presentations, get everybody up to speed. The moment we launched, I assumed my position in Mission Control.

I had a corner of a little console, because that's all we had, that I reported to every day for seven months. Every day at six thirty a.m., seven a.m., I would report for duty. I would go home whenever the last meeting ended, around five thirty. I would take my computer, I would plug it in

During STS-133, Flight Director Ginger Kerrick monitors events as they unfold from her console in the Flight Control Room.

at home, I would log on, and I would work in the evenings. This is what you do as a lead flight director.

The evening work is really just to put out any brushfires that could be brewing so that folks aren't doing extra work. Ultimately, I had the working knowledge of everything that I needed to do. The on-console team can crank away and handle things, but if they ever have a question about what this means long-term or how it impacts the overall increment objectives, it had to go through me, so I had to stay engaged the entire time.

There's always someone on the flight director console, but the increment flight director has no on-console responsibilities. I was there every day because things change every day. Although I was sitting in the back of the Flight Control Room, I wasn't working on things occurring on orbit that day. The on-console flight director did that. I'd be working on other stuff—for the next week, the next month, longer lead items.

Things come at you all of the time. It is busy-busy-busy nonstop. That's not even dealing with what's happening during the workday. They encourage you to take time off. I never did. I didn't have any time off during that seven months. What I did do to maintain my sanity is run. That's how I would de-stress. I'd call the console and say, "If anything weird happens in the next hour and a half, I'm going to be running. I'll call you when I get back." I actually wound up training for and running a marathon. It kept me sane. We encourage people to take off a week here and there. We do have flight directors that'll come in and step in, but most people don't take that time. It takes over your life for an extended period of time, but it's fun.

We had lots of challenges on that expedition. We did the first triple EVA (extravehicular activity).[21] You go in knowing that you're going to do three EVAs in the span of nine days. That was a challenge, but our crew came through with shining colors.

There was one event, however, that I was not happy with. Progress vehicles come and bring supplies.[22] During my mission, this Progress vehicle is coming along. It's supposed to retract its antenna because there's a physical interface between the antenna and the handrails on the module. If the antenna doesn't retract, it's supposed to retreat per our flight rules.

It comes in. The antenna doesn't retract. The on-console flight director calls over to the Russian counterpart. "Per the flight rules, I assume you're going to back out and station-keep."

"No. We're going to press with the docking."

I was working the management console during that mission. The flight director kept trying to get them to station-keep. They're continuing to come in. So I pick up the phone, and I call my Russian counterpart over the phone. In Russian I say, "What are you doing? Our flight rules say to retreat."

He says, "No, no, my engineers tell me that the structural integrity of the antenna will yield before the structural integrity of the hull."

While I am having this conversation on the phone, the on-console flight director is talking with her flight director. They kept pressing with the docking, so we got them to remove the crew from the module port that the Progress was docking to and seal the hatch in the event that something did happen. Sure enough, the antenna gets wedged underneath the handrail. We wound up having to do two spacewalks to saw that handrail off so that we could move the antenna and undock the Progress.

That was memorable. I could live my whole life and not see that again. It did cause us to have some pretty serious discussions afterward with our Russian counterparts to get them to acknowledge that they should have abided by the rules. The rules are there for a reason. That's my most memorable adrenaline-flowing moment, I think, from that particular increment. We had a lot of good memories, though.

L-A used to have music trivia. He'd play a song and each Control Center got to play along. The CapComs had to call up and guess the song title and artist and then you won something, although we were never awarded any prizes. I think it was just to bond with the ground, but he made it a good seven months.

I took a couple weeks off after this expedition. When I got back I asked to be assigned to be the flight director for the Generic Joint Operations Panel. That is a panel that the Flight Director's Office owns where you have new concepts for operations, or new rules, or anything that will affect how we do things generically day-to-day for ISS. That board needs to approve it before it goes off and gets implemented in the Ops environment.

Having had things like this Progress incident and several other lessons learned in that increment environment, I thought there were a lot of generic improvements that we could make for future increments so they wouldn't face some of the things that I faced. I asked if I could be placed in charge of that board so I could help implement some changes while it was still fresh in my mind. I was allowed to do that, I think, for about a year and a half before I got assigned lead to STS-126.[23] I had a good time.

I had a great crew. Fergie—Chris Ferguson—was my commander.[24] Originally, we didn't have anything defined. Then 10A, STS-120, discovered the solar alpha rotary joint (SARJ) on the starboard side of the vehicle was eating itself away.[25] I liken it to having a problem with your shoulder's rotator cuff. That's what this solar array joint was doing. We found particles out there. We wound up having to park the SARJ, which greatly reduces the amount of power that the arrays outboard of that can generate. We realized that if we did not fix this, we wouldn't be able to power our international partner modules that were coming.

We threw a whole bunch of folks at this problem: a lot of engineers, a lot of Ops folks, a lot of materials researchers, safety folks. For the better part of a year, we met twice a week. Folks just needed to understand the mechanism that was causing this, then figure out how to clean up the damage and prevent it from happening again. It took a long time. It was new techniques, new hardware, new everything.

I think that's what made it the most fun, because we didn't know where we were going in the beginning. We got to plot our course for ourselves. The crew in the Neutral Buoyancy Lab would help, saying, "I don't think that technique is going to work; I think I have a better technique," and work with the engineers.[26] The crew participated in this whole experience.

On my mission we had four spacewalks dedicated to fixing the SARJ. Three of them were dedicated to cleaning and fixing the starboard SARJ. The last EVA was dedicated to applying some lubrication to the port one so that it would not incur similar damage in the future.

It was a pretty memorable flight, not only for the SARJ, but we also were bringing up the regenerative ECLSS equipment that would allow us to go to a six-person crew. We were tasked with installing certain pieces of that equipment and generating water with it during the docked timeframe so that that water could be flown back down and tested to give a go/no-go for the six-person crew. While we had all the EVAs on the outside, we were tearing up the inside of the vehicle installing all this equipment and starting the test so that by the end of the mission we could get this water sample.

It was down to the wire because we had a failure or two here and there, but we wound up getting the samples that we needed back on the ground. They were tested successfully, and we went to a six-person crew when we were supposed to. It was great. Same thing with the SARJ; we were able to clean out the joint, lubricate it, and it's still rotating today. I think we've had to apply a second set of lube on one mission since then, but now we have a routine for how often we need to do that. It worked perfectly.

MOD has changed quite a bit, I think, since 1995. When you look at the old pictures of Mission Control, the men in the white shirts and skinny ties and buzz cuts, the mix of folks has changed. I think they're still just as young as they were. Back in Apollo days they were pretty young. Today we still have a lot of youth in there, so that's good, especially since the average age of our NASA employees is starting to get a little bit older. It's good to have those young folks in there still enthused about the program.

I remember one briefing about women at NASA and how we needed to increase the number of women supporting NASA. Then I went to Mission Control later that week, and we had a loss of com so we all went to the restroom.[27] There was a line in the women's restroom but no line in the men's. I thought, "We've officially made it. I'm waiting in line at the women's restroom in Mission Control, and I am so happy about this!"

I was named to be the lead for the Space Station flight directors in 2011. When we were flying the Shuttle, we had the chief of the office, the deputy chief of the office, then we had a lead for Space Station and a lead for Shuttle. The lead for Shuttle, of course, has gone away, but we still maintain that lead for Space Station. We also have leads for Commercial Crew and for multipurpose crew vehicle and our visiting vehicle program now.

I'm responsible for ensuring we have the right folks trained and supporting shifts in real-time and in simulations. I also provide guidance to folks working current and upcoming increments. I used to be the technical person. I would get up, and I would defend my technical truth of the day to the program or to whoever I needed to. Now I'm at a point where the folks under me in my office, it's their turn to be in the spotlight. My job is to make sure they have enough of those moments to establish themselves in this program, to defend them, and just help guide them.

I like that aspect. I have enjoyed making that switch from being the lead technical person to being the manager that enables employees to do what they do and to get better and to be the bright shiny example in MOD. The last week has been a busy week with the spacewalk that didn't go exactly as planned.[28] We had one of our young flight directors leading that spacewalk. We had another one of our young flight directors leading the associated increment. They've been working together and solving problems, and we've been providing guidance as required. "You might consider this. You might want to talk to this person about that." But they've been doing the work. I sat there in the IMMT when we did the go/no-go, and I was just so proud of them both. I've only done it for about a year, but I think I'd like to keep doing it for a little while longer.

I still get on the floor now and then. I pull a shift probably once a month and then I sit at the management console for dynamic operations involving the Soyuz and the Progress. So I still get in the Control Room, and I'll do a few sims. I'm not at a point where I completely miss it because I still get my feet wet. Now I enjoy the management aspect of it more than the day-to-day.

I've had a lot of girls—when I go to conferences for women—ask if I've encountered any roadblocks here in either category, women or minority. Since I've been here I have never felt like an opportunity that was out there wasn't made available to everybody. It's never even been a question. When I talk to some of the kids, I explain that if it's not an expectation in your mind when you come here, you'll never see it, whether it's minority or female. I've honestly never run into a problem.

My greatest accomplishment—I was walking down the hall one day, probably about four years ago. This kid came up to me. He asked, "Do you remember me?"

I said, "Oh, I'm not sure."

"I'm from New Mexico State University, and you came to talk to me once. I just wanted to tell you that you're the reason I'm here."

I had to leave the hallway because I was going to cry. You don't cry in the hallways of MOD!

I think about all the technical things I've done here. Sure, I've helped to solve a problem here and there, but if you ask me about my greatest accomplishment, if I can get two or three more of those moments in the hallway, that's what I'll walk away from here with. That made my year. I know it sounds dorky and sappy, but it's true.

My biggest challenge—the hardest one—was the internal shift when I had to give up on the dream of becoming an astronaut. I was thinking I'd send in my application letter and get the coolest rejection letter ever because I was only twenty-six. But to get the call and to make it through interviews, the disappointment that I felt was incredible. When I talk to co-ops now I talk to them about this. I could have very easily taken a different turn. I was miserable. I could have given up on this place because I didn't get what I wanted. I had to convince myself that I was going to look at it from a different perspective.

Initially I said, "I can't even continue to teach. Yes, I like being an instructor, but now I'm going to be face-to-face with the astronauts every day—the people that hold the job that I wanted to have."

I made myself stop thinking that way. I said, "No, you're going to wake up every morning, and you're going to say, 'I can't be an astronaut. There's nothing I can do about it. But if I teach these folks, each one of them can carry a piece of something that I shared with them up into space. Each one of them will carry a piece of you up into space.'" Every morning I said that to myself when I woke up and I didn't want to get out of bed because I was miserable.

They tell you this in these self-help books. You say it enough, and you start believing it. Until you do it, it's just a line that somebody feeds you. After a few months I didn't have to tell myself anymore. I actually started believing it, and then I started having a really great time. It was right around that time where all these other doors started opening then: Russian training integration, CapCom, flight director.

I sit here seventeen years after my astronaut interview, and I can't imagine myself being anywhere else. This is exactly where I was supposed to be. I think that was my biggest challenge early on, making that mindset change, but it has brought so many rewards, and I encourage folks to take themselves out of that pool of misery long enough to have an objective assessment of the situation and see if there are other directions you can go.

Sarah L. Murray[29]

Official NASA portrait of Sarah Murray.

I got a phone call from RSOC, Rockwell Space Operations Corporation. They wanted me to come to Houston for an interview. My husband was in the military at Fort Gordon in Augusta, Georgia, so I was there. That call happened because I had applied for the co-op program at NASA. While I was in school, at the University of Texas at El Paso, we neededw money. NASA was offering six dollars an hour. El Paso Natural Gas was offering ten dollars an hour, so I co-oped at El Paso Natural Gas, but they still had my résumé at NASA.

I didn't know what Rockwell was or what they were involved in. I didn't really do any research. Typically you do some research about the company so that you mind your p's and q's during the interview. But my husband had orders to go to Germany, and I knew we were going there. My husband said, "It's a free trip to Houston. You can brush up on your interviewing skills. Go on, do the interview." I got here, and I was taken to Rockwell's red brick building. I was interviewed by a few folks, still not realizing that RSOC had something to do with space.

During the interview they kept saying that we will take you "onsite" later. Onsite? People don't normally use that term unless you're associated with something that uses the terminology. They put a group of us on a bus and drove us over, and we were all yakking. I missed going through the gate. I didn't know where I was. We drove up to this building, they took us inside, and there were all these console positions. The tour guide is talking to us and explaining, and my eyes started getting big. "This is Mission Control!" I just couldn't believe it.

Afterward when I got to the hotel room, I called my husband. I hadn't planned on taking the job because we were going to Germany. I said, "Dear, it's the Space Shuttle Program." Oh, my gosh—this is awesome!

We decided, well, we do need the money. If they offered me a certain amount, then I'd do it. He'll go to Germany, and my son and I will stay here in Houston. This tells you just how naïve we were. I had no idea how much they would offer, but it was more than we thought. That's how I ended up here the first time around.

Going into the space industry has been awesome for me. It just fell in my lap. It's been great because there are different types of engineers. There are the engineers that have to know every little bit that's in a circuit. That's not necessarily me.

What we do, at least in Mission Operations, Mission Control, is real-time operations. That was a great fit for me because I can relate it to the emergency room where you needed to make decisions fast. You needed to triage. During Space Shuttle missions you have things that are going on; you have to triage, prioritize the work, and figure out how to do it. You didn't have a whole lot of time to get it done, especially during ascent, when there's just eight minutes to work any failures. So there was something about that aspect of the job that was good for me, and it fit me.

I told my husband that there would be three things that I would love to do. If I did any of these I would be happy. One would be an astronaut, another would be a flight director, or a stay-at-home mom. Any one of those would be awesome. I didn't get any of those, but I was so close. It's that old adage of you shoot for the Moon and you land among the stars. I applied twice later in life for the astronaut corps. I spoke to Astronaut John Casper about applying. He said, "I applied six times. You need to keep on."[30] Had I started earlier I would have continued, but I think if I were selected now they'd really only want me to do experiments on menopause.

I started in the communications groups, a position called INCO (instrumentation and communications officer).[31] I was at JSC for two years while my husband went to Germany. We thought he would be back in two years. Something happened and he had to stay longer, so I ended up having to leave and join him in Germany.

After my husband's tour, he was stationed at the Pentagon. I'd just had my second child and we were living in Springfield, Virginia. I started looking through the newspapers. It was time for me to go back to work. I saw an ad from a company called Computer Sciences Corporation (CSC). I said, "Let me go check this one out." I got on the road, started driving, and it was a lot further than I thought. I started thinking, "Sarah, why are you going? You don't like computers; it's cold."

I finally get there. There are hundreds of people waiting to be seen. They tell you to turn your résumé in, so I turned my résumé in. We're in this huge cafeteria. I sit down. I'm there for ten minutes and they called my name. They didn't call my name for the job that I applied for. They looked at my résumé and said, "I know someone who's looking for a flight controller." I had had the flight controller experience here at JSC, and it got me right into the flight dynamics area there. I became a flight controller at Goddard Space Flight Center in the Flight Dynamics Facility.

I was hired as a mission coordinator. We had ground stations all around the world, and we had satellites and antennas that would point to the Space Shuttle or ELVs (expendable launch vehicles). The mission coordinator's job was to coordinate with all the different entities that had to provide data to the ground facilities. The coordinator gathered all of that data from Goddard and the different entities at the ground stations so they had the correct data—pointing data primarily—to point to the Space Shuttle or the expendable launch vehicle.

My supervisor told me a couple months after he hired me that I did not belong in his group. He said, "I hired you because the group that you do belong in, I don't think they would have hired you." What he meant was that the group that I should have been in was an all-white, male group at the time. He just knew that they wouldn't hire me. "I'm getting your foot in the door." The group that I hired into were not degreed. I didn't know, didn't really care, at the time.

In about eight to nine months I transitioned to the mission manager group, which is the group that's actually over the mission coordinators and the entire mission. I moved into the mission manager job. If I could compare it to our setup here at Mission Control, it's like the flight director in the Control Room during missions. The mission manager was the lead for each of the missions and pulled everything together. The mission coordinators had a smaller scope than the mission managers did, and the mission managers had to report to the program managers. I did that job for a few years, maybe two years with CSC. I was later hired by NASA to work as a network manager in the Network Control Center.

The transition was good. I don't think I gained the rapport with the supervisor that we should have had. I just didn't feel as comfortable as I felt with other supervisors, I should say. As far as the actual fellows that were in the group, we had no issues whatsoever. Typically, it is, I would say, abnormal to run into issues. There are going to be onesie-twosies that are out there, but for the most part folks are doing their best to treat everyone as equals. Also, there's an element of not even being aware that you're treating someone different, and I know that's out there, too. As far as folks being blatantly unfair about it, I think that it's more abnormal, at least in my case, in the environments that I have been in. It's still there, but it's probably not as bad as it was when I first arrived at JSC.

I remember a comment being made back in '88 at JSC. I hired on, and my girlfriend and I were in the hallway, and she was an African American, too. Some guys walked by us talking to each other. They said, "Oh, it's starting to look like the ghetto around here." What's unfortunate—well, it's fortunate and unfortunate—is that you learn not to let that deter you. You don't want to totally ignore it because you want to make sure that when it needs to be addressed, it's addressed. You can't let it deter you or cause you to behave in a manner that's inappropriate or unprofessional, but it's still there and you just have to be smart and know how to handle it.

To get back to Goddard—I worked as a mission manager maybe a year or so. The network director at that time was Gary Morse.[32] He heard that I had been in the INCO group at the Johnson Space Center. He worked in the Network Control Center, the NCC, at Goddard, and they are responsible for the TDRSS (tracking and data relay satellite system).[33] He asked if I was interested in being a network manager. I worked side-by-side with him.

From there my husband got orders to go to Hawaii. I was fine with going to Hawaii. Then we started thinking. My husband was about three years away from retirement—if we both go to Hawaii, when he retires we'll both be looking for jobs. We'll probably come back to the continental United States. It'd probably be wiser if we had something established here.

Not only that, I was really tired of the cold weather at Goddard. At the time a friend of mine at Goddard had gone down to Johnson Space Center, applied for a job, and had used me for a reference. Someone from my old group called and said, "We found you! Would you like to come back to the group?"

"Yes!" All of that happened at the same time. I ended up coming back to the exact same group I was in before I left.

I was with RSOC the first time. When I became a network manager I became a civil servant, so when I went back to the INCO group I was a civil servant.[34] That was the only difference; I was doing the same job. It was great because I was already familiar with INCO. Some of the folks that I had known were still there. I left in '90 and came back in '96.

My husband went to Hawaii. He took the teenager and left me with the two-, three-, and four-year-olds. I think they were there for about two, three years. I came to Houston with the three little ones and started getting recertified in my communications position.

Within communications, at least while I was there, there were four positions. There were three back-room positions and the front-room position in Mission Control. Those are the folks that you typically see on TV. Those people in the front room have a host of people in the back room that are supporting them. You work your way up through the back room to the front-room positions.

Certification usually starts out with reading and then the single-system trainers. They're really mockups of the Space Shuttle cockpit. Crewmembers' training and the flight controllers' training are different but also, in a lot of ways, the same. Because we are on the ground and are commanding the Space Shuttle or we're telling the crewmembers what to do, we do a lot of the same training, so that we know — "Yes, they can do this," or "Yes, they can reach this from where they're sitting" — because we have to tell them what switch to flip. In the single-system trainers we either sit in the commander or pilot seat, and we go through the procedures just like the crewmembers would. We flip switches, and we make mistakes so that we understand what all of that means.

After you get a certain amount of training under your belt, you are ready to sit in Mission Control in a training environment in a simulation, where it's integrated with all the other systems and flight controllers. Mine was communication. Then you've got your environmental specialist, which we call the EECOM (electrical, environmental, and consumables manager).[35] You've got your prop.[36] You've got booster.[37] The simulation starts out nominally; things are going smoothly. You're watching your screens looking for any indications of problems, or malfunctions, as we call them.

If you see something that's abnormal, you report it to the flight director. We call it FIW (failure impact work-around). What's the failure? How does it impact you? What's the work-around? That's really what you want to learn to do. When you have malfunctions, you see issues, you understand what the failure is, you understand what the impact is to your systems and even to other folks' systems. You're impacting a lot of folks when the com goes away. Then there are other systems where a piece of the com fails.

For instance, the instrumentation position, which is a back-room position for the INCO group, was responsible for the communication between the payloads. You may have a payload on board, and it may be connected through a hardline or via an RF (radio frequency) link. The payload signal processor—we call it the PSP—is typically how we communicated with the payload if it was hardline, meaning attached to the vehicle. You can command to it. You could get data from the payload to make sure it's healthy. Then you have the payload interrogator. Some payloads would start out hardline, and then that payload is deployed into space. It's no longer hardline, but you have an RF link. Then we turn on the payload interrogator. It's still talking to the payload, we're still getting data, and we can still command to it.

Certification required learning all of that. The fun part is ascent. In ascent certification you typically learn what you can do quickly to recover. When you're on orbit, depending on the failure, you may have a little time to work some issues. During ascent you just want to recover. I don't really want to know exactly what's wrong, just recover until we get on orbit then we'll fix it. So that's the difference between the ascent and the on-orbit piece.

I started in the middle with instrumentation. The back-room positions were different levels. We had our recorders. That was typically an entry-level position, because all they did was command into the recorders. They recorded data on board, then we dumped it when we got to a ground site or had the TDRSS satellite. It was a good entry position because you couldn't hurt anything. It got you comfortable with commanding to the Space Shuttle. You could make errors in your commanding, and it wasn't detrimental to the mission. Instrumentation was responsible for all the telemetry coming from the Space Shuttle, to include the payload data, the PSP, and payload interrogator.

I certified as instrumentation first, then ascent/entry. We call it orbit instrumentation or orbit inst, so I certified as an orbit inst. Typically what comes after that is the ascent inst. That's a little bit easier. The orbit inst—because you're on orbit and you need to understand the systems well and understand the malfunctions and know how to work through them—takes a little bit longer. When you get to ascent inst, you already know this stuff, now you just have to learn all the quick remedies to getting through your malfunctions. It wasn't quite as difficult.

From there I went to the next position, which is RF com. I started training; however, that was delayed because I saw another position I was interested in. There was a position called RIO, Russian interface officer.[38] I said, "Wow, that sounds interesting." So I applied. I stayed in INCO, but I gave 25 percent of my time to the RIO group. I started with Mir, then I went on to the Space Station.[39] I always had an interpreter that was sitting next to me on console. The interaction that I had with the interpreters was interesting. RIO was a good job; you didn't have to know the details about every system, but you had to know enough so that when there was a failure you went to the procedure, read through the procedure, and followed along, because you needed to understand what needed to be communicated to the Russians.

At the same time I went through training for the RIO position, I started working console in the blue FCR (Flight Control Room).[40] I learned all about the Space Station. I was certified as ascent/entry inst, and at that point I think I had also gotten my ascent/entry data com—that's the other position for the recorders. There was a time when they combined them, so the ascent/entry inst also did the recording for ascent/entry. When we moved to Station I'd work my ascent/entry phase for INCO. Then I'd run over to the blue FCR and work the RIO position. That was pretty cool.

I didn't complete the RIO certification because I was selected to be group lead for INCO in 1999, and my branch chief said, "No more RIO. You can't do it. We need you to focus on this group." So I had to give it up, but I learned a lot from it, and it actually helped when I was eventually moved to Space Station.

I worked the night shift as the instrumentation officer. I think one of the things they liked about me was that I enjoyed working the night shift. Many of the group leads did not work console, but I felt like I needed to work console so that I stayed in touch with what the groups were doing and I understood the difficulties they were facing. Also, it allowed me to be available during the day if I needed to do office work as a group lead. The other thing that I liked about working the night shift is that it didn't take away from my home life. I'm at work at night when my kids were sleeping, I'm sleeping when they're in school, and then I'm up during the time that they were home. The night shift worked out for me, and I didn't mind being up at night.

I was the group lead for INCO until the Space Shuttle *Columbia* accident in 2003.[41] I had been on the night shift, and I handed over to the team that was going to bring the *Columbia* crew in. My husband had retired by then and was also working in the training area. I had gone home and was sleeping on the couch. My husband called; he was on a console. He said, "We don't have com with the crew."

I looked at where they were. "Well, that's typical when you get closer in. You start getting some interference." But usually we have some com. It's a little ratty, and it's got some noise. He said, "I don't know. We don't have any com at all." Then we discovered that the vehicle had broken up.

Some interesting things happened. I believe that things happen for a reason, no matter how good or how bad. We can always learn from them, and something positive comes out of it. When we got back in the office, I went to my branch chief and I said, "I'd like to have a prayer. Can I do that? Is that okay?"

He said, "No, I don't think you should do that."

I said, "You know what, I'm going to do it." So I sent an email out. I specifically said, "I don't want to offend anyone, but I would like to have a prayer. We will meet in this office. If you want to join us that's fine." I'm so glad I sent that note out. Everyone in the group showed up. Everyone. I thought, "This is awesome." Then we started pulling together all the data that we needed to pull together, certifications, and paperwork.

At the same time, I was trying to think of something the group could do to keep us pulled together. There was one guy in our group, Gary Horlacher, who did a lot of cycling.[42] I thought why don't we all start cycling and practicing for the one-hundred-mile Hotter'N Hell Hundred.[43] Some of the group said, "Okay, let's do it." So we started working on that together. I was just trying to do something to get their minds off of the accident, because we weren't flying. I wanted to keep morale up. We would have cycling 101 at Horlacher's house. He'd teach us about changing flats and all of this sort of stuff. Before you knew it, I was in the one-hundred-mile Hotter'N Hell. It was awesome. I think that was a great accomplishment. I rode my bike one hundred miles in one day.

It was good for the group, too. That's why I say something positive came out of that accident; I probably would not be riding my bike today. That may be minor, and I'm sure there are lots of other things that came out of it, but I love cycling now. I ride my bike to work every chance I get. I think it keeps me healthy. Some of the group members are still doing the same thing.

Also after the accident, I was sent to Barksdale Air Force Base in Louisiana. I was sent there to help collect debris. We were working with the Air Force, the Civil Air Patrol, FEMA (Federal Emergency Management Agency), the US Forest Service, and the Army. That was such a huge operation.

I worked with the Navajo. The Navajo consider themselves their own country. They have their own government and so they had their own FEMA point of contact that I worked with. When we were looking for debris, their tribe members were going to their summer camps, I guess that is how they refer to them. We had to drop flyers that had been translated into the Navajo language telling them that if they found debris, here are the people to contact. That was interesting, working with the Navajo.

I didn't know that I would meet so many different people in so many different fields. The NTSB (National Transportation Safety Board) was there too; they worked very closely with everyone. I started out at Barksdale Air Force Base, then they decided to start trying to consolidate some of the camps, and I ended up at Lufkin, Texas.

The NASA folks, the FEMA folks, and the military folks were all in one room. We'd have meetings at six in the morning and at six in the evening every day to figure out where everyone was, and everyone reported in. I would say that it didn't seem very different from the way we operate here at JSC. Everyone's trying to provide whatever information they had and provide ideas they had for solving any of the problems that we had. I remember one of the search helicopters crashed. Of course, everyone came together for that because those people were out there helping NASA, so everyone felt really bad about the accident.

As far as coordinating with different entities, it proved that we knew how to come together and make it work. Things just fell into place, and everyone knew where they fit in the entire effort. It almost seemed seamless, now that I think back on it. Everyone just knew what to do and who to go to if you needed help. It was a good thing, but if I say I'm surprised, it was because it just seemed to work so well with so many different organizations. No matter how different those entities were, they all had different resources that the others didn't. The surprising part was what resources they had.

NTSB helped us map out the areas that would be searched. They would receive results from satellite imagery. "We think that there's some debris here." They would give us a quadrant to search and what the square footage was. We'd work with Civil Air Patrol so that they could fly over and determine what kind of terrain it was before we sent people out there. A lot of times it ended up being old refrigerators and things like that, sinks that folks had thrown out.

Mission Operations sent us there for about ten days, then we would rotate someone else in. When it was time for someone to rotate in for me, the person that I was working with asked, "Who is replacing you?" I told him. He said, "Well, I don't know that person. When they get here, we're going to send him there, and you stay here." My ten days turned into three months. My poor family didn't see me for a while. Sometimes my husband would bring the kids on the weekends to see me.

I ended up leading a group of folks at Lufkin. We took a lot of phone calls from folks that thought they had found debris. We'd try to get them to send photos or describe it to us so we could determine whether we actually needed to go out and pick the debris up. The recovery was just a huge operation. We had teams in Arizona, California, Utah, Nevada, New Mexico, and Florida. I would contact one of these teams, depending on where we got the phone calls from, to go out and take a look at the debris and determine whether we needed to send it in. A lot of the search was concentrated here in Texas because that's where a lot of the debris was. The heavier pieces of the Shuttle landed in Texas and Louisiana.

All of that debris had to be cataloged and tracked. We worked with Weston Solutions, Inc. When we received calls for debris, I'd get maybe four or five pieces on the list before I'd call Weston Solutions and have them pick up the debris. They would take it to a facility in Palestine, Texas, the National Scientific Balloon Facility. They would take it there and let it collect until we had a certain amount, then they would put it on a truck and drive it to Florida. All the debris had to be put into a database. After it was all entered into the database, I flew out to Kennedy Space Center and handed over everything. I worked with the folks, trained them on the database, trained them on the procedure for gathering the information from the callers, who they needed to call to go pick up the debris, and making sure that all the right information was entered into the database. I don't know where it is now, but I can tell you that it was pretty huge. There were probably thousands of entries in that database.

We probably didn't start the database until later in the effort. We were concerned about everything not being in the database, but we haven't had to go back into the database for very much. At least, I haven't, and no one's come to me. But at least the information is there if someone feels they need it.

As you know, we had camps looking for debris. The flight crewmembers went to the different camps and thanked the searchers. In Nevada and Utah we actually had prisoners that were doing the searches. I think for each day that they participated in the search, they got a day taken off of their sentence. Those guys were out there searching, and some of them were getting hurt and breaking legs, because the terrain was pretty rough. I said, "We need to send a crewmember out there." Doug Wheelock and I flew out to Nevada.[44] We had to get on four-wheelers to make it up where the searchers were. We went up there and talked to the prisoners, and those guys were just thrilled. They had good questions. We also went to a school in Nevada while we were there and talked to some kids. For me that was pretty fulfilling because they were really intrigued and interested in what we were doing and what they were doing. I was glad to be able to get Doug to go out there and talk to them, and Doug and I have been good friends since then.

In East Texas the locals were so welcoming to NASA. My oldest son and his fiancée were living in Nacogdoches at the time. She was going to Stephen F. Austin State University. There was debris in Nacogdoches, and a soldier would stand and watch the debris until someone came and categorized it and did whatever they had to do. My son said, "Mom, that guy must have stood out there all day long, but he was really fed." People were coming by, and they would bring food; they would just do anything to help the folks that were involved in the effort.

We were in a hotel for a while, but apparently the hotel had already prebooked another conference. It had been on the books for months or a year, so they had to move us out. There were cabins in the woods, and the owners were offering up their cabins. It was nice. I got to go out and stay in a resort cabin for free. Everyone in the community was trying to help. They would bring food and drinks. You didn't have to ask for anything.

One night I actually had to drive home for something. My child had something scheduled at school. It was dark, the middle of the night. I was driving from Lufkin to Houston, and I got stopped by a police officer. He pulled me over. "Oh gosh, I'm going to get a ticket."

I had my *Columbia* recovery baseball cap on, and he asked, "Oh, are you with that team?"

I said, "Yes."

He said, "Well, let us know if you guys need any help and take care, make sure you get some rest." He shuffled me on my way.

That entire community was out to do nothing but help and assist us. That's all I saw. It was just amazing. It was encouraging that humankind is kind, and there are a lot of good people out there. A lot of people know that it's important to come together when there is a crisis, and to help folks out and help them get through it.

As the effort started to wind down, we were moving the *Columbia* Recovery Office to JSC. John Casper and Dave Whittle at the time were over the Lufkin efforts.[45] They said, "Sarah, we would like you to lead the rest of the effort." They were going to go off and do other things, but they needed someone to finish out the recovery effort. They said, "We need you to lead it at JSC."

I thought, hmm. There was another gentleman that I felt should have led it because he was out of Florida and he was the first person onsite at Barksdale that pulled everything together. I told them I thought he should lead it.

They said, "Well, we need it at JSC, and he's in Florida. You're at JSC. We need you to lead it."

I said, "Okay."

They said, "You need to find your own group of people also, because everyone else is going back to their own jobs, but we still need a small group of folks to finish this out."

I started just looking around Johnson Space Center for folks who were interested in working with me in the *Columbia* Recovery Office. We were located in Building 30. I found a small group of folks and we continued to work taking phone calls and getting debris. As a matter of fact, I got a letter from a little boy, he may have been ten or eleven years old. If he had not told me he was ten or eleven, I would have sworn he was an engineer because he sent a picture of the debris and he had it next to a ruler so that you could gauge the size of the object. He was in Arizona. It turned up at the same time my father-in-law had passed away—he lived in Arizona—and we had driven out to Arizona to attend the services. I said, "You know what, I'm going to go see this little boy."

I called him and said, "I'm going to come by and see you. We appreciate the information that you provided us." I didn't think about it at the time, but it was in the middle of the day, and his dad was home. They were dressed up. They even had a camera set up, so I was really glad to be able to do that. I spoke to the young man and I thanked him and told him that we needed more people that were interested in the space program. I just wanted it to be a good experience for him and his family. I love kids.

When I was running the *Columbia* Recovery Office at JSC, there were a couple of folks who would call on a regular basis because they felt some alien had destroyed the Space Shuttle and we weren't listening. Then there was the one that felt the Space Shuttle was destroyed because we weren't meant to go into space. They were interesting phone calls. We had to be very careful about what we did with some of those phone calls. There was a process. If they sounded as if they could be threatening or harmful in any way, then we had a phone number that we were supposed to call and report them. I think we only did that for one phone call. It was a different type of communication than I was used to dealing with on console.

The team of folks that I worked with was small. We actually became very close, though we're off doing different things now. One has retired. One was an SES (Senior Executive Service) at the time; I learned so much from this guy.[46] Another works in the Astronaut Office. The team that I pulled together for the effort became very close. They were really, really good.

When I got back to my group after working the *Columbia* recovery, my division chief said, "Sarah, we want you to move to Space Station." He asked me to move to another group that was struggling a little bit in the Station world. I probably hadn't been back more than two or three days when my division chief said, "We want you to move to the PHALCON (Power, Heating, Articulation, Lighting, and Control) group," which is the group that is responsible for the power on Space Station: solar arrays and batteries.

I said, "When do you need a decision?"

"I'd like to make the announcement tomorrow."

It was a hard decision, but what finally made me say yes was that I'd been the group lead for the INCO group for probably three to four years. It's always good to move on and open up opportunities for someone else.

I knew that there were at least two people within the group that could run the group given that opportunity. So I said, "Okay, fine," and I moved to Station.

One of the folks who was going to be my branch chief was trying to convince me to move over also. I said, "I know nothing about Station. I did a little bit of RIO."

He said, "Oh, that's okay. All you have to be able to do is spell I-S-S." I learned that moving can be very rewarding because you have different challenges. I had forgotten how much I enjoyed challenges and learning new things. So I became the PHALCON group lead. Moving to the PHALCON group, I had to learn about the solar arrays, the BGAs (beta gimbal assembly), the SARJs (solar alpha rotary joint), and the hybrid FET (field-effect transistor) failures that typically the RPCs (remote power controller) would have.[47] At first, I thought I was drinking from a fire hose, but in about four months I started to get very comfortable and I was happy with where I was.

The group needed some leadership. I remember having a really candid discussion with the senior people in the group, telling them that the group had a reputation in Mission Operations. "I don't know what's true and what's not, but we have to fix that. This is what I expect of you as the senior folks." It didn't take long before we were doing just fine, and we got those solar arrays up and going. That was before we had our solar arrays up with 12A, 12A.1, and all the issues that we were having with the solar arrays.[48] We were working really hard with the engineering folks because we were learning as we went. That was a tough time for the PHALCON group, and I think we did a good job marching through all of those issues. There are still some challenges with the solar arrays, but in general they're generating energy like they're supposed to.

I then applied for the branch chief of the Electrical Systems Branch. I got that branch chief job in 2005. At the same time we were about to put six crewmembers on board ISS instead of three, and our budget was being reduced. One of the USA (United Space Alliance) managers and I had been talking about how we were going to handle this.[49] We were going to have to find a better way for supporting twenty-four-hour ops. That's where the OSI (operator/specialist/instructor) concept started. We went to the division chief, and he said, "Well, go off and figure out how you're going to do that."

I got a few folks within our division from each of the different disciplines. We went off on a three- or four-day retreat. We pulled a flight director in and started thinking about different ways to do operations. One of the goals was to decrease the amount of time it took to train flight controllers to get on console so that we could start using them sooner. The way we had been training, they knew almost every detail about their system. The operator concept was to teach them enough to recover. They don't have to know all the details, but we'd teach them enough to do the routine work that happens during quiescent periods. If there is an emergency, then they'd know enough to recover from the emergency or "safe" the vehicle and call in a specialist. It would reduce the amount of knowledge they have and reduce the amount of training they need, and we can get them on console.

When we first started the Geminis, as we called them—they were the folks that were on console during the quiescent periods when there's not a whole lot of activity and one person can handle two systems. Actually, I think we had one person handling three systems. They were our more experienced people. It didn't seem right to have our more experienced people working the bad hours. With the operators, we're giving them a little bit less knowledge, and they're fairly new. They're just getting out of school, so they can handle those off-hours better. When you're right out of school, typically you're not married and you don't have a family, so these off-hours are okay. As you get older and have been in a discipline for a certain amount of time, you really would like your job to work with your lifestyle. Now you have kids and you're doing things, so we were trying to make it where we kept folks in the organization versus making it even more difficult as they became more experienced but they were getting worse hours.

We envisioned operators being the new fresh-outs and working the off-duty hours. The specialists would be the next level, where they had a little bit more training. They would be called in if there's a malfunction, and they would work the dynamic periods, like EVAs and reboosts and things like that, so they're not working console all the time.

OSI was also called Top Gun. The reason it's called Top Gun is because in the military, they use their most experienced people to do the training. The people who do the training have actually done the job. That's not the way Shuttle worked. We hired folks to be flight controllers and instructors, but those instructors had never actually done the job. There were many times we'd be on console doing a simulation, and the instructor would put a malfunction in, and someone on console would say, "That signature is not correct; that is not what we see in the real world." A lot of times the instructors wouldn't get respect because they didn't know what controllers saw in the real world. It made sense to us to have done the job before you start teaching. That's one of the goals of the Top Gun, to make sure our instructors are more experienced and they've actually done the work and can speak from experience. That's the "I" part of OSI, the instructor.

Around 2007, MOD started moving on their plan to reorganize. Part of the reorganization was to implement OSI, the Top Gun concept. That meant locating the instructors in the same areas as the flight controllers so that they're all working together. We felt it would be more efficient because some of the things the flight controllers were learning were the same things the instructors had to learn. Why were we teaching these things in different areas when we could have one? Then you teach them all the same thing at the same time. Where they need to diverge, then we start diverging.

We came up with what we called boot camp. Boot camp is for all new hires. This is all the information, all the training that you have to have whether you are a flight controller or an instructor. That way we prevented the training organization from teaching the same thing at the same time that the flight control organization is teaching it, and you're using all these resources separately when you don't need to.

That also gave us some standardization and consistency in what was being taught. Say everyone had to learn how to command, but ODIN (Onboard Data Interfaces and Networks) was teaching their folks how to command, CATO (Communications and Tracking Officer) was teaching their folks how to command, and the PHALCONs were teaching their folks how to command.[50] Just have one command class rather than having eight disciplines do it. Now it's just one class where everybody goes and they learn. I think that was pretty huge, because we reduced the resources used and we made sure that the training was consistent.

Opportunities for women and people of color have changed quite a bit. I would say for the better. There's still room for change. Within MOD, for the most part, I have not seen any barriers, but I do know that overall, just talking to other African Americans that work with NASA, and even some females, there's still some room for improvement.

One of the things I've talked about with MOD's director is the opportunity to reach out to minorities, women, and African Americans, or Hispanics. How much are we reaching out to the historically black colleges? When we do our recruiting, if we're going to predominantly white schools, then that's what we're going to get. Today, typically when we hire, we are hiring from our pool of co-ops, so if we don't have minorities, then we're not going to hire minorities. But after you get your foot in the door and you prove yourself as a co-op, then your chances are pretty good that you're going to do well in MOD.

When I talked about the OSI concept, I explained that we recognized that flight controllers had families. My experience as a woman and a mother did not play into my management style or decisions. I was taught to always work hard, period. For years—probably decades—I didn't care what was going on at home; I never mentioned it at work because I didn't want it held against me. I grew up in the age where they didn't want to hire women because they'd get pregnant, or they'd get married, or they'd have to take care of their children. So I never mentioned that my child had an appointment, or I had to go to the doctor. As a matter of fact, I was pregnant four times during my career and never took a day off until I went into labor, period. It was because I didn't want to fit that stereotype. I didn't want that held against me when there were opportunities to move or be promoted.

What I had to do was make myself a little bit more sensitive to the needs of our senior folks and their families. Even though I didn't apply it to myself, I knew that it was important. When I thought in general about how we were working folks, it just didn't sound right. It just wasn't right at all. These are our best people and we're shoving them off onto a night shift and the weekend shifts. We never see them. They could be working on projects and things to excel us, but the paradigm that we were in didn't allow that.

As far as being a wife and a mother, I would say that my family has been pretty neglected. I look back on that and I think I could have done a better job. I probably should have devoted more to my family. I was just so afraid that if I missed a meeting or a conversation or a telecon, then someone else was one step ahead of me, or someone would say, "Well, you should know that. Where were you?" I was just trying my best to stay on top of everything. I didn't want to represent the black community or the female community in a negative way. I didn't want someone to point to me and say, "Yes, she worked for us, but she didn't do a very good job," or, "She was out taking care of her kids." In hindsight, I should have devoted more time to my family.

It's hard to change, even doctors' appointments—I might get to the doctor every two years, maybe. My daughter-in-law said something very interesting last night. I had a telecon this morning at six. Depending on whether I'm conducting it, I'll either come into the office or I can just call in from the house. I said, "I think I'm going to call in," not realizing that most folks think "calling in" means I'm going to call in sick.

My daughter-in-law says, "What? You're going to call in? I want to see that! You would never call in."

I said, "Oh no, I'm not calling in sick. I'm just going to call in to the telecon." So even my family sees that I'm that type of person. I've never called in sick and just not made it to something. If I'm on vacation and there's something I can tie in via phone, I'm there. I need to stop that. There were times when I didn't take vacation. I'm one of those people that have weeks of "use or lose" leave. One time our division chief went on a three-week tour during one of our very first Station missions. I thought, "If he's expendable then I am too, and I can take vacation." It's hard to do, but I have to think of it like that.

One thing that's different now is men. I was recently interviewing for a job, and I went to talk to some folks in one of the organizations. I asked them about their supervisor. The guy that I was talking to was an older gentleman. He said, "He's always doing stuff with his kids." I see this younger generation, especially the guys, who know it's important for them to spend time with their family. The older generation is like, "What's up with that?" They're always out; the kids have a recital. Even the guys see that it's okay. It is most definitely different than it was years ago.

NASA's *flight directors, including the first woman flight director, Michele Brekke, gather for a photo outside the Mission Control Center.*

Technicians

Female technicians are a rarity at NASA, though their numbers are gradually growing. Thanks in part to the women's movement, federal legislation, Supreme Court decisions, and the support of JSC management, women technicians are more visible than ever. The Upward Mobility Program instituted by NASA in the 1970s provided secretaries the skills and training to develop new ideas, concepts, and products for the agency and its workforce, and the opportunity to move into positions with higher pay and greater responsibilities.

S. Jean Alexander was one of those women. She became NASA's first female suit technician in 1980, just before the maiden flight of the Space Shuttle *Columbia*. The selection of women astronauts two years earlier had opened the door for women to fill other positions across the center, including that of suit technician. Having worked as a center secretary for more than a decade, Alexander sought a new challenge. She applied to the Upward Mobility Program and was one of a handful of applicants accepted. Alexander went on to become an equipment specialist with the Crew and Thermal Systems Division.

Sharon Caples McDougle was the first woman to become a crew chief in Crew Escape Equipment Processing, which oversaw the suits and equipment astronauts wore in case they had to evacuate the Space Shuttle during launch or reentry. Before working as a NASA contractor, she served as an aerospace physiology specialist with the SR-71 and U-2 reconnaissance aircraft and pilots at Beale Air Force Base in California.[1] Those pilots wore pressure suits similar to the launch entry suits the astronauts began wearing after the *Challenger* disaster. In 1990, McDougle moved to Houston to work as a suit technician for Boeing Aerospace Operations and eight years later became crew chief. In 2004, she became manager of Crew Escape Equipment Processing at United Space Alliance.

A suit tech assists Astronaut Tracy Caldwell Dyson with her launch entry suit during emergency egress training in the Full Fuselage Trainer.

S. Jean Alexander[2]

I started working at NASA as a secretary. Went out to Ellington Air Force Base and was out there for a couple of months.[3] They sent me to different offices that needed people because they were shorthanded. There was a hiring freeze at the time. I finally got into an office and stayed in the Crew Systems Division. It was interesting because I worked with the human resources group that interfaced the suits and the helmets with the people, making sure that the people part of it worked and that it was going to be okay for them. That was really interesting. Gemini was going on then.[4]

My husband was in the service. He came home from the military, and we had to go to Washington State for six months, so I transferred and worked for the Army up there in a food lab, testing ice cream and dairy products. Had to eat all of the winners! When I came back, I went to work for SR&QA (Safety, Reliability, and Quality Assurance) in the director's office.[5] I worked for the special assistants up there for a while and then ended up working for the deputy director of SR&QA, Bill Bland.[6] I had a lady that came to my house and took care of my kids, so it wasn't a problem until she retired and moved. I left and stayed home until my kids started school. I want to say I was home for a year and a half.

I went out to lunch one day in Building 45; the Personnel Office was in that same building. When I was getting on the elevator, one of the ladies from Personnel saw me. "Are you coming back to work?"

I said, "Oh, no, I'm just out here for the Christmas luncheon."

She said, "Oh, we need people. I really need you. I have an interview you could go on today."

"No, I'm just here for lunch." I went upstairs and had lunch with the girls.

She called upstairs to the girls and said, "If she's even thinking about coming back to work, tell her to go over and talk to Glen Brace," and she gave me a room number.[7] I was bored at home. I'd worked since I got out of school. She said, "Just go talk to him—you don't have to take the job."

I went over there and I talked to Glen Brace. He was in the Awards Office. I had to pick up my son from kindergarten, and it was time to go. I literally answered the questions with "Yes. No. Uh-huh," and, "Okay, I'll see you later." When I got home, there was a phone call: "He wants you; you've been hired."

When my husband came home that day, I said, "I have a chance to go back to work," so I did. That was interesting, working in the Awards Office, because I really developed myself and my writing skills. It helped me later down the line doing write-ups for other people for awards. I knew what they wanted and what they were looking for. That was fun.

Then my old job with my boss from SR&QA, Bill Bland, opened up again. He wanted me to come back and work for him, so I did. He worked on the Three Mile Island project and had to go to NASA Headquarters in Washington, DC; he would be gone for two or three weeks and then come in with stacks of paperwork.[8] We'd go blind for three or four days taking care of everything that he had to do, and then he'd be gone again.

During that time they put out the announcement for the equipment specialist job. There was another position, too. I don't remember what the other job was, but it was like a fabrication person. Phyllis Morton ended up getting that job.[9] It was when NASA was doing upward mobility.

A lot of the women were transferring over from secretary positions into contracting jobs, administrative assistant jobs, and budgeting. None of those interested me at all. I'm not a people person. When this job came open it just sounded like me, so I applied. I understand that over two hundred applicants applied, because it sounded like a really good job. Walt Guy was the division chief at that time.[10] They narrowed it down to twenty candidates, and then the twenty of us interviewed with him. Then he got to pick. Walt Guy picked me.

They were planning on Phyllis and me cross-training. She worked in the special projects lab where they made all the covers for the cameras and stuff like that. Her job was a lot of sewing. I could sew, but it's definitely not my favorite thing to do. We cross-trained a little bit. I went up there to the special projects lab for a few days. They sent us both up to ILC (International Latex Corporation) and had us change out wrist bearings on the EMU (extravehicular mobility unit), which I didn't work with at all, but we got to go on a trip.[11] I got to eat soft-shell crab. After we came back from that trip, Phyllis definitely felt like she was out of her element and didn't want to do that type of work, and I definitely didn't want to do what she was doing. It was just too hard. We were barely learning what we were supposed to be doing, much less trying to figure out the other job at the same time.

I don't remember if Walt asked us, if he gave an option, or if we went to him and just told him, "We're happy where we are; just let us stay here." It was agreed that we would no longer try to cross-train. I don't know if he was doing that in the beginning just in case somebody in their group hated the one that they had, as a way to get rid of them; I'm not sure. We both liked where we were and the people we worked with were satisfied with us, I guess.

When I got the job, I had never met the guys, any of them that I was working with—even Jim Schlosser.[12] He was the boss over our group. I was expecting not to be welcomed, exactly, being the first woman up there in the office, and them not having a say in who they picked. I thought that was not right, but that's how NASA does things, and Walt in particular. There were only four NASA suit techs when I got selected: Ronnie Woods, Al Rochford, Troy Stewart, and Joe Schmitt.[13] We also had a couple of contractor support people.

I never felt any pressure being the first female suit tech just because

I was a woman. I just didn't want to mess up. I'd never seen a suit much less put one together. On one of the first preflights that we did at the lab with the first suits, the brown ones, the launch entry suit (LES), had a vent tube that came up to the neck ring to blow air, and it could go to the back or it could go to the front.[14] I can't remember to this day if it went to the front or to the back, but whichever way it went, when I put the suit together I put it in the wrong position. The quality guy caught it—I didn't—but one of the guys was there, Troy Stewart. I was so embarrassed. You have paperwork that you've done. There's not a drawing, really, but there's paperwork. I changed it, and we re-torqued everything. I know I must have looked like I was upset because Troy said, "Don't think anything about it. Everybody in here has done that same thing or something similar, so don't worry about it." That made me feel a lot better. You didn't mess up often, but every now and then—because it's life support equipment, you don't want to. Just like that, they were always there, very supportive, and I did not learn until later that they were apprehensive about getting a girl. I assumed that they had been in on the decision, but I was kind of forced on them. They never acted like they were upset with me, so they were very generous people.

The guys were really good. Joe Schmitt took me in, taught me all about tools. I knew how to use a hammer and screwdriver and what some of the tools were. They had written up a training program and covered all the tools that I would use. He would take me in there and show me how to do things. Ronnie Woods worked with me on the commercial sewing machines a little bit. I learned like that. At the time, they were flying the brown full-pressure suits. Of course, there was a lot of training on those— taking them apart, putting them together, and supporting the training sessions. After that, we went to the blue coveralls and helmets.[15]

I was hired when they had the first women astronauts. I can remember when the women astronauts were selected. I wasn't a suit tech when they were first picked, but I don't remember it being that big of a deal. You knew the selection was going to happen. You were proud of them and you were glad that it was happening. It turned out not to really matter as a suit tech if you were female or male. I think they thought the female astronauts would want a woman there for modesty's sake. They obviously didn't know these women, because they were there to prove they were as good as any man, could do anything a man could do and not complain about it. During our first suiting session with Sally Ride, where they were just wearing the blue coveralls and the helmets, we were in Building 9, where the mock-ups were.[16]

We had their coveralls laid out in this one dressing room, but we had Sally's in the ladies' restroom around the corner. There was a sofa in there, so I'd taken her stuff there. When they all came to this room and had their briefing, they said, "Okay, all right, we got it." The guys turned around, started taking off clothes. Sally went in the corner and started taking off her clothes even though there weren't coveralls or anything there. I remember the guys and I looked at each other. I was like, "Oh! Oh, my gosh," but she was already stripping. Evidently, she hadn't gotten the memo that we weren't going to make her do that. I went and got her coveralls, I brought them back in there, and she put them on. On orbit, they might have had to do that, and I'm sure that's what she was probably preparing for, but we didn't have to at that time.

Afterward I said, "We were planning on keeping you separate."

She said, "Oh, okay." It really didn't matter to her; she was just like one of the guys and she was going to do whatever it took to be one of the guys. That's the way it was. To me, that was an example of her saying, "I'll do whatever it takes. I'm here to do whatever."

It looked like the rest of the crew was comfortable with that. They all had their backs turned, they were getting their own stuff, and they didn't make a big deal out of it. No one said anything. They might have later, I don't know. They might have already talked about it and said, "We're not going to let this be an issue. You're one of the crew." That was the only time that that happened. I don't know what they did on orbit. I do know they had a privacy curtain they could have pulled shut. Before long, they built the dressing rooms, and we had two separate dressing rooms, so it didn't matter.

Later, when they went to the launch entry suits, they wore long underwear.[17] We would have their underwear and their diapers and all that laid out. They would come get them, they'd go into the dressing room, they'd each trade, and then men and women would come back out in their long underwear and sit down and get dressed. But underneath the flight coveralls, they just wore the bra and panties or their boxers and a t-shirt, and that was it.

I always thought that when we got a female crewmember, I would be

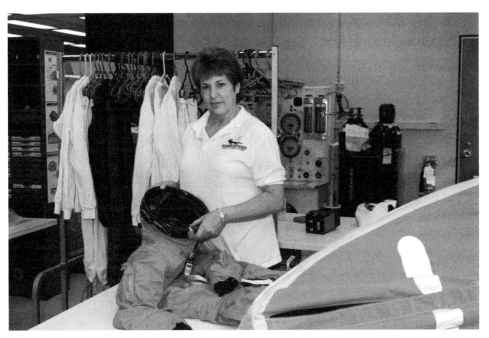

Jean Alexander works with the launch entry suit.

assigned to her. Sally Ride's flight came along and I didn't get assigned to her—well, I did go and support her. The first flights, there were so few of us suit techs that we worked them all. You went to training and to launches and to landings because we had to split up and do it. I didn't do her strap-in.[18] I'm not sure who did, but I got to work with her during training. I honestly can't remember. That's how unimportant some of that stuff was. We were so busy. You had four and five crews training at one time; it just all blurred, especially after the guys left. I worked every flight after the suit tech guys retired, and they both retired at the same time.[19]

When I first heard they were retiring, I just laughed at them. I said, "Yes, there's no way NASA'll let them do this," but NASA was turning our jobs over to the contractor, so it was right up their alley. They wanted to get rid of us civil servant types, in a sense, so they didn't really care. It was really hard. I not only had to work every flight, but I was having to train the contractors to take over our jobs. I was on the road six months out of the year. Home two weeks, gone for a week—because I was either at launches or landings, training somebody, or doing something—up until the time I retired. The other day somebody asked me what my last flight was, and I said, "I think it was STS-110, but I can't even tell you for sure about that."[20] I think it was the last one that I actually went out to the launch pad and did the strap-ins.

I got to go to the landing of STS-1.[21] Al Rochford and I went out there. I was on the lakebed when they landed, and in the Astrovan with them when we picked them up.[22] That was definitely a moment to remember. We were out in California for a long time because the mission slipped. They sent us out there a week ahead to make sure everything was ready. I just remember it was forever. I missed my son's birthday because the mission ended up slipping. I think we were gone sixteen or seventeen days before we got back home. I hardly ever traveled, so it was a long time for me. My kids were around ten and twelve during that time, so it was hard, but gosh, to get to support the first landing of the Shuttle — it was awesome.

Suit techs had to take the equipment to the Cape and then support the launch, and we went wherever they landed and recovered everything.[23] We were the first ones to be there when they opened the hatch and helped gather everything, making sure everything got back. We trained for all of that. We supported STA (Shuttle training aircraft) flights when they would fly suited or just used the helmets.[24] When they wore the clamshell helmet, we would take their coveralls and their helmets and fly on the STA with them.

I got to fly in zero-G; that was part of my training when I first started.[25] I supported an EMU flight. I remember holding a helmet for them. That was about the extent of what I did during that flight. When we had to certify the launch entry suit, that you could don and doff it, Pinky Nelson was our subject and I had to fly with him.[26] That was where I really worked. We had to suit him up in between parabolas. That poor man. My job, basically, on that day was to hold the barf bags for him, me, and the doctor, and then try to get him into a harness. It's so hot and the gyrations that he had to go through to get into the suit — he earned his pay that day. Certifying all the equipment, we did all of that. It was fun.

Body dragging was another part of our job.[27] We're out there at the pad. If something goes wrong, the suit tech and the ASP (astronaut support person) are the first two in to start un-strapping the crew.[28] Different men in the VITT (Vehicle Integration Test Team) Office and the astronaut crews — you could tell some of them weren't too keen on the idea of women out there because they didn't think we could do the job. We had to prove that a woman could go and do it.[29] When we did body-dragging tests, we'd use each other, the suit techs, but the ASP and different people would also be subjects. Sometimes we had real crewmembers because we had to have their input that it was acceptable and tolerable. For most of the stuff, we just used each other because we'd do session after session after session; each team had to do it repeatedly.

Anna Fisher and I went down and did body-dragging classes.[30] We were paired together. I think she was going to be an ASP, the one that went out to the pad and helped do strap-ins, so she was going through classes, too. Either that or she was just from the astronaut side, proving that the women over there could do it.

When we got back to the suit room that day, I turned to the guys and I said, "Did it look like Anna Fisher had a baby bump going on today?"

They were all like, "What? No."

I said, "I swear, I think she's pregnant."

They just looked at me like, "Yes, whatever."

A week or two later, NASA announced that she was pregnant. I said, "I told you!" She went out there and you'd never have known it, with what work we did that day.

Back in those days it wasn't as bad because it was just coveralls and the helmet. You still had to lift and tug and pull, but later, when we got to the launch entry suits and we did it out at the pad, it was really hard. We would have to take them from the Orbiter, from the White Room out to the baskets, load them in the baskets, then stop, go down to the other end, they'd release the baskets, and people would get in them again.[31] We would have to get them out of the baskets and get them to the bunkers. It was always August or summertime when we did it, one-hundred-degree weather or something. It was hard.

Shannon Lucid and I worked together; I think when she and I did it, though, it was coveralls.[32] Marsha Ivins and I were at the pad for one of those exercises.[33] We had to be able to go in and get them off of the flight deck, down and out. We did it. Did we do it exactly like the guys would have done it? No. We actually did it smarter, in our own way.

The way the men did it was very difficult because they would pick up the head and the feet and then try to walk with them, and with all that gear, it was really hard. She and I, each of us got on one side, grabbed a harness strap, and we moved them to the basket. It may not be the most comfortable for the crew, but they're unconscious anyway. The way some of the guys handled them couldn't have been much worse than what we did.

I remember, even after I got older and had done it for years, a new group would look at me and think, "She can't do this." You'd have to show them that women can. Sometimes you had to do it more often than the other guys did, but it was good.

Usually it was a two-man crew. If you had a middeck crew, the suit tech would go to the middeck and the ASP would go to the flight deck. They would do the straps for MS-2 (mission specialist-2), and then just turn them and get them to the hatchway.[34] Meantime, the suit tech is on the middeck doing the same thing. All we did was get them out of the seat and to the hatchway. The close-out crew would reach in and take them out. At that point, it was basically gravity helping you do that. As soon as I finished with the middeck, I went upstairs. Usually it was a suit tech that went up and undid the commander and pilot and then lowered them to the ASP. Then the ASP turned them and scooted them on toward the hatch. It was a joint operation, though parts of it I did by myself. The close-out crew was there to do the tugging and the pulling. That guy really earned his money.

The first time I went down there and saw the mockup we would train on, I was like, "Seriously?" They had a little wooden, very low-fidelity mockup out there that they used for this. The seats weren't right; straps weren't right. Some of the seats didn't even have headrests. You had somebody in there and their neck's not supported. They had a little wooden walkway. We would just get them down from the vehicle and out, get them to the end of that walkway, and that was basically it. You didn't load them in baskets or do any of the rest of the stuff that we had to do later.

Over the years, they worked hard trying to make it better. That's when we started bringing the close-out crew to Houston and doing it in our trainers because we had the right seats. We'd do it up to that point, getting them out of the Orbiter. Then we'd go to the Cape, pick up from the White Room, and go to the baskets, go down to the bunker area, and pick up from there and get them to the bunkers. It was pretty realistic training.

But my thought always was, "If something goes wrong and the crew goes down, what makes you think we're going to be standing?" It was one of those things you just had to fill a square, I think, because NASA had to have all these redundancies and plans in place for what they would do if there was an emergency.

In fact, STS-41D where the engine caught fire—that was my flight.[35] I remember the fall-back area where we were gathered, and my boss Jim Schlosser was down there. I remember looking out the window, saying, "Seriously? You're sending me out there with the Orbiter on fire?"

We got out there. Judy Resnik was the first one out.[36] We opened the hatch and the deluge system was going off, so there was water coming down on the pad. As soon as we opened that hatch, she was coming out. I put my hand out to stop her, "No, my word was that we were taking helmets off; harnesses off."

She said, "Are you sure?"

Everybody was there, so she stopped, and we took all their stuff off, got everybody out. We got in the elevator, went downstairs, and water's still just pouring off of the thing onto the Astrovan, right there. They just stood there, and I said, "You're going to have to go through it to get to the Astrovan." We didn't have umbrellas for them at that time. I think umbrellas were added to the Astrovan after that. They ran and got in the Astrovan.

Then I went back up. Mike Smith, I believe, was the ASP on that flight.[37] I remember the close-out crew said, "Go in there and get him and tell him we've got to go."

I went in there, and he's flipping switches. "Mike, we got to go." I'm picking up kneeboards.[38]

He says, "Okay, give me a minute."

And they're out there, "Tell him we've got to go."

"Mike, really, I'm not sure if this thing's fixing to blow or not, but we need to go."

"Okay, give me a couple more minutes." A few more switches and we left.

Later when we went back out there and I saw where it had been burned, I was like, "Holy moly!" because I'd finally decided, well, it must not be bad or they wouldn't let us be out here. Then when I saw it, I was like, "That was bad! Those people are nuts!"

They had the slide-wire baskets. When it happened, why didn't they just tell the crew, "Egress, go get in the baskets, and go to the bunker?" That would have been the safest thing. But they hadn't certified the baskets—they'd never had live people really ride them down. They had put dummies in them and done it, but they had never really done it. They were afraid somebody was going to get hurt, so they wouldn't use them. I don't know that officially, but I promise you that's what happened. Then, later, Charlie Bolden actually got in a basket and rode it down and got live certification on it.[39] Why would you risk sending people out there? I just always thought that was weird, but you don't know what their thinking is. It worked out okay.

I remember another time we were out at the pad—this is where being a female suit tech was important. When we first started wearing the LES suits, they had to wear the long john underwear under them. That with a harness on top of it rubbed your breast. One of them, and I don't remember which one it was, called me over. I don't remember if she was doing something that made me think she was uncomfortable, but that's where it came out: "This thing rubs me. By the end of an exercise, I'm raw."

I said, "Oh, my goodness, no one's ever said that." We went out and found the thickest double-layer sports bras, and that's what we started providing from then on for suited training. It helped the problem a lot. They probably wouldn't have told their guy suit tech about that. They would have just sucked it up and kept dealing with it.

Our job changed when we moved from the coveralls that they wore pre-*Challenger* to the LES. Of course, it was a lot harder because they're a lot heavier. That was one hundred pounds of equipment they would have on between the parachute and the harness, the suit and the boots, and all the equipment that went into the pockets. There was a lot more testing. You had air bladders, so it was a lot more complicated and a lot more uncomfortable for the crew; there was a lot more complaining. The partial-pressure suit squeezed the astronauts like a blood pressure cuff, so any time it was pressurized, they were hurting. You'd pressurize them basically just to stretch the suits out and then you would not do it any more until you had to do a test. They were very uncomfortable, very hot. We finally got thermoelectric cooling units. In the beginning, they just had a fan that blew cabin air on the crew. It was like having nothing. We had to swab the suits after landing. We took swabs before and after, for bacteria growth in the beginning.

When the full-pressure suit came along, it was so much easier to don.[40] The partial-pressure suit had double neck dams that the astronauts had to squeeze through. It was like going through the birth canal. The full-pressure suit just had a wrist and neck dam that you had to go through, so it was a lot easier to get through. It was a lot more comfortable. But when it was pressurized, you couldn't do much in it. It was a bad day if you had to pressurize. You were along for the ride, basically, till it depressurized. It was a lot more work, a lot heavier, a lot more hassle.

In the beginning when we first started flights, we worked right out of Building 7 before the suits and crew equipment were turned over to the contractor. We would test the suits, pack them into black suitcases, take twenty of them out to the airport, get them on an airplane with us, check them in, get them at the other end, and haul them to the suit lab. Once a contractor took over, they shipped them. Then we always had to worry when we got to the Cape—was the equipment there or lost?

There were days we got there and found out that the suits or the equipment weren't there yet. We were always there days before the crew got there because we had to get everything tested. We'd have a day or two to recover. A few times, we had to really track things down and find out where they were, but we never missed a deadline. After we would get to the Cape, everything was checked, their pens and their pencils, that they had lead in them, that there were batteries in their flashlights. We checked everything. Everything was double-checked and triple-checked. After they came back out for the dry count, we went through it all again and checked it all again. There was a lot of stuff that you usually don't think about that's time consuming.

Equipment development came through the NASA subsystem manager.[41] We got to work on a lot of good things. I got into making things better for the astronauts, like the kneeboards. That's one of the things I did. They had this big clunky, metal kneeboard that they used. It was uncomfortable and hard to use, and they wanted a softer version. I found a fabric version of it that commercial pilots or military used, I'm not sure who it was. It was in a little shop at Ellington Field.[42] We went from there and made the modifications so that it suited what they needed, to carry all their pens and pencils and flashlights. We made a double kneeboard and we did single kneeboards. At one point, we were trying to come up with the board for their laptop, but that only flew on one flight and then they never pursued that.

At that point, we did the drawings for the changes. We literally drew them there in our office, where Schlosser signed off on them. We walked him through the signature process and then went back to the lab and sewed them up. It was that way in the beginning, and then later, of course, it got to where we couldn't do anything without everybody at the site signing off on it, and it made everything a little bit harder.

We had to have a parachute cover for the LES. We developed the prototypes and took them to exercises until we found one that worked and stayed on the seat and did what it was supposed to do. It wasn't a matter of just covering the parachute; we had to be able to cover them and then leave them in the seat. The astronauts didn't want to have to take them out. We had to be able to strap it to the seat somehow. We did a lot of different things like that.

We also did water testing. Took them over to the swimming pool. The crews had to be able to get into their life raft and bail water and use their radios. In one of the training exercises there, one of the VITT guys was a test subject. He went in, he came back up, and he floated right on his back without the life vest inflating because the suit bladders held air, so he just floated up. The anti-suffocation valve on the helmet was in the back so it wasn't long till he was sucking water into his helmet. That opened up a whole new can of worms. We'd always dropped them with their life vests inflated so that they popped right up like a cork, no problem. Since this could potentially happen, we had to design a new life preserver.

We did that right in our lab. Two or three of us went in there and worked. We actually took bladder material, cut it, glued it together, made different prototypes, went to private swimming pools, and we checked it out until we got to a point where we thought we had something worth testing in the WETF (Weightless Environment Training Facility); it was so expensive to get divers and do everything.[43]

The new LPU (life preserver unit)—I worked on that from day one, me and two or three of the contractor guys. Literally, when you went into the water, it popped you right back up into a reclined position. If you went face-down for some reason, which was a concern, we had the frontal lobes designed such that it would bob you around and wouldn't let you stay face-down. The crew, a lot of them didn't like it because it was bulky once it was inflated, but you could let the air out and still get into the life rafts. When you'd have someone that would complain, I used to always think, "If you were out in the Atlantic somewhere and you needed that sucker, you'd be wishing it was five times as big as it is now." That was a very challenging thing and one of the most rewarding pieces of equipment

I worked on. The company that got the contract to actually build them eventually made a smaller version for women that would work. I heard that a lot of the guys switched, too, because it was easier.

Now, would it work for them when they're out there in the ocean? I don't know if anybody cares, but you just can't imagine the things that people think of. One day someone had said, "Well, what if I landed face-down?" You could be dragged by the parachute like that, I guess.

I said, "I don't know, let's see what happens." Of course, they stayed face-down, and I'm like, "Oh, no." We had to go back and start all over again.

I have a patent with Dr. Frederic Dawn on a helmet pad; he was our materials guy.[44] I went down to him one day and told him, "I need something that's comfortable, some kind of foam that doesn't burn, and that we can make for helmet pads." When an astronaut is on his back, he has to prop his head up inside the helmet, otherwise he can't see his kneeboard or anything. For entry, they would take the pads back out because they would be pushing their heads forward. Dawn had some new material he was really happy about and he had been looking for a use for it, so I made the drawing of what size it needed to be and we put leather over it to protect the rubber or whatever it was. He told me, "I put your name on this patent."

I said, "No, that's your stuff."

He said, "No, I have to put your name on it," so there's a patent out there somewhere with my name on it. I remember him sending me the paperwork after he got it through. Little things like that were just some of the comfort things I worked on.

There was an external pad that raised the seat for visibility, and in the beginning they were made out of bubblegum foam. After a while, with the indention from use, it would go down and the crewmembers could no longer see this switch over here. We had to come up with another foam or pad, something stiffer that was still comfortable enough, something that kept their head up and wouldn't make them lose their line of sight.

I started developing these things out of necessity, like the pull-tabs in the partial-pressure suit. The crews, the first couple of times they flew the LES, came back with little slits in their neck dams because they were so uncomfortable and hot. Then the suit would no longer work. One day I heard one of the crews talking about it. They were laughing, "Don't worry.

We've got so-and-so, he's working on the answer to that in his garage at night. He's going to fix something." He was making an S-hook that they were going to hook into their neck dams and then hook over the ring of their suit. First of all, they could damage the neck ring so that their helmets wouldn't even lock, or they would rip their neck dams. The others, they were taking their lanyards that they used on pencils and pens, wrapping it around the button on their underwear, and then pulling it out and tying it off. These were the types of things that the crews were doing to make the suits comfortable enough.[45]

I went back to my subsystem manager one day, and I said, "We've got to do something about this," so we started making elastic tabs that Velcroed to the inside of the suit and came out and Velcroed to the outside.

I remember there was all kinds of grumbles that, "We are not going to be able to do that; that's not going to seal the suit."

I said, "Well, we've got to do something, so let's try it."

Sure enough, it worked. The elastic was thin enough and snug enough up against the other. It aggravated the contractor to no end because of the upkeep of the Velcro, but it was a fairly simple solution. We said, "We'll just do this until you can come up with something better," a zipper on the neck dam or something to make it more comfortable. Of course, they never did. Eventually they came in different lengths because some people needed smaller ones. That was an example of something that out of necessity, you had to go back and come up with something new or put up with the consequences. The crew was going to fix it themselves, one way or the other.

In the beginning when I was working on all these different ideas, one of the suit techs would be in the suit. Eventually, when we got to the real testing phase, it was all done by the book. It was done with paperwork and cameras and documented in every way. We had astronauts and real test subjects that tested the suit and life preservers. Other times, it would be VITT people if we needed more bodies, but we always had to have astronaut participation and approval, showing that we got the piece of equipment to what they wanted and what they needed and what they were expecting to see.

I was always the only female NASA suit tech. There was another female contractor, Valerie McNeil. She worked for Lockheed, and I was originally thinking she was going to be my replacement, but she went back to being just a regular suit tech, and then went over to USA (United Space Alliance).[46] It just wasn't her deal. She didn't like the body dragging and having to lug the equipment around. She had younger boys at home, and the travel was hard.

Sharon McDougle—when we turned the work over to Lockheed, I tried to get her to apply and take that job because she came from the Air Force and knew the equipment inside and out. She was the best they had over there as far as a woman that could have done the job. At that point, she was shooting for being head of their lab at USA, and that's what she wanted to do because she didn't like the training either. She was a very, very, very good suit tech, and knew that equipment from front to back.

In general, the contractor didn't like me very much. If we had a discrepancy on a microphone or earphones or if the suit leaked, we documented it. You're supposed to document it on a DR (discrepancy report).[47] I kept records. I turned over to the subsystem manager what we had written DRs on, what had happened. I sent them to everybody. The contractor got them; everybody got them. It's not like I came home, snuck around, and told on them. They had an engineer at each of the flights, so they had a copy all of the DRs. It wasn't like I was doing anything sneaky, but they hated that. The other NASA techs—I'm not going to say that they didn't follow through, but if the guy said, "I'm just going to put a new microphone in there," because microphones are cheap, they wouldn't document it. That doesn't help the overall problem because we had these things, especially these sensor switches, that were constantly going out. When you have a constant problem, you fix it. You get a new one. That's how you do it. You trace it by DRs. The contractor didn't like me because, invariably, if they get enough of those things, it affected the company's award fee, which it should. If they send crap down there, then something should happen. They didn't like that. I wasn't mean about it. That was my job.

I didn't have to test the suits and I didn't have to know exactly how to test them, and I think that was one of the things that some of them resented as contractors. I knew how they were supposed to work in the suit. I knew when the suit didn't work right. I could tell you what part wasn't working right, or what the problem was. I think they had chips on their shoulders—I think they felt like they were so much better because they could take that suit apart and test every component on it. That was their job. If I had been trained to do that, then I could have done that. But I wasn't trained to do that. I had to know everything about it and how it worked. I can't tell you how many times our persistence paid off. This comes back to a contractor versus a NASA person. I was so worried when it went over to the contractors because I can't tell you how many times they were willing to just say, "There's nothing wrong with it, it's fine."

We had one suit in particular that kept having a leakage, that had been on my flight. The engineer kept insisting that it was the exhalation valves or something like that. The next flight, Troy came home one day and I said, "Well, how'd things go?"

He said, "Good. Had trouble with Pinky's suit."

I said, "Oh, really?" He started saying what it was doing and I said, "Was it suit number such and such?"

He said, "Yes."

I said, "That's the same suit I had, and the contractor engineer kept insisting that it was this instead." We flew it, but I knew it had a problem. I came back and told my manager even though it passed the test in the labs. It was the suit controller. It was a leakage. When I told Troy, I said, "That's the same suit—you make them fix it because there's something wrong with that suit."

Another time, we had a communication problem during one of the TCDTs (terminal countdown demonstration tests). One of the crewmen on the middeck complained that his communication line was really weak.[48] When we got back to the lab, we were testing and checking everybody's lines. In the lab, it checked out just fine. I went back to him and I said, "Could you be more specific? Exactly what happened?" It was a crewman I trusted—it wasn't a payload specialist that didn't do it but once or twice in his life.[49] We made an issue out of it, and we set up a session because we checked everything fifty times and couldn't make it do anything back in the lab.

We went out to the vehicle and plugged them in, all three of the helmets from the middeck. We plug them in and nothing happened. They were fixing to call it quits and say, "Well, it's just one of those things, the crewman just must not have been hearing right."

I said, "Well, the only thing we haven't done is plugged all three of them in at the same time." We went from seat to seat. I said, "Let's plug them all three in at one time." Plugged the first one in, no problem, it was fine.

Plugged the second one in, it was fine. Plugged that third one in to that middeck and the com dropped off; his just blacked out. It was a problem with the Orbiter in that com box, but the contractor had been so willing to say, "It's not our equipment, there's nothing wrong." I worried about that afterward because of that. If it's a problem, then that just looks bad for them; they don't see it as an opportunity to make something better. It all comes back to money, in a way. I worried about that, but it worked out fine, I guess.

We had a lot of fun, too. When the Dog Crew flew—you remember when the little Beanie Babies were popular?[50] We went out and got little Beanie Babies. My Lockheed suit tech and I traipsed to every store in town till we found five Beanie Baby puppies. On their tags, we put the astronauts' color codes and we wrote a little verse that fit with each one of them. Then with the Rat Crew, we changed their color codes into little rat faces to put on their knee boards.[51] We would try to make signs for the room sometimes. We did some fun things like that.

On the first Shuttle flights, because of all the stuff they put in their pockets, every time we'd get all their survival gear together, the suit tech guys would ask, "You sure you don't have a small child you want to put in there or something?" to make fun of them. Al and the guys picked up watermelons and put on the color codes red and yellow and had them out on the table to put in their lower pockets before they left that morning. That was funny. That was John Young and Bob Crippen. Joe Engle and Dick Truly, they changed their eyeglasses every week.[52] We had to order new eyeglasses. Right before they left the suit room, Al opened up a trench coat, and he literally had sunglasses and eyeglasses hanging inside the coat. He said, "You sure you don't want to make one last change before you go?" Just little things to lighten the mood before they went out.

The other thing relating to being a female suit tech—one of the first times I rode out with the crew to the pad, oh, I don't remember, it wasn't my first time to ride out, but maybe it was my first time to ride with George Abbey.[53] We would stop and let the managers off at the O&C (Operations and Checkout) Building.[54] Right before he gets off, Abbey usually turns and looks at the crew, and he says, "Don't f— up," and gets out. He started to say it. He looked at me, and he said, "Don't screw anything up," and got off. It was just funny, and the guys all looked at each other. Later I was telling the guys and they said, "He really wanted to say it, but he wouldn't."

I thought that was funny. I was surprised that he would be that careful, because he wasn't normally.

I never thought about leaving the job. I loved it. I got to do the design and development, so that was really fun, which I didn't think about when I first went over there. Other than the traveling and the physical hardships of it, it was a very exciting job. The quality of the people at NASA—I hear other people talk about their jobs and the jerks they work with day in and day out, and the environment they work in—it just doesn't get any better than what we had, with the opportunities, and the classy people, not just suit techs, all of the divisions and all of the center. SR&QA, some of the best people in the world over there that were working in that area. It was just really good.

People these days, they're so into job-hopping. When you tell people that you worked at the same job for thirty-eight years, they look at you like, "Oh, my goodness." You wanted to. It was a career, and you could do it. It wasn't a job—it was a career. Kind of like going and getting a haircut: sometimes you get a haircut and sometimes you get a hairstyle. That's kind of what this is. Instead of a job, it's a career, and NASA was wonderful about that. To all women, not just me, but to all women. Phyllis Morton, she loved her job up there and she stayed in that job until she retired. Sometimes I think about how they made a big deal out of that, a woman being up there in the survival lab. Why would they? Women sew; that's all they did, basically, up there. It was fantastic sewing, but that's basically all they did. I don't understand why that was such a big deal that it had been a male environment all that time.

I'm glad I was in the right place at the right time. I worked hard at every job I did, no matter what I did. I know that was the difference, when you get to that point. How do you look at twenty women and decide this one's going to be able to do it and that one's not? It came from your recommendations, I really feel. The people that backed you up, that said you were willing to do whatever it took. I've got to credit the guys—they were so good about training me and working with me. They never made me feel ill at ease whatsoever. I was one of the guys.

Sharon Caples McDougle [55]

I didn't have a learning curve when I came to NASA. In the Air Force, you learn the process and you do it. If you break it, you fix it and you keep moving. Here you have the engineering team, you have quality inspectors. In the Air Force, you were the inspector/engineer/tech. When you put your initials on the paperwork, you were saying the equipment was good. Here you have other people who have to look over your work. That was a culture shock for me because it made me feel like they thought I didn't know what I was doing. I understood that wasn't the case; it was just a whole different culture and I had to get used to it. It wasn't a learning curve, but a culture change from military to the civilian world. That was hard for me, but, like I said, I got over it. It was just a slower pace than I was used to.

In the Air Force we loaded the aircraft, we suited up the crew, we tested the suits. We strapped them in, we worked on the suits, we did everything. You didn't just perform a portion of the task. You got to be a jack-of-all-trades in the military. When I got here I assumed we'd suit up the crew, load the Shuttle, and strap them in. "No, you just work on the suits and suit up the crewmembers."

"I don't go strap them in? I don't load the Shuttle?"

"No, you just do the suits." Lockheed Martin and NASA strapped them in at that time. We have that contract now, of course, at USA.

We suited up the astronauts, took care of the crewmembers, waved at them when they walked out, and then we recovered them on landing. We would send people to each landing site: the prime, which is Kennedy Space Center (KSC), and backup is Edwards Air Force Base in California. If they had to call up White Sands Space Harbor in New Mexico for some reason, we'd send someone out there, but only when they called it up, not immediately like at the other two. We'd send suit technicians and an insertion technician, who are the guys who strap them in.

Something else I was surprised about was, in the Air Force—and I figured out why—the crewmembers had their own assigned sets of gear. This was their assigned suit, their boots, their helmet in their own bin assigned to them. When I came here I was looking for the same setup, and it wasn't that way, because, if you think about it, these guys may fly once on the Shuttle. So of course they're not going to get assigned their own gear but in the Air Force they were assigned gear because they flew several times a week. No such setup here. The suit an astronaut wore, tomorrow somebody else could be wearing it.

STS-37 was my first flight.[56] I suited up Linda Godwin on that flight, and she was a Mississippi native, too, so I was excited about that.[57] We wore white uniforms then. Now the suit techs wear a tan uniform, and the insertion techs wear white uniforms. We wear these suits when we're suiting up the crew and flying on the Shuttle training aircraft. When we fly on the STAs we wear a Nomex uniform in case of a fire. It's not fireproof, it's fire-retardant for sparks. Also, it looks sharp for the team to be dressed alike.

Every person on the crew has a suit tech assigned to them, though sometimes a suit tech will have two crewmembers. You can be assigned two crewmembers because they come in to the suit room at different times. The commander, pilot, and mission specialist-2, who is the flight engineer, come in last because they're attending the weather briefing. The other crewmembers come in prior to them by about ten minutes or so and get suited up and tested and are done when the other crewmembers come in. Then the suit tech can suit up the other crewmember. I had two people most of the time. It takes approximately five minutes to suit them up; it doesn't take long.

STS-47, that is my most memorable flight, the one I cherish the most, with Dr. Mae C. Jemison.[57] I wanted to be her suit tech as soon as I saw her name on the flight. We had a board out in the lab here, and they listed all the astronauts. I had already heard the first "sister," as I say, was going to space. So when my lead put the names up, I placed my name by Mae's; she's my crewmember. Everyone already knew the first black American woman in space would be assigned to me. That was the only time I've ever done that.

I wanted to make sure she was taken care of, she was comfortable, and I wanted to be the one to do that. I didn't want anybody else assigned to her. I'm sure they would have done a fabulous job, too—I'm not saying nobody else was good enough to take care of her. I just wanted to be the one. You know that kinship, that family feeling. I was proud to be the only black woman in my department. I know my stuff and I'm really good at what I do.

I had never met her, but it didn't matter. I still would have taken good care of her because I wanted her to do well and feel safe. I felt her suit and training should have been her last worry. I treat all my crewmembers the same way, but I had never felt so passionate about an assignment.

My first meeting with Mae was for her fit check. They all have to come and get fit checks so we'll know what size equipment they'll need for everything from their diaper size to their harness size. We receive general information that says they wear size nine or whatever, but we still need to measure them for our specific gear. Helmet is one size—no sizing with the helmet, just for helmet pads. Boots are easy. The boots are usually a size and a half bigger because they wear thick socks and the bootie part of the suit takes up more room.

We use anthropometric measurements. We perform all the measurements, and it's just like if you get an outfit tailor-made, same thing: the trunk, around the waist, the girth, up the back, everywhere. You measure everything because you need to get as good a fit as possible for the crewmember. With the suit you can adjust the arm length, the leg length, the waist, and the girth.

Sometimes crewmembers can wear somebody else's harness. We have a few crewmembers that have to get special harnesses made because nobody's harness would fit them, even though you'd think you'd be able to have small, medium, large. The harnesses are actually serialized with the crewmember's name, but you do have some crewmembers that can fit another crewmember's harness, so you won't have to make one for everybody. The suits are usually small, medium, large. You have small and short, medium and long, large long, extra-large long, and so on; it's not like size eight and ten. The gloves are from A through L, with some customs.

The launch entry suits weren't made for women. Even a small woman can have hips, so it can make you wear a larger-size suit. The breasts can make a difference sometimes. Usually the hips and butt area for women can be a little harder to zip down because the suit's made for men, without curves.

A female ACES (advanced crew escape suit) was made later.[58] They actually went to the company that makes the suit and measured to make a suit with a better fit for the smaller female astronauts. The suit, mind you, is not made to be formfitting at all. When you see them in it, it looks frumpy.

Before the crew launches they get the fit check first, of course. Sometimes if they got delayed or scrubbed for too long—like STS-114 took forever—they'd have to come back in and get fit checks again.[59] Maybe they'd gained weight or lost weight because it'd been such a long time. Now, if it's just been a little while, a few months, then we usually won't have to refit them, but we have had instances where they may have worked out and put on more muscle, had some body changes. You may have to change a harness size or make other adjustments. We have final fit checks two days before the actual launch, and we've actually had changes up to that point. And you just want to say, "You've been wearing that all this time, and now you tell us?" But we have to be prepared for that. It's why we take spare equipment down to Florida.

We have a really good rapport with the crewmembers. We work with them sometimes for more than a year before they launch because they have to do all their training events. They'll have about twenty to twenty-four training events they perform before they even launch. Things like ascent simulation, where they go to the simulators over in Building 5.[61] Building 5 has the motion-base and the fixed-base simulators. In Building 9 there are the crew compartment trainers and the full fuselage trainer.[62] They can tilt up vertical like launch or stay horizontal.

In the motion-base simulator, they can simulate flying and also having issues. Sometimes the Mission Control folks throw problems at them, and they have to practice what they would do in that scenario and practice landing. They practice landing and flying in the Shuttle training aircraft also. They do some training fully suited; they do some in just a flight suit. Of course, they want to practice in a suit at least once because they want it to be realistic when they practice their Shuttle landing. The commander and pilot do that, and the suit techs are there to suit them up and strap them in. When they fly the STAs and they put their suit on, we have to inspect it and test it again when they come back because anything could happen between going out to the plane and coming back to the lab. So it's tested again to make sure they didn't damage it.

They also have bailouts.[63] They have the one over at the Neutral Buoyancy Lab, which is probably the most realistic, the closest to setup for actual flight because we have to have everything working.[64] They have to inflate their life preserver units, activate their oxygen bottles, the whole nine yards. They hoist them up, they drop them into the water, then they practice going under the parachute and getting into the life raft. That's one of the most intensive training events they perform.

Then they have TCDT, where they go down approximately three weeks prior to launch and practice their emergency egresses in a real-time launch flow, starting with suiting them up, timing everything, and getting them strapped in. Then they practice as if they had an emergency, where they would hold hands, go to the slide wire basket, and pretend to slide down.

The other practices are normal. They'll do the postinsertion operations. Sometimes they'll do one or two between TCDT and launch. It depends. Sometimes the commander may want to practice one more time because they want a refresher.

Wherever they go, if they're getting into a suit, our team has to be there. Of course they can suit themselves up, because they have to do it in space, but it's less pressure, less stressful for them, to have to deal with that—and less damage to our gear—if we help them while they're here on the ground.

Crew Escape Equipment utilizes Class 1 and 3 flight equipment for training. Class 1 is the flight equipment. It can go into space. Class 3 will not ever go into space and is strictly for training. Then you have the Class WIF (Water Immersion Facility), which is the equipment that's used for the water training. The crew may use some of their Class 1 gear that's actually going to space to train in, so we have to perform very thorough inspection tests. It's called a PIA (preinstallation acceptance) when we inspect and test everything to get ready for launch.

When the crew is assigned, technicians are assigned. You have a crew chief, which is the lead of the team, then three or four other technicians. The crew chief is in charge of all the equipment processing. Then there are the insertion techs. NASA and Lockheed used to do the strap-ins, but USA has the contract now, so we have insertion technicians. Two are assigned per mission, one for flight deck, one for middeck. You have a backup that goes to KSC the night before launch just in case something happens to one of our guys that straps them in. They're certified to be insertion techs, but they are also suit techs. The suit techs cannot do insertion work; they're strictly suit techs.

We have to do an unpressurized and pressurized leak check before they go on to the Shuttle. When they get strapped in they inflate the suit one more time to make sure nothing happened on the way to the pad. We check to make sure that it's going to pressurize and hold pressure when it needs to. That's when the suit gets hard and stiff but they can still move their fingers. The test also makes sure the helmet, gloves, or anything is not leaking and that the crewmember is getting oxygen before they leave the building. The helmets and other equipment are taken out to the van, then they'll walk out just with their suit and boots on. The test takes maybe five minutes. They walk in wearing their underwear, then they have a seat, don their gear, and the test is really quick. It doesn't take long at all.

For landing, we have to sit on the crew transport vehicle and wait for them to come off the Shuttle. We're out there hours ahead of time.[65] For most landings I've participated in, when the crewmembers come off they're not feeling well or are fairly weak. Sometimes we have wheelchairs available; sometimes they drape over our shoulders. It's because they've been in space ten, eleven, fourteen days sometimes. They're just woozy; they're getting used to gravity again. Everybody's different. It's really hard on their bodies, coming back into the Earth's atmosphere. Most times they're not very strong.

But my girl, Mae Jemison—I'm ready to assist her and she comes walking out like she just came from the mall. Girl, she's like, "Hey, Sharon! How you doing?" I was thinking to myself, "Isn't she supposed to be weak?" I couldn't believe it, once again I was in amazement of her abilities. That also shows you how strong she is. It didn't affect her at all. I'm looking at her like, "You know you just came from being in space for two weeks." It didn't faze her. I was so glad she was my crewmember. She was so easy to work with, no problems. She's one of the few people that I've seen like that.

Back on the launch morning of STS-47, she was still her normal self. She didn't seem afraid or nervous. She was always interacting with the other astronauts. Her walking off like nothing happened was the most memorable thing about that landing, but also on launch morning it was just like another day in the office. She came in, I suited her up, and we tested the suit. Everything was good, there were no problems.

I was assigned with Joanie Higginbotham until I got promoted.[66] I was going to be her suit tech, too, because I wanted to keep making history. She was the third African American woman in space.

Suit techs have a broad base of knowledge. It's not a mandatory requirement to be able to sew, but the majority of them can sew using the sewing machine. That's just for the orange cover layer and sometimes sewing labels on the harnesses and things like that. We have about three techs that are great at sewing. One in particular is a master fabricator. A lot of us are prior military here—the majority of us are Air Force—so we have a very solid foundation. A lot of us retired from the military, so we each have twenty years or more experience with suits before we began working here.

Let me tell you a little bit more about the tests that we conduct on the suit. When we first get to Florida, we unpack the gear. We have to inspect and test it again. You never know if something may have happened during the shipping process, so we perform the tests over again, one of which is called a vacant test. During the vacant test we test the dual suit controller to make sure it's functioning properly. There are exhalation valves in the suit and also a breathing regulator. We have to put the ACES ensemble together. It's lying on the table. You have to attach the gloves and the helmet to the suit because if you don't have them attached it's just like having a hole in the suit. You hook it up to the tester, pressurize it, and perform checks. It gets really firm. We test it to see if it'll hold pressure when it should, what the leak rate is, and check the Magnehelic differential pressure.

We test the anti-G suit's pressure control assembly, which controls the G suit's pressure.[67] They wear the G suit only when they're coming back into the Earth's atmosphere, so we test it to make sure it's working properly. Then the oxygen manifold is tested to make sure it's working properly. The communication checks are performed ahead of time before the vacant test, because you don't want to use a helmet that might have been faulty.

Then we test the ensemble once they've donned the suit and during their fit check. We have to perform those tests. They're called the manned suit testing, of course, because somebody's in it, unlike with the vacant suit test. We have to ask them to hold their breath to get their leakages because if they're breathing we can't get a good reading. Of course, when nobody's in there you don't have to do that because you just control the switches, but when they're in the suit their movement and their breathing will affect the readings.

Usually we take a breath with them so we don't have them sitting there holding their breath forever because it's easy to forget, especially because we get a little nervous even though we perform this test all the time. On actual launch morning it's like, "Okay, this is it. Take a breath and hold."

Of course, not at fit check. Sometimes people get nervous at fit check. It depends, but I tended to shine under pressure.

When you pressurize the suit you dial in the controller, which is the heart of the suit. The dual suit controller is the silver circle. The harness contains their emergency oxygen green apple. That's how they would activate the two bottles of emergency oxygen, about ten minutes' worth of oxygen. When they bail out they would activate that, not just here at home. The parachute harness attaches to the two frost-fittings, and this is also where their life preserver units are stored. If they submerge in water it will automatically inflate for them, but if it doesn't they just pull the tabs and activate them.

When performing the manned suit test, we only test System 1. There's a System 1 and a System 2, which is the backup. After we test it we take all the helmets, gloves, and CCA (communication carrier assembly) caps down to the Astrovan.[68] When they get out to the pad and get strapped in, the insertion techs will check the gear again to make sure nothing happened on the way out to the pad. There hasn't, that I can recall, been a case where anything was bad once they got out to the pad. Well, we've had communications cap concerns. We take spare helmets and CCA caps and gloves to the pad. So far so good—it's been a really good run. We've been very, very fortunate not to have had any incidents.

People tend to think, "After they launch, you all just sit around." No, you constantly have crews in training, and we have to prepare all of their equipment. It's not as extensive as for the equipment that's going to flight, but you still have to prep it, get it packed up, and support the different events. There's always a lot of activity going on.

We could have two to four crews in training at one time. Our schedule is usually full. Right now we only have STS-133 and STS-134 training.[69] With only two launches there's not a whole lot of training going on. Then we have presentations we support. We go out to schools and the community to do presentations on the ACES and our department.

All the equipment has a usage life and/or a shelf life. We have a whole fleet of suits and helmets. The suit is overhauled—inspected and tested. We're constantly testing and repairing. The equipment can have discrepancies, and we work to repair all those discrepancies, to get them ready for the next use. We have a lot of repairs and overhauls we perform on gloves, the helmets, the harnesses, everything. There's constant work here.

Even Class 3 equipment has to be repaired because we use it for training, so it still has to be operational even though it's not going to fly.

We have a team of nineteen people. There can be five to seven people assigned to a mission. There can also be two training events in one day. STS-134 could be training at Building 5, STS-133 could be at Building 9—so you could have a majority of the team out because they're supporting these events. If it's an event that's going to last all day, we'll leave one person over there in case the crew needs help and everybody else comes back to the lab and works until it's time to recover the event. Now for the bailouts where they go out the top of the mockup, practicing in case they have to escape from the Shuttle, everybody stays in place because it's very busy. You have to reset their oxygen bottles and assist them with a variety of things because they're in and out of the Shuttle mock-up.

You can also have people on travel. An example is STS-133. Our tech team is down there with them. If the astronaut crew is done, they come back to Houston that same day. My guys have to stay a couple more days to finish processing the gear, inspecting it again, and testing it again after it's been used for TCDT. Meanwhile, the STS-133 crew came back and

Sharon McDougle helps STS-106 Commander Terrence W. Wilcutt with his launch entry suit.

they have a training event the next morning, but their team isn't here, so techs here are going to have to go and support. It gets crazy. Of course, people still take vacations and have appointments. They're not all here all the time.

This is probably one of the slowest times besides the Return to Flight when we were waiting for STS-114 to go up, and there was no training or anything going on. That was bad. We thought we were going to get laid off then. It was about two years before Return to Flight. I was so shocked they didn't have layoffs, because it was really, really slow.

We also clean the suits. We have a Stericide, a disinfecting-sterilizing solution, that we use to clean the suits. We take the suits into the drying locker, hang them up, invert them, and spray them down really well. After they dry, they're inspected and the seals are cleaned and lubed. There is some spot cleaning involved on some equipment.

If urine got in the suit, which could happen—even when they wear their diaper, it can leak out sometimes—we use a baking soda and water solution. You have to be careful with what you use to clean, of course. It has to be approved because you don't want anything that's going to damage the equipment or cause any problems.

The engineers check everything out before we use it. Remember the culture shock I told you about? When they're going into space it's different than just flying a plane at high altitude. So they really have to check, check, and recheck. I got used to it, but I was still in shock with how things worked. It's a lot more procedural-based; you have to make sure you read everything and follow procedures.

Now I'm in management. I'm always tied to my computer, meetings, and things of that nature. It's not as fun as being a suit tech and being out interacting with the crewmembers. I'll pop up at an event every now and then just to say hi to the crewmembers, but so many that I knew are gone now. There are so many new astronauts that I don't even know that many anymore. I just go and show my face so they'll know who I am when they see me down at launch.

I started at the bottom as a C technician even though I had eight years' experience in the Air Force. I knew once I got here and the management saw my work ethic and what I had to offer that I would move up. And I did. I was the only black woman for a long time, then we had another girl come in for a little while. I was at the bottom of the totem pole. I was brand-new to Boeing. I finally became an A tech, which is crew chief, and I was the only woman to achieve that at that time. A crew chief is over the team that goes down for launch and has to make sure all the gear is getting processed.

STS-75 was my first time being a crew chief.[70] That was February 1996, the day before my birthday, February 22. Then STS-78 was the all-girl suit tech team.[71] That was the first time and the only time that's ever happened. All girl suit techs and I was the crew chief. That was another highlight of my career, the first time we had all girls.

As far as the people that make decisions, my management, I think I've been so fortunate and treated so well since coming in the door. USA, they are taking such good care of us and offering so much. I moved through the ranks. I think everybody is treated fairly and judged and promoted on what they bring to the table, on their merit and credentials, and how hard they work. That's why I wasn't really concerned when I came out here. I don't think gender and race has affected my career at all.

I've just been treated well. I'm not going to say they gave me anything, because I earned my raises and promotions. I don't think they had to be as good as they have been to me, and I try to do the same thing. I try to be very fair across the board. I think that's one of the reasons I was promoted. Twenty people interviewed for the position after the manager of Crew Escape left. I told them I bleed orange, and they knew I actually love what I do.[72] I love Crew Escape. I felt I could make a difference because I had been a technician, I knew everybody, and I'd grown with everybody. I felt like I could put together a better team and just really do a good job—do right by the company and for the team, and, of course, provide good support for the crew. I was just very happy.

That's what I tell people. All that work from when I first got here up to this point is what got me in this office. If I had a bad history over that time they probably wouldn't have even considered me, but because I had been so ethical and fair—as a crew chief, you're managing a team of five people. You have to make sure they're doing what they're supposed to do and they're at the training events and processing the gear. You report up to your manager. It was a big responsibility, but it never seemed like work because it was so much fun. Even as a crew chief I was still a suit tech. I was still suiting up the commander, enjoying myself, learning a lot along the way as far as people skills.

I used to get upset because sometimes you would hear people try to blame things on "Oh, it's because I'm black," or "because I'm a female," or other excuses, and I don't agree with that. I don't know everything behind whatever their issue may be, but I refuse to believe that. It's hard for me to see that when I haven't been treated that way. I'm always wondering, "What's the whole story? Are they really doing what they're supposed to do?" Don't just try to use race as the reason all the time.

Of course, growing up as a black woman, you do have to work harder. Because back in the era when I grew up—I don't know if it's as bad now—but you did have to work harder. I grew up in Mississippi, so I had to work harder and be better than the white kids. That was just part of life. And I knew that, but I didn't let that deter me from trying to do well and do the right thing.

I just have to say, we have an awesome team here. Flight Crew Equipment as a whole, but especially Crew Escape Equipment, which is our department, is the face of USA because we're the ones out there suiting up the astronauts. We're the ones that you see in all the pictures. We have the coolest job, I think, at USA. We're working with astronauts, we're suiting them up, we get real close with them. My team even straps them into the Shuttle. We're the first ones they see when they step off the Shuttle when it lands. It doesn't get too much cooler than that.

We have a great, great team. We're very knowledgeable, and I'm very proud. If you talk to anybody out in the lab, they'll tell you they're proud of what they do. And most of them are sticking it out to the end of the Shuttle Program. Ninety-five percent of the people said they're going to stay till the end because they want to. I know if a good opportunity comes along, of course, you have to really weigh your options, but everybody's so proud to be a part of this team, including me. I'm not going anywhere; I'm not going to even look for another job right now, because I know I'm going to stay till the program is over.

Every time the Shuttle goes up I think of the lyrics "Proud to Be an American."[73] I get emotional, even if I didn't work that launch and actually suit somebody up. I still get a little chill of *pride*.

5

Nurses and Scientists

Opportunities for scientists and medical personnel abound at NASA. Scientists perform experiments in space, operate the Space Station's robotic arm, and walk in space. Physiologists and other researchers, called principal investigators, or PIs, design spaceflight experiments, define the procedures for them, and then write reports about their findings. Others conduct research in labs scattered across the center, coming up with innovative technologies for spaceflight missions, from new fabrics for crew clothing to recipes for astronaut meals. Flight surgeons and nurses monitor the crews and their families in space and on the ground.

In November 1959, Dolores B. "Dee" O'Hara became America's first aerospace nurse. Historically, nurses help to heal the sick, tend to the wounded, and assist physicians during surgeries. Aerospace nursing is

A researcher uses a microscope in one of the Medical Sciences Division Laboratories.

quite different. Her patients were the United States' first seven astronauts. O'Hara remained with these men through the conclusion of Project Mercury in 1963 and then resigned her Air Force commission to stay with the space program. In 1964, she accepted a position at the newly built Manned Spacecraft Center in Houston, where she set up the Flight Medicine Clinic. She continued to work with the astronauts through the Skylab Program and returned at the request of the flight crews for the Apollo-Soyuz Test Project, the world's first international spaceflight, and STS-1.

Vickie L. Kloeris was known as NASA's "Top Chef." She grew up in Texas City, just down the road from JSC, but had never considered working for NASA. After earning her master's degree in food science from Texas A&M University, she moved to Houston, where she attended a local meeting of the Institute of Food Technologists. After talking with JSC food scientists, Kloeris decided it was where she wanted to work. Technology Incorporated, a space food contractor, offered her a job in 1985. Four years later, after working for a series of NASA contractors, Kloeris became a civil servant and the subsystem manager for the Shuttle food system. In 2000, she also began managing the Space Station food system and the Space Food Systems Laboratory. She retired in 2018.

Dolores B. "Dee" O'Hara [1]

My roommate came home one day and said, "Let's join the Air Force and see the world."

I said, "No, I don't think so. Nice girls don't do that." We're talking several centuries ago, you see. Anyway, we mulled it over for a while and thought, "Well, why not?" It's a way to travel and to do something different.

We went to downtown Portland, walked into the recruiting office, and said, "Here we are. Where do we sign?" Of course, the recruiter was a bit stunned at that point, because females didn't just walk in off the street and ask to join the Air Force.

We went to officers' training at Maxwell Air Force Base in Alabama. From there my classmate, Jackie, went off to Mobile, Alabama, and I went to Patrick Air Force Base at Cape Canaveral, Florida. This was in May 1959.

The first seven astronauts were selected in April 1959, and in November 1959, I was working nights in the labor and delivery room at the hospital there at Patrick. I had a message that the "old man," meaning the commander of the hospital, wanted to see me when I got off duty the next morning. Naturally, I was terrified because I'd only been there six months. I knew that when you went to see Colonel George Knauf, it was for two reasons: one, you were in trouble; or, two, it was for a promotion. [2] Well, I knew it was not for a promotion because I'd only been there six months, so I kept thinking, "Oh boy, what have I done?" I didn't remember harming anybody or harming a baby.

I gave my morning report the next morning and went to his office, and there sat his executive officer, the chief nurse, and all these people. I really was terrified because I didn't know why I was there. I literally sat on the edge of the seat. He started talking about Mercury, and I thought, "Well, there's a planet Mercury and there's mercury in a thermometer." Then he mentioned astronauts. That, of course, didn't mean anything to me. I didn't know what they were. He mentioned NASA, and I thought he was saying "Nassau," because of the island of Nassau. I had just been there, and I thought, "How in the heck did he know I was down there?" I was quite confused.

He turned and asked me, "Well, do you want the job?" I turned around, because I didn't think he was talking to me. He said, "You haven't heard a word I've said, have you?"

I said, "No, sir."

And he asked, "Well, do you want the job or not?"

I didn't know what else to say, so I said, "I guess so," absolutely not knowing what I had committed myself to. Of course, the chief nurse, who was there, was furious with me afterward because she was losing me. Also, she thought NASA was crazy because they were going to be putting a man on the top of a rocket. That's how I started.

In January of 1960, I went out to Cape Canaveral and set up the Aeromed Lab.[3] I felt that this position really was an opportunity. It was never an order, because the choice was certainly up to me.

Colonel Knauf had been tasked by NASA to put together medical support teams from all of the military services—the Army, Navy, and the Air Force—and these support teams were to consist of surgeons and nurses and people of all disciplines. They would go aboard recovery ships and set up little hospitals should there be a problem upon landing. They were stationed all over the area in case the capsule wasn't quite where it was supposed to be. We had to put together medical kits and everything that people on board the ships would need to treat an injured astronaut. Colonel Knauf decided that he wanted a nurse, and NASA said, "We don't want a nurse."

He said, "You're going to have a nurse." He wanted someone that would get to know the astronauts so well that she would certainly know if they were ill, because, as we all know, pilots and astronauts are not about to tell a flight surgeon when they're sick. That's understandable because pilots are so afraid of being grounded, and the flight surgeon's the only one that has that authority over them, so they're not usually very friendly with their flight surgeons.

People have said to me, you must be very special, you must be whatever, but no, it was the luck of the Irish. I was in the right place at the right time. I did have a varied background. My understanding was that when he decided to have a nurse out there, he had gone to Washington, DC, to the Department of the Air Force, and went into the Air Force Nurse Corps Office and looked through files there. Apparently, he did not find what he was looking for and thought, "Well, why don't I go home and see what nurses I have at Patrick Air Force Base."

He came back and apparently pulled my record after looking at all the other nurse records. Since I had a varied background in surgery and lab work, he apparently said, "Well, what about her?" Unbeknownst to me, he observed me for two weeks and said, "She's going to be the one." So I guess I was in the right place at the right time. It was just as simple as that.

This hadn't been done before, so it was all new. Putting a man on top of a rocket was a bit strange. There was no one to follow. All of us, physicians, engineers, me, and everybody associated with the space program, we were all marching forward and not with a lot of guidance, just making up the rules as we went along.

I just went along with whatever was going on, and, obviously, it was a great learning experience. I didn't know quite what to anticipate. Of course, my involvement was only from the medical side of the house. I set up the crew quarters and the Aeromed Lab there in Hangar S at the Cape. We had an exam area and a laboratory area and then my little office. Next door was a spacesuit area where all the suits were checked out and tested. Then just past that was the crew quarters, where the crew slept just before launch, or, if they were down at the Cape for training, that's where they would spend the night. It was a very small area, but everything certainly functioned very well.

Back during the Mercury days there weren't a tremendous lot of people. It was a very small core group of people, mainly engineers supporting the Mercury Project. The first seven astronauts were all stationed at Langley Research Center in Virginia, as were most of the engineers.[4] They would come to the Cape for their simulator training, their suit checkouts, and whatever testing needed to be done in the spacecraft. We would do their preflight physicals, and that's where I got involved, in the medical preflight exams.

One of the fun things was going down to Grand Bahama Island (GBI). This is where we set up a little fly-away hospital so that if an astronaut, particularly during the first suborbital flights of Alan Shepard and Gus Grissom, was injured upon landing we could take them there.[5] That is also where we would take them for their postflight examinations. I went down and helped set this up. We set up a surgical suite and an exam area. This little hospital was collapsible, and it could be flown anywhere it was needed. That was a fun thing to do. Fortunately, we never used it except to see them after the flight for their postflight exams.

Surprisingly, maybe it was because of the times, but back then, whatever

NASA wanted, they got. The space program was new and it was much favored by everyone, particularly all of the politicians. We had a wonderful NASA administrator, Mr. James Webb.[6] He was just fabulous. He was very astute politically, and politicians were falling all over themselves to be a part of this wonderful new adventure with this brand-spanking new agency. In my case, all I had to do was go out and find the furniture and tell procurement that's what I wanted. I don't recall any hassles, certainly not like you have nowadays. I was out there by myself, and I kept Colonel Knauf informed of my work, but I really worked independently.

In January of 1960, the first seven astronauts started coming down to the Cape to start training for the first suborbital flight. At that time, the astronaut that was going to fly first had not been chosen, or at least it certainly hadn't been announced. They would come down, as I said, to do their simulator runs and work in the altitude chambers. The Mercury spacecraft had to go through all sorts of tests because it was a brand-new vehicle and certainly untested. Since I was there on a daily basis, more or less doing everything to support them, I just did whatever was necessary or whatever needed to be done.

The first time I met them, I was terrified of them. I was very intimidated by them. But they were extremely nice to me, and they made me feel very comfortable and welcome in this all-male world, if you will. During the early Mercury days, I remember opening the door to the conference room, which was between the suit room and the crew quarters, and there they were. I just froze. I said, "Oh, excuse me," and I slammed the door shut. I remember racing back up to my office.

John Glenn came up.[7] He said, "Come on back. Meet the guys."

He introduced me to all of them, and I was so nervous that I blurted out, "Would you like something to drink?" Of course they all wanted something, which meant I had to go back and get it and then bring it back. I thought, "Why did they want something to drink?" Fortunately, I had enough soft drinks in the refrigerator in the lab area, so I took them all back the soft drinks.

They said, "Oh, stay and talk."

I said, "No, that's okay," and boy, I got out of there. Little by little, that went away. I don't think it was instant bonding, but it's like anything. Friendships develop over time. They were the cream of the crop and being around them daily, when in the area, we just became good friends. It's like that with anyone that you meet and you're around daily. You become more familiar with them and you work it into a very nice friendship and relationship. I think the friendships just grew as they would normally under normal circumstances, because I'd help out in the suit room, or I'd do the lab work, or I'd do some communication checks. I would do whatever was required. Back then you did whatever it took to get the job done.

Occasionally their wives would come down, so it was nice for me to get to meet them, although I didn't get to know any of the wives really well until we all moved to Houston. They came down once in a while for brief periods of time, because they all had small children and the kids were normally in school.

At the Cape, it really was a man's world. I was the only female in Hangar S—maybe there was a secretary here or there—so I stood out like a sore thumb. I wore a white Air Force nurse's uniform and an Air Force nurse's cap. It was all men, and never once was I ever discriminated against or made to feel uncomfortable. It was a wonderful, absolutely wonderful, environment. Of course, I wasn't really the sensitive kind, if you will, or looking for every little nuance, that I think is prevalent today. I was treated beautifully by everyone and nicely and with respect, and we had a lot of fun. We joked and kidded around, but we were also very professional and we each had our job to do.

After Alan Shepard had been chosen for that first flight my day-to-day routine didn't change, per se. It just became more medically oriented. We had twenty-four-hour urine samples to collect and a number of other medical tests. So just prior to flight, there was more emphasis on the medical exams, and I was the one person that was there all the time that they could either call or contact, and I would go out and get whatever was needed.

Bill Douglas was their flight surgeon, and he was a wonderful man.[8] He was the kind that wouldn't allow them to go on the centrifuge or do a chamber run or any test unless he did it first. When it came to an astronaut's health he would ask me, "Is everything okay?" or, "Have you noticed anything?" or, "Any problems that you've observed?" I would certainly have a verbal input as to the status of the individual at that time.

One of the things I probably should mention about the medical aspect is the business with the flight surgeons. I'd made an agreement with the astronauts, actually each group that came in, a long time ago, that they could come to me with anything. It didn't matter what it was, and that I would never betray them. There was one condition. I said, "You have to understand that if, in my opinion, it would jeopardize you or the mission, then ethically I will go to a flight surgeon with this. So don't come to me with anything you don't want them to know." That was the understanding, and that's the way it worked all these years. It was always that way. I would never betray them, but they had to know that if it was serious, we had to tell someone.

I don't know if the flight surgeons knew. I just never discussed it with them. It didn't matter because it was my bond and my word with the astronauts. It was my agreement with the guys, and that's the way it was. On two or three occasions, they did come to me, and we would check it out. If it wasn't serious, then nobody knew about it. That's just the way I operated, and it worked very well.

The days prior to Shepard's first flight were probably the most emotional, excruciating time for me. Since we had never launched a man on top of a rocket, it was very nerve-wracking, and I think it was nerve-wracking for everyone simply because we didn't know quite what was going to happen. As you know, they were always testing missiles, and they'd go up and explode, or they'd nosedive into the ocean, and now we were going to put some guy on top of one of these rockets. So it was very scary.

The night before the flight, I was out there all evening. One of my jobs was to call and wake the various support people and get them up well in advance of the launch. Then I would go and wake up the crew. The crew would come in to the exam area, and we took his temperature, his blood pressure, heart rate, and got his weight.

The weight was very important. You always need a weight before the launch and a weight after landing. We'd make sure that we checked his skin and his ears, particularly the skin on his chest area, to be sure there was no irritation. We checked to see so the headphones and electrodes didn't become an irritant for them during the flight.

I could feel the anxiety building up, particularly in myself. It was kind of a scary time, and I'm sure it was for a lot of people. You could feel the tension in the air.

After the physical, he would then go have breakfast and then go back to crew quarters, which was about fifteen or twenty feet away, and get into the undergarments he would wear under his flight suit. He would then go into the suit room to be suited, and once that was done, he would walk down the hall and downstairs and off to the launch pad.

I would then go over to what was called the forward med station. This was an old Snark hangar that I set up as a trauma room in case there was a pad abort or a launch disaster, and this would be where they would bring him.[9] I had a med tech and myself. We were stationed there during the preflight hours, and that's where we would watch the launch. It was very emotional for me to watch.

I remember John Glenn's flight. Any time we would see a cloud, maybe over Orlando, coming our way, they would scrub the launch. The least little thing would cause a launch delay or a launch scrub, simply because we had not done it before. This was the same with Alan's and Gus's flights. Every time anything out of the normal would happen, why, everything came to a screeching halt.

As soon as Alan launched I don't think I breathed for quite some time, and even then, his flight was, what, sixteen minutes? Those were very, very

Dee O'Hara talks with Astronaut John Glenn prior to his first spaceflight.

long minutes until we got the word that he was downrange, and he had landed exactly where he was supposed to.

Then I was taken over to the Cape Canaveral Air Force Station Skid Strip, and Dr. Douglas and I flew down to Grand Bahama Islands for the postflight physical exams. Even though he had been recovered and was aboard the ship, he was brought to GBI for his major physical exam. The astronauts were seen aboard a ship very briefly, but the more extensive medical exams were conducted there at GBI in that little fly-away hospital.

I remember I was so relieved to see him. We all were, and there were so many people talking to him. I think I just smiled. I was just so glad to see him; I don't know that I said anything to him. We were just so happy he was back, and it was a great accomplishment. Really, it was a great feat. I think we tend to forget that. With all the exotic technology we have now compared to what was done then, it was really an engineering feat that was unbelievable. The talent that NASA had back then, and really still has, but the talent back then was awesome. The best and the brightest wanted to work for NASA, and rightfully so, because it's a wonderful agency. The opportunities were unlimited—to be a part of something as historic as these first flights or any of the flights, but particularly the first ones.

It never really got easier as we flew the second Mercury flight. I think we all had just a titch more self-confidence, but not much, because, again, it was still all very new and very experimental. The same type of launch preparation was done for Gus as was for Alan, and then, of course, Gus got into trouble upon landing. Afterward we heard the details. We were on GBI and didn't have instant TV like they have now, so we did not know what was going on.[10]

John Glenn launched on an Atlas rocket, and there was tremendous anxiety, at least, again, on my part. We had watched those blow up right and left. I don't remember how many times John's flight was scrubbed because of the weather, which was usually one of the determining factors, and in Florida, as you know, the weather can change instantly. We'd see a cloud coming, so they would reschedule the time, or they'd put it into a hold, and that was simply because they just weren't going to take any chances.

Every time that his mission was put on hold we had to start some of the procedures all over again. We really didn't do a lot of in-depth exams on them. We just made sure that the heart, lungs, and skin, and the ears were clear before they left.

The preparation before the later Mercury flights was very much the same. I don't know that anything changed except probably their simulator training. During the debriefing, you learn what things need to be changed, and they would try to modify and either add or take away from, say, the flight plan. But there weren't a lot of drastic medical changes.

All through the programs I was involved in, we really didn't change the way we operated. We did not usually allow anyone new, a new face if you will, because it really upset the crewmen. I know that sounds silly. You wouldn't think it would bother them, but it really did. The same team was assigned to a mission so that when the crew came in, they saw the same faces and were familiar with what we were going to do to them, and we tried never to change that. We really worked as a team. Of course, the team kept growing as the flights became longer and more sophisticated.

All my records, everything was handwritten. We didn't have computers. I'm sure there were some around, but we didn't have any. I think they were just coming out. We had paper checklists. You went and followed your routine, and you didn't vary much from that.

We collected all of the data from the physicals and brought everything back to Houston, and it would be analyzed and filed. We soon started a computer database. One of the fellows there set up a wonderful system for collecting and storing all of the information to archive all of the medical records. Whenever I would go down to the Cape for a flight, I would take their medical records with me. I was always terrified the plane would go down and the medical records would be gone, and they'd never forgive me.

I really just took their medical records on board with me in the aircraft. The rolls of data, rolls of EKGs, and a lot of the other medical information was shipped back. I didn't bother to bring that, but I would certainly make sure that I got the medical records back. The investigators that were doing certain medical tests on them, they were responsible for their own data and to cart their own data back. I just took responsibility for the medical records, the pre- and the postflight medical information.

The workload increased with the Gemini Program. They started selecting more astronauts. The second group was selected in 1962, and I think there were nine astronauts. The next group was fourteen. They all started arriving with children and wives. Well, we had a couple of bachelors, but it was a pretty heavy workload.

When I came to the Manned Spacecraft Center in Houston in 1964, I went from being an Air Force employee to a NASA employee, another government employee. My duties expanded because we had wives and children, and that entailed things other than just flight-type activities or preflight medical activities. It became busier because the population kept growing and growing.

It had been decided that the Flight Medicine Clinic would not only see the astronauts, but the charter for us was to see and care for the families. The reason we were designated as family physicians was that the astronauts were usually on travel and rarely ever home, and at least this way they knew that they could go away from home, knowing that if anything happened, we were there to take care of their families. If it was above what we could do in Flight Medicine, they were referred to a specialist, but the flight surgeons and I became the family physicians for them, if you will.

I think it was a comfort to the families and to the astronauts, because many of them were military wives and used to military outpatient clinics. They would say, "You have no idea what a luxury it is to come here and be seen by friendly faces." Military wives with a sick child would go and sit in these outpatient clinics and wait and wait and wait and never see the same doctor twice. In Flight Medicine, it was someone they were familiar with and that they could call. That gave not only the astronauts assurance that their families were taken care of, but it was a place for the families to come that was familiar to them, and they were usually seen right away. We had great capability in the Flight Medicine Clinic to do this, so it worked out very well. It became a family medical unit for them. In fact, I think it still is today.

At the time, there was just me and five flight surgeons. I was very busy because I'd do all of the preflight stuff, then go to the Cape, where I'd stay until the crew was recovered and brought back to the Cape, and we'd do the postflight exams. We'd all fly back to Houston. In the meantime, everything that I normally did in the Flight Medicine Clinic was still waiting for me. I did all their EKGs, eye exams, all of the administrative forms, the hearing tests—you name it—and, of course, the immunizations on the families, plus the guys. So, I kept pretty busy.

I did not have much of a personal life. Particularly during Gemini and Apollo, I really had no life of my own because I was at the Cape for two months. I'd be home for two or three weeks, then I'd go back down to the

Cape. It was back and forth and back and forth. I have to confess that it really got old after a while because I was constantly living on the road. I had the same room at the Holiday Inn in Cocoa Beach for nine years. I've often joked that we need to make a shrine out of it one of these days.

What's that saying? "I'm a girl that can't say no." It's just the way it was. I think at one point I had a patient census of five hundred, and, about the third year or so, we finally did get a secretary to at least answer the phone. She took care of some of the administrative duties, but I maintained all the medical records, the Federal Aviation Administration records, and typed up their physicals and histories. In addition to taking care of the astronaut families, we took care of the NASA Aircraft Operations pilots and all the crew personnel—anyone that flew NASA aircraft at Ellington Air Force Base.

There was also a lot of fan mail, which always surprised me, at least during the Mercury Program. Various groups all over the United States asked me to speak. I hated it with a passion. I absolutely hated it. When I was an Air Force nurse I did it because I was ordered to, and everybody always wanted to know things about *them*. Well, I don't talk about them. I was very uncomfortable doing this because it just isn't my right to talk about them. What a lot of people don't understand is that they allowed me to be close to them, and you just can't betray that. I never could, and to this day I wouldn't betray that. I took it as a sacred obligation, and that's why I would never write a book about them.

I was invited everywhere, and I was treated like a celebrity and accorded so much. I know it wasn't me, per se. It was because of my association with them, and I have never really lost sight of that, because you can certainly become very self-important and go on and on about your so-called accomplishments. But it was simply because of my association with them.

I was called and referred to by all sorts of titles, astronauts' nurse, aerospace nurse, what have you, and I never called myself any of those. Other people have, and I guess astronaut nurse is probably the proper term. I don't know what else you would call it, but I don't use the term. I was always embarrassed by that title, "astronaut nurse." The "nurse" part is kind of a misnomer, in a way. Here were seven guys who were in top physical condition. They were all top fighter pilots.

I was just lucky to have a very elite group of patients, and they really weren't patients, of course, because they were certainly never ill. I'm not really big into titles. You have a job to do, and the point is to just do it, and it doesn't matter what they call you as long as you do whatever it is you're supposed to be doing.

There were tragedies in the astronaut corps, and, no question, the Apollo fire was horrendous.[11] That really was horribly difficult, I think, for everyone, but a lot of lessons were learned that needed to be. Fortunately, a lot of things changed for the better.

I think one that hit me particularly hard was when C. C. Williams was killed in October 1967.[12] His wife was my best friend and still is. She was two months pregnant with their second child when he was killed. Their first child, Catherine, was ten months old. And his death was extremely difficult for me to deal with. The deaths were all very painful. It's not only losing the guy, but then you go through the tragedy with the family and the children. There was a lot of emotional—well, I don't want to say turmoil, but there were a lot of emotional issues to deal with, and you can't be that close to that many people and not expect it. They were doing very dangerous work. Spaceflight is certainly not for sissies. Spaceflight is dangerous, and I think we lose sight of that because we've been so successful, very successful.

The Moon landing, now, that was a spectacular engineering feat. On occasion you would hear, "Oh, they were so lucky." Well, it was not luck. It was hard work, and it was brilliant planning and engineering. Like with Apollo 13, we were so lucky to get them back.[13] Yes, we were lucky, but it took a lot of teamwork and brain power to figure out how to get them home, not only on the part of the crew, but also the Mission Control people and all of the ground support people. The ground support people should never be sold short. I know the astronauts certainly don't sell them short. They're very complimentary about the people that have supported them throughout their missions. And, boy, if you want to see teamwork in action, the space program is the place to watch. It's not any one individual. Nobody can do it by themselves.

The early days of the space program weren't routine, and yet, in many ways, they were. For example, a couple of times when the astronauts were out training in the Lunar Landing Training Vehicle, in what we used to call the "flying bedstead" out there at Ellington, a couple of them happened to land in chiggers.[14] This one guy came racing into the office, dropped his pants, and said, "My god, I'm dying! Do something!" Of course, that was a wonderful sight.

I said, "God, the blinds are open!" They finally ended up closing the door, and we pulled the blinds. But wonderful, fun things like that fortunately didn't happen all that often. Nothing surprised me—believe me—or very little surprised me.

The practical jokes played by the astronauts were unbelievable. They were quite naughty at times, and they were always, of course, extremely funny. The jokes and practical tricks, those went on all the time, particularly when the crews got to the Cape. They'd come over and say, "Hey, we want to do such-and-such to so-and-so, and can you do this?" I'd do it, and we had great fun.

Then they would put these things on board the spacecraft so that the astronauts would find them midflight somewhere, and, of course, that was always a surprise for them. So I got to be a part of a lot of that, and it was great. They were deadly serious about their work, but there was also a tremendous sense of humor, and they were very, very clever. I think you have to have a great sense of humor, particularly in that kind of work. Oh, boy, did they have one.

Vickie L. Kloeris [15]

In 1985, I was hired primarily to begin planning the food system for Space Station Freedom, which was going to be the US space station.[16] Of course, it was all on paper at that point. We were just planning. When I first started we looked at Skylab, of course.[17] Skylab had the most sophisticated food system that NASA has ever flown, before or since, because it included frozen food. We went into Space Station Freedom assuming we would have frozen food. We actually went into the International Space Station (ISS) thinking we would have frozen food. Of course, Freedom went away and morphed into the International Space Station.[18] Originally, on the International Space Station, we were supposed to have the US Habitation Module. The US Habitation Module was going to include freezers and refrigerators for food.

We were planning on developing frozen food, planning how we were going to package it, how we were going to heat it on orbit, etc., then they canceled the Habitation Module. That was a budget thing, and it was also a power thing. They determined that they weren't going to have enough power to support all that. They had already put a significant amount of money in the budget over a several-year period to develop this frozen food system. Dr. Charles Bourland convinced them to leave that money in the budget and let us use it to develop shelf-stable foods for the Space Station.[19]

If you go back to Mercury, Gemini, Apollo, those food systems were all custom. There was very little commercial off-the-shelf (COTS) food used in those systems. Things were made either by Natick or Pillsbury.[20] It's very expensive to do. Going into the Shuttle Program NASA made a conscious decision to use as much commercial off-the-shelf food as they could. When I came to work here we weren't doing any product development. If a commercial product that we were using got discontinued, we would go out and identify another commercial product to take its place.

During the Shuttle Program we utilized the pouched entrees that they use in MREs. We were buying the Meals Ready to Eat from the same vendors who made them for the military.[21] That worked very well for the Shuttle Program, but we knew going into long-duration spaceflight that the MREs were too high in salt and fat for what we wanted. That was one reason we wanted to morph into this frozen food system, so we could provide foods that were lower in salt and better quality.

For Shuttle, our food list was fairly short. We didn't have a huge amount of variety because the flights were so short. You didn't really need this long list of foods and beverages to support a ten- or twelve-day Shuttle flight. Going into Space Station they were talking about three-and-a-half to four months at that time on orbit. We knew we would need more variety. We knew we needed to try to get away from the MREs as much as we could.

In the late 1990s, we started true product development, where we were coming up with our own formulations. Our food scientists started developing thermostabilized pouched products to use on the International Space Station. Over that time, we developed about sixty food items that are unique formulations we make just for Space Station, so we have expanded our food list. On the US side we have about two hundred different foods and beverages. There are still commercial off-the-shelf items in there. All of our beverages are COTS, because we use beverage powder. We're buying Kool-Aids, the instant coffees, and the instant teas. We're not in the beverage production business. We still utilize commercial cookies, crackers, and candies.

Up until about two years ago, most of our freeze-dried foods were commercial frozen items. We would bring them in-house to cook and then freeze-dry them so they were further processed. That all goes back to the philosophy that we're going to utilize as much COTS stuff as we can. In the long haul that came back to haunt us. Now that we've got years of Station under our belt, we have come to realize that there are medical issues that occur in some of the crewmembers. They were experiencing increased intracranial pressure on orbit. They think the high salt exacerbated it.

In 2010, we were directed to reduce the sodium in the foods. What that meant is we had to get rid of all those commercial off-the-shelf products, especially the freeze-dried items. Everything that we had been further processing from COTS we're now making from scratch, because commercial products tend to be very high in salt. It's a very cheap way to make things taste good, so food companies utilize salt quite a bit. We have come up with new formulations on about ninety of our products.

We have completed that process now. We've successfully reduced that sodium content of the standard menu by about 43 percent. Even with that we're still above the recommended dietary allowances. We can only get it down so far without refrigerators and freezers. One of the big problems that we face is the fact that these crewmembers get little, if any, fresh food. When you and I eat fresh fruits, vegetables, and salads, it lowers the total sodium content of our diet. Because they're not getting that on orbit, all they're getting is the processed foods, that means that their sodium content is going to be higher overall. We're trying to get that reduced sodium food to orbit as fast as we can.

We tried to reformulate what we already had rather than coming up with a whole new food list. We just tried to modify the ones that we had, because we already had a nice balance of different types of foods: starches, fruits, vegetables, meats. Previously we would bring in a frozen commercial macaroni and cheese, cook it, and freeze-dry it, but it was loaded with salt. Now we are making macaroni and cheese from scratch, so we have macaroni and cheese that's much lower in sodium because we're using a reduced-sodium cheese.

When the ISS Phase One Program came along and we were going to send our crewmembers to Mir, I was managing the Shuttle food program.[22] Dr. Bourland was managing Station food at that point. When the original agreements were signed, there was not going to be any US food on board Mir. Our crewmembers were going to eat all Russian food. When that word got back to the Astronaut Office they were like, "No!" This would have been 1993, 1994.

Dr. Helen Lane, who was head of the Nutritional Biochemistry Lab at that time, and Dr. Bourland both started negotiating with their Russian counterparts to try to get US food on board Mir, because we knew we wanted to at least get some up there.[23] It was interesting the way that worked. Russia was being paid to provide 100 percent of the food. We basically said, "We'll give you some food, and we're not going to charge you anything for it." What that meant was, "We're going to get paid to provide 100 percent of the food, but we only have to produce 80 percent of the food." We had to make it profitable for them. That's how we originally got the Russian Space Agency to agree to accept our food and send it to orbit on the Progress.[24]

Our biggest challenge for the Space Station from the start was getting stuff into Russia; it has always been very difficult. One of the first shipments was rejected. Russian customs was a nightmare. They would change the rules every time you would go to ship. The next time you'd see the shipment, they wanted some other document. This went on for a long, long time, even once we were on the International Space Station. It took quite a while. Now it's basically a smooth operation.

Norm Thagard was the first US astronaut to fly on Mir. Norm and the two cosmonauts who were with him were participating in medical experiments.[25] During those medical experiments they would have to record everything they ate because they needed that information for the data. Scott Smith, me, and a couple of the people from his lab made trips to Russia in late 1994 and early 1995 to do baseline data collections (BDC) for Norm and the two cosmonauts.[26] During those baseline data collections they were running a lot of medical tests on them, getting blood work, all kinds of stuff. During those days that they were medically tested, we had to weigh and record everything they ate, so we prepared the food. His dietician and I were the two chief cooks and bottle washers. We went to Russia and the first BDC was in November 1994. It ended up that one of the weeks was Thanksgiving. We missed Thanksgiving at home.

When we went over, we flew, obviously. I had a suitcase, and I had four cardboard boxes of food with me, same with everybody. We took all this food as checked baggage because we had no other way to get it over there because of all the issues with customs. We had to get it there if it was going to make it for Norm's flight, so we hand carried it. We were just so unbelievably afraid that we were never going to make it through customs to get into Russia. It was bad enough getting out of here but trying to get into there—I remember the science liaison from the US embassy met us at the airport to help us get the stuff in. It was nerve-wracking because you thought you were going to end up in jail.

Around this time, from 1995 to 1997, NASA did a series of chamber tests in Building 7's high bay where they were testing air and water recycling technologies for the Space Station.[27] They did a fifteen-day, a thirty-day, a sixty-day, and then they had planned a ninety-day test. We got involved in that because we were providing some of our flight foods.

They were looking for volunteers, so I volunteered. Much to my surprise, I was selected to be one of those crewmembers. I became a prime crewmember; four of us became prime, and the other four became backup. Then a decision was made to do it for ninety-one days. They chose ninety-one days because they wanted to break the US record for chamber tests. They knew they couldn't break the world record because Russia had some huge record, I think. This was in 1997. It was planned for September 19 through December 19. We did team training for all the eight crewmembers during the months leading up to that period.

I was the science coordinator. Space Life Sciences had agreed that they were going to do a series of experiments during the chamber test. My job was to coordinate between the investigators on the outside and the crew on the inside, making sure that the crewmembers knew when they needed to participate, what they needed to do to fulfill the requirements for participating in those experiments, making sure we had all the equipment, supplies, whatever we needed. We did have an airlock where we could transfer trash out and equipment in as needed.

We did, I think, a series of about thirteen different experiments. Part of what we did was to help develop and validate a food frequency questionnaire, the FFQ. We did a weighed diet record where for a period of time we weighed and recorded everything we ate. The FFQ was a quick and dirty way to collect dietary information. Principal investigator Scott Smith wanted to compare how the quick and dirty method was to the manual way. The FFQs are now used on orbit on the Space Station, where US crewmembers fill this out on a weekly basis.

We also did exercise protocols that were developed for us; these were precursors to the protocols that they use on Station. We also did things like microbiological surface sampling. They were trying to validate how often they would need to sample on the Space Station. We did questionnaires that were developed by the psych support team. Again, they were trying to develop questionnaires that they were going to use on ISS with the crewmembers on a weekly basis to try to keep track of their mood and how they were doing on orbit.

Nigel Packham was also a crewmember; Nigel and I were trained as medical officers for the crew.[28] We went through some of the training that they were using at the time for nonmedical Shuttle crewmembers who were trained as medical folks for the flight. We also did the simulation of remote medicine. We had a monitor inside the chamber, and we telemedicined to the outside.

Basically, we were there as biological loads on the chamber. We were testing air and water recycling equipment. That was really the main thing we were doing. Some of the earlier tests had done biological systems and physical-chemical and/or mechanical systems separately. Our test was a combination of the two. We were using part of a biological system to recycle our air and a physical-chemical system for some of it as well. We had ten thousand wheat plants growing in an adjoining chamber. We were providing CO_2 to the wheat and the plants were providing a fourth of the oxygen to our chamber. The rest of it was being provided by mechanical and/or physical-chemical means. We were also testing water recycling. This was all part of the development of what they're using on Station to recycle water.

We did a lot of educational outreach. Twice a week we would do a briefing for the Space Center Houston mission status room of what was going on in the chamber.[29] I remember we had one outreach event that we did for the Education Office. They teleconferenced in all these schools, so there were several schools that were able to talk to us directly and then there were others that were just listening. It was a lot of fun. I was a little bit apprehensive about what it was going to be like to live in close quarters with three other people for ninety-one days.

The chamber had three levels. The first level of the chamber was the main living area, that's where we had a bathroom and what served as a kitchen. We had a shower. We had a washing machine. We actually had to wash clothes because we had to push a certain amount of water through the system every day. We spent most of our time on that level.

The second level was mostly all equipment, so it was very loud up there. That was where all the equipment was processing the air and the water. We didn't spend very much time up there at all. We did have an exercise bike up there, because we had to exercise. We would do tests of our aerobic capacity. Throughout the ninety-one days we were trying to see the efficacy of the exercise protocol. You always wore ear plugs or listened to music while you were exercising so you could cut all the noise that was around you, because it was really loud.

The third floor had living quarters, which were very small. It's basically where you slept and dressed. There was a bunk bed. Actually, the top bunk bed was in one room and the bottom bunk bed was in another, so you were sharing a bunk bed with the person next door to you. You can visualize how small the rooms were. We did have a desk, a chair, and a workstation in our rooms so we could telecommunicate. We could do videoconferencing with our spouses, with our significant others, but the walls were really paper-thin. If anybody needed to have a private discussion we would leave and vacate to the first floor.

We were drinking iodinated water, just like they did on the Shuttle. That's the way they guaranteed the purity of the water, so we were drinking iodinated water and lots of it during that test. We were sending out blood samples for analysis periodically throughout the test. They observed that our TSH (thyroid-stimulating hormone) values, which is one of the things they monitor, were going up. There's a broad range of what's normal, and we were still within that normal range. This was not the first time they had seen this in these ground-based tests. It wasn't the first time that they had seen it on orbit either, which we didn't know about at the time. That was all private medical data. They didn't really discuss that.

As a result of data that came out of our tests, the Shuttle crewmembers were no longer drinking iodinated water on orbit. They installed a point-of-use filter to remove the iodine. They did the same thing in the chamber for us. After the test, they continued to monitor us to see if we would return to normal.

We were eating a combination of flight foods and foods that were plant-based, like what we thought astronauts would be eating on a Moon or Mars mission. But there was quite a bit of the flight food in there. It wasn't the first time I'd consumed it.

I had done ground-based studies before where I spent a month consuming nothing but Shuttle foods. That was for an experiment Helen Lane was doing when she was head of the Nutritional Biochemistry Lab. She was really trying to validate a process on the ground before she used it on orbit, so we were guinea pigs to work up the process for her experiment. Part of the time I did that. When I went to Florida to support a launch, I had to go into restaurants with my scale and weigh everything I ate. I had to choose things off the menu that were very pure, not with sauces, so that I knew what the ingredients were and could give that information to Dr. Lane. Or I would have to ask for recipes or ingredients from the restaurants so that they could formulate the nutritional content of what I was eating. I have to admit when I was at home it was easy, but traveling and doing that was pretty tough.

We had some Shuttle crews, like Spacelab Life Sciences-1 (SLS-1) and other flights, that did in-flight medical experiments that required nutritional monitoring.[30] On orbit they would scan the packages with a barcode reader. We instituted barcodes on all our foods so the crews could easily track what they were eating on orbit. When they would come back after flight—maybe for a week or two after flight—they would weigh and record everything they ate. Usually that meant we were cooking for them so that we knew what was in their food.

There have been times when I've gone to postflight parties with a scale in hand and followed crewmembers around weighing and recording what they were eating. I've done that in crew quarters, because we've had several flights where the four crewmembers who were science on that flight would have to weigh and record everything they ate during the quarantine period. We were cooking for them in quarantine, but we would have to make sure that we weighed everything that went out on the plate to them. If they left any plate waste we would have to weigh what was left so that we knew exactly what they consumed.

At some point the crewmembers get a little cranky about having to do it. So you're having to deal with that as well, because they do get tired of it. After doing it for a month I certainly know why they get tired of it. The nice thing is, in most of the cases, we were doing it for them rather than having them do it themselves.

Part of the Shuttle food contract responsibility was to feed the crews at JSC and Kennedy Space Center (KSC) during quarantine. Typically, it was a seven-day preflight quarantine. They would divide that time up between JSC and KSC—it was three to four days here, and the rest down at KSC prior to launch. It varied from flight to flight. We would work with the crews ahead of time to try to plan the meals for that period.

We weren't giving them flight food. It was just regular food. The only stipulation was that it couldn't be restaurant food. It had to be ingredients we bought and prepared. The two ideas behind quarantine were to keep them away from very young children at home, who might make them sick, and to keep them away from restaurants, where they might get foodborne illness, because they didn't want them getting sick prior to flight. That's the whole reason we were cooking for them in quarantine. The same people who did the food processing for the Shuttle and the packaging and stowing of the food, those were the same people who did the food preparation and the quarantine work in Houston.

At KSC we did it a little differently. We would have one or two food people from Houston go down to supervise that operation. Then we had part-time employees who worked the preflight period. So when the crew went down there for TCDT (terminal countdown demonstration test), which was usually two days, those same part-timers would come in and prepare the food for them during that period.[31] That was just more for expediency than anything else, because during TCDT you would have the prime crew there but then you'd also have the astronaut support people, the C-Squares (Cape Crusaders), the astronaut crewmembers who were down there supporting the launch.[32] You'd have people like the flight surgeon in crew quarters too. There was a fair amount of people who needed to be fed, so we would use those part-time people to do that.

The part-time food folks also worked during the preflight period when the crew was in quarantine, and they would also work postflight. On landing day, they would provide a meal not only for the crew coming back but for the families of the crewmembers who had come to view the landing. If the crew was staying overnight for any reason, they would provide breakfast for them before they flew back to Houston. The part-time people were the main backbone of the food service activities down at Kennedy.

These were your grandmother types. They'd been trained just like restaurant people would be, but these were your genuine cook-from-scratch, real cooks. These ladies loved cooking for these folks. The crewmembers really appreciated it. Some of those ladies did it for years down there, so the crews knew their specialty. They would come in, "Oh I want some of this or that while I'm down here." It was almost like it was mom or grandma cooking for them.

When I was with FEPC (Flight Equipment Processing Contract), I was required to go down there for every launch.[33] When I came over to the NASA side, I went down there for every launch for quite a while. At some point they turned that over to the contractor, but I spent many a quarantine in Houston and in Florida. During that time period I basically didn't have a life. Well, I did have a life, but you couldn't plan vacations, because all of those launches kept shifting. Many of the folks that work on the Space Station food system also worked on the Shuttle previously, so they were used to all the launch slips. Now that Shuttle is over that's all done in Russia so we don't have to deal with all those expediencies now. It makes your schedule a lot more predictable.

For SLS-1, that flight landed out in California (Edwards Air Force Base) and had a lot of dietary information that was gathered on orbit. We were feeding them postflight as well. We had to weigh and record everything they ate postflight. I can't remember the exact duration. There was a dude ranch about fifteen, twenty minutes away from Edwards, and that's where they stayed, so we stayed there too.

As part of the on-orbit data process we had to evaluate the trash that came back from the flight. It had been on the Orbiter for quite a while, and it was a few days after it landed before we could actually access it because they were getting other hardware and other more important things off the vehicle. At some point, they allowed us to go out there on the flight line. They had taken the trash off the vehicle and put it in barrels. It was sitting out in the sun, out in the desert. It stunk!

We had to go through it. We compared the serial numbers on the packages to what had been recorded by the barcode reader to make sure that nothing had been missed. Then we would look at the empty packages to see if there was residual food in there, and if so had the crewmember recorded that they hadn't eaten 100 percent of it. We were going through item by item. We wore gloves and a face mask to cut the smell and bunny suits to keep from getting nasty stuff on our clothes. This was a hot and sweaty and stinky job that we did. That was probably the worst case of that.

We also did it in Houston a few times when I worked at FEPC. They would go through the trash just to be sure that the crew hadn't inadvertently thrown away something they shouldn't have into the trash. We rotated that responsibility. A few times when I was there I would go through the trash as part of my duties. We would do it on the loading dock out back before we threw the trash away. Usually before we'd go through the trash we'd get a list of what we were looking for. "Here's what we're missing. See if you find it in the trash." Usually it was because they had stowed it in the wrong place, and it would turn up later. Occasionally we did find stuff in the trash, so we'd have to pull it out.

Let's talk about Shuttle and Station food. Shuttle food logistics were a lot simpler than ISS. On the Shuttle, we were able to do a personal preference menu, meaning each crewmember could do a custom menu to their own liking. On Station, we're not able to do that. On Shuttle we were because the food and the crewmember went up on the same vehicle, so they never got separated. It was easy to do and it was a small amount of food. It was only for a week or two. We went into the Space Station era trying to do the same thing. It was an abysmal failure. It was a huge challenge to get all the food for a crewmember there at the exact time that that crewmember was there.

What ended up happening was that for a certain percentage of the time they were on orbit they would end up eating somebody else's food. Believe it or not, even if it was only like 10 or 20 percent of their time on Station, it became a huge psychological issue. They would come back and complain. "You promised me this personal preference menu and part of the time I had to eat Joe Blow's menu and Joe Blow has horrible taste in food." They really would complain about it. That was one of the ways the psychological importance of food on long duration spaceflight became very apparent.

When we got ready for the first Automated Transfer Vehicle (ATV) to launch, it kept slipping.[34] We didn't really know which crew would be on orbit, so we went to the Astronaut Office and we proposed a standard menu. "We would like to propose this on the first ATV because we don't know when it's going to get there. We don't know whose preferences to use." We used virtually everything we had, provided as much variety as we could without repetition. Much to our surprise, they came back and said, "Not only do we want you to use it on ATV, we want you to use it from now on, on all the increments. Use the standard menu." So that's what we do.

On the Shuttle, they could choose exactly what they wanted. If it was a nine-day mission, they could choose nine separate menus: breakfast, lunch, dinner, and a snack every day. It was strictly up to the crewmember. Some crewmembers, for a nine-day flight, would choose three menus repeated three times. We had one crewmember who, every time he flew, had a one-day menu that he repeated. You should have seen that menu. He loaded it down with stuff so it gave him a lot of variety. Regardless of the length of the flight, his was that same menu over and over again. That made it very easy to plan for him.

For the Shuttle, we prepared the food in Houston. We used middeck locker trays, the blue trays that had Velcro nets on top. We'd fill them with food and they would get stowed in the lockers—two trays would fill the locker. The lockers got boxed and shipped down to Florida. When they unboxed them in Florida they were ready to be stowed in the middeck. They were all labeled. They already had the decals on the front that would show where they went. The folks who received the lockers on that end would just take them out at the appropriate time and put them in the Orbiter. About a week before launch they would get installed. The Orbiter was already on the pad in the upright position when they would install the food.

We did have typically one locker's worth, or two trays' worth of what we called fresh food that was late stowed. It was typically one of the last things that got put on the Orbiter before they started fueling the vehicle, because at that point they didn't want people out there. We would stow that in crew quarters down at Kennedy. One of the responsibilities of the person who traveled from Houston to Florida was to stow those trays, because the part-time people down there did not have experience.

We would work out beforehand with the crews what they wanted in those fresh food trays. Obviously, volume was limited. The whole crew shared the volume of those trays. They would make their list, and 99.9 percent of the time the list was longer than what would fit in the trays. It had to be stuff that had a long enough shelf life to last through the Shuttle flight without refrigeration. Typically, we would send things like oranges, apples, citrus fruit, things that would last a while without refrigeration. We'd also put in special request items, like if crewmembers wanted their favorite candy bar or favorite commercial cookie. As long as it had sufficient shelf life and met our microbiological requirements, we could put it in there.

The fresh food tray came into existence prior to my coming to work here. My understanding from Rita Rapp and Dr. Bourland was that it came into existence because the Astronaut Office wanted it.[35] They wanted some way to put fresh items on there.

We would put tortillas in there. Tortillas became the bread of choice in space. When I came to work here, for the first few fresh food trays that I worked on they were sending sliced bread. They would pack it in a Ziploc bag, put it in the fresh food tray, send it into orbit. Then one time we had a payload specialist from Mexico fly.[36] He wanted to take tortillas with him in fresh food, so he took a limited number. When that crew saw how easy it was to take a tortilla and roll something up in it, how it was like having a sandwich without having to deal with two pieces of bread, from that day forward we had tortillas on every Shuttle flight. Flour tortillas became the bread of choice.

We had to go through the process of determining whose tortillas we were going to use. We started going around Houston trying to figure out who had the best tortillas. We ended up with La Espiga de Oro in the Heights because they had the cleanest tortillas as far as yeast and mold.[37] When bread spoils it's usually either yeast or mold. So we would go to La Espiga de Oro and buy the tortillas before every flight. They would be on the NASA plane down to Florida.

Early on in the Shuttle Program you couldn't get tortillas in Florida much, at least not in that part of Florida. Eventually there were some close-by sources of tortillas. So, at one point we took samples of those to see if we could buy them down there. They didn't come near to being as clean as what we were getting here in Houston, so throughout the Shuttle Program we continued to ship tortillas on the NASA plane from Houston down there.

When we bought the commercial tortillas, those would last through the shorter Shuttle flights without a problem. Even though there was no refrigeration they would last eight to ten days or so, and then they'd mold. Crewmembers wanted tortillas that would last all the way through without having to worry about them molding, so we had to develop an extended shelf-life tortilla. That was one of the projects we did in the lab.

You can vacuum-package bread products, and they'll last. You remove all the oxygen so you package them in an anaerobic environment, and they can't mold. They just can't because yeast and mold need oxygen to grow. When you anaerobically package and take away all the oxygen, you create an environment where there's a small probability for anaerobic pathogens to grow, like *Clostridium botulinum*, which is very, very nasty.[38] In order to eliminate that altogether, you reformulate the tortilla to lower the water activity of the tortilla. Water activity is a measure of free water in a product. If you get the water activity below a certain level (they have equipment for measuring this), there's not enough free water for the bacteria to grow.

We had to come up with a formulation of tortilla that we could vac-uum-package, so we got into the tortilla manufacturing business in the food lab. We had the dough divider and the dough press so that we could make tortillas, and we did that for quite a while. We wouldn't make many. We were making enough to pick up the end of the flight, the part that was too long for the fresh. We would send a combination of fresh and vacuum-packaged ones, so between the two they had tortillas for their whole flight. We did that for quite a while. The tortillas we made in the lab would last about four months and then they would develop an off flavor. They would become somewhat bitter.

We were about to undertake a project to research how to get rid of that bitter flavor. As we got to the Space Station era we wanted them to have an even longer shelf life than four months. About the time we were going to start that, Taco Bell started marketing a kit in the store that had flour tortillas in it. That kit, every time we would look at it in the store, had at least a nine-month shelf life on it. We're like, "That has got to be a low water activity tortilla." So we got samples of it and analyzed it, and sure enough it was.

We went to them and said, "Who makes this for you?" We started buying from that company, and we were able to get out of the tortilla manufacturing business. Then all we had to do was buy them and package them. That's what we're doing now. We buy those and package them to send to orbit. We were doing that at the end of the Shuttle Program and of course for the Station. Those tortillas we have tested out to a year. They would probably have gone longer. We just terminated the test at a year because that's really all we needed for the Space Station.

Besides development, we also do food sessions with the crews. We do it a little bit differently for Station than we did for Shuttle. For Shuttle, typically we would do one food session for a crew. That was because they would see the food again during the simulations. They had more opportunities to taste the food. With the amount of foods that we have, they couldn't try everything at one food session.

We would make them come at lunch so they would be hungry. It would get scheduled just like any other training class they had. Typically, they would have forty to fifty food items at a food session. That's food and beverage, so it wasn't all foods, but most of it was. They would score these foods. We had a rating scale that they used. They had room for comments so that when they went to choose their menus they would know what they liked, what they didn't like, why they liked it, why they didn't like it. We would keep copies of all that so they would have it when they got ready to choose their menus.

With ISS, they don't get any food during training. They have a galley training class. They might consume food. Typically, they only rehydrate the food and learn how to use the rehydrator, but there are no simulations

Vickie Kloeris conducts a food tasting session with Expedition astronauts at the Johnson Space Center.

for Station where we're providing food, so our food sessions are their only opportunity to try the foods. For Station we do multiple sessions, but it depends. If a crewmember has flown on the Shuttle, they're familiar already with the food, so we don't have to do as many sessions for that crewmember. If this is someone who's never flown on the Shuttle, like a cosmonaut or one of the newer astronauts, then we'll do three to four food sessions so they try everything. They'll go to Russia, and they'll try all the Russian products as well as part of the training. Again they'll score it.

A lot of people ask, and even the crewmembers will ask, "Why am I doing a food session if you're using a standard menu?" It's because you get to augment that standard menu with bonus containers. You get to choose what goes in those. What you choose can be more of our flight foods, so if there's something you particularly like then you can include that. That's the purpose of having them try all the food. Plus, it gives them familiarization ahead of time before they get to orbit with what's going to be in the standard menu.

We try to get them to try everything before they ever leave the ground. They come at lunchtime. They sit at the counter in the kitchen in the food lab. Depending on how many people we'll have there, we'll rehydrate one to two servings. They'll spoon out a serving on their plate. They'll try a bite of everything. They don't eat a full serving of forty items. Obviously, they couldn't!

I don't do that much hands-on stuff anymore. Hands-on stuff is done by the technicians in the lab. I'm doing more paper pushing. These days the biggest thing is manifests. We've been playing, for the last month, the SpaceX-1 game.[39] What food is going to be on there?

They'll tell us, "This is how many kilos of upmass you have on this flight for food." We have to look at our plan, see what needs to go up when, and figure out what we're going to put on each vehicle. We sent what we thought was going to be the food for SpaceX-1, then they asked for more. Then they came back and said, "No, we're going to take everything off, because we need to put something else on." Then they said, "No, we're not going to take anything off." Two weeks ago they took half of it off. Last week they put all but about one-third of it back on. It's just been one thing after another. That's so typical of all these flights. The planning is just horrendous.

When we're trying to support the science experiments and we know food for those experiments has to be there at a certain time, we've got to be sure it's there when they need it on orbit. It's the same way with their bonus food. The bonus containers obviously have to get there at the same time the crewmember is there, or it doesn't do any good, so I spend a lot of my time tracking the food that's on orbit, the usage rates, how quickly they're going through the food, planning the manifest. It's a big deal trying to figure out what upmass I need in order to get stuff there at the right time and looking at the shelf life of the foods on orbit. How old they are. How long it's going to take the crew to get through them.

This week they've been after me. "What are you going to put on SpaceX-2?" SpaceX-2 is launching in January 2013.

"Until you tell me what I'm getting on SpaceX-1, I can't tell you what I'm putting on SpaceX-2. Every time you change what's on SpaceX-1 it changes what I'm going to put on SpaceX-2."

Today Space Life Sciences had the operational readiness review for Expedition 33.[40] We go through this process where we say we're green, yellow, or red, depending on what issues we have. One of the things I had to discuss was the food that we have on orbit is expiring. Our expiration dates, they're really best-if-used-by dates. They're not true expiration dates, so the food can be consumed beyond those dates.

But the quality starts to degrade. It's not a safety issue; it's a quality issue. One of the things I talked about today was the fact that by the time we get to the end of the year, when Chris Hadfield, Roman Romanenko, and Thomas Marshburn get there, they're going to be eating food that's past its best-if-used-by date.[41] It's a psychological issue for the crewmembers. They're opening up a container of food that plainly says best if used by November of 2012, and it's December.

That's a big fight that we have. We know that we have twenty containers of soups and cereals, and they reached their best-if-used-by date on August 30. Well, at that usage rate, it's going to be another almost six months before they get through with all those containers and move on to the next best-if-used-by date. They're going to be eating that stuff potentially for six months past its best-if-used-by date. We have a control set on the ground that has those same products made at the same time, packaged at the same time. We're going to be monitoring on the ground. When we see that the quality of certain products has degraded, then my job is to go to Mike Suffredini, to the program manager, and say here's the stuff we want to throw away because it's not good anymore.[42] He doesn't like that, because we've paid to get it to orbit.

At the same time, it's a catch-22 because we know ATV-4 is coming. And HTV-4 (H-II Transfer Vehicle-4) is coming.[43] They are loading those vehicles up with food, because they don't have enough other stuff to put on them. We are the ballast of choice. That's what happened on ULF7 (STS-135).[44] ATV-3 was the same way. We added a bunch of food to ATV-3 because they didn't have enough to fill the vehicle. But Mr. Suffredini said, "We will not launch an MPLM (Multipurpose Logistics Module) to orbit that isn't full.[45] So fill it up with food. Fill it up with clothing." You end up constantly kicking the can down the road so that they're always eating old food, which is not good. At the end of June 2012, they probably finished all of the food that came up on ULF7. That food had a best-if-used-by date of July 31, 2011.

Now some of the categories they had finished off earlier, but some categories they were continuing to eat all the way till early this summer. So, for a two-and-a-half-month window they've actually been eating food that was within the expiration date. Now we're starting down the path again. At the end of August, we had some expire. Another category will expire at the end of September, another two at the end of October. By the time they get to December they're going to be eating all food that's past its best-if-used-by date.

Today I had to brief Chris Hadfield, Tom Marshburn, and Roman Romanenko. The three of them will be launching in December. I had to sit at their consumables briefing and say, "You guys are going to be eating old food the whole time you're up there." They understand the situation. They know why it is what it is. It still pains me, and it pains the folks in my lab because we know if they were eating it when it was fresher they would enjoy it more.

When they come back we will have a food debrief with them. We ask them, "Were there products that you particularly disliked? What were they?" We've had so many debriefs on Station we can almost predict what they're going to say.

They'll name a product like our teriyaki beefsteak. "Oh, that stuff was horrible."

"If it hadn't been five months past its best-if-used-by date it would have been fine. You would have liked it had you eaten it five months earlier." So it pains us because we know the quality degrades when it's that old.

With the Shuttle ending, until they're more confident about the commercial vehicles, they're hoarding food.[46] They're hoarding everything, because they're worried they won't have enough. But they're just condemning them to eating old food the whole time.

The shelf life for most of the food we have for Space Station is about two years. We have two hundred different products. To completely produce everything we need, it takes about three months or so.

We have to ship food everywhere. We send it to Russia if it's going to launch on the Progress. We'll send it to Japan if it's going to launch on HTV. We'll send it to South America if it's going to launch on ATV. Of course, we were shipping to Florida whenever it launched on the Shuttle. Now we're shipping to Florida to launch on SpaceX because that's where they launch from. We have to ship all kinds of places. Life got very complex when we started sending food to Russia, and the ESA (European Space Agency) and the JAXA (Japan Aerospace Exploration Agency) started sending food to us.

We have to ship food out the door six months before the ATV or HTV flight launches. So you're looking at food that's about nine to twelve months old before it ever even leaves the ground, and then it's got to sit on the Station for eighteen months, twenty-four months before they eat it. ATV and HTV, they can carry a lot of food, but they're not really good vehicles for the food team because we have to do it so far in advance. Progress was better because we only had to ship about two months (roughly) in advance of the launch. The Shuttle of course was the best because we didn't have to send the food nearly as far in advance. SpaceX is not too bad. It's about two months in advance of the flight. We can handle that. That works much better for us than six months.

When food goes overseas it has to go by commercial shipper. The Cargo Mission Contract (CMC) handles shipping to the international destinations, so we send our food to CMC. CMC ships it to Russia, Japan, or wherever it's going. They're responsible for boxing it up, but we have to do all the paperwork that goes with it. With food there is a ton of paperwork. Every time we do an international shipment, you're looking at a stack of paperwork. We also have to get a permit. When we ship to ATV in South America, believe it or not, it has to go through Paris. Apparently, you cannot ship directly from here to French Guiana. Because it goes through Paris, we're dealing with the European Union and their import requirements. The European Union and Russia both require a USDA (United States Department of Agriculture) permit. Basically, the USDA permit states the food is safe.

We had to set up an arrangement with USDA to provide us with the export permit, because we are not a USDA-inspected facility. Because we are a federal facility and not a commercial facility, the USDA has no jurisdiction over us. They don't want the responsibility, but they did agree that they would provide these permits if we paid them. I have to write a letter that says, "You're not responsible if Russia or Europe rejects this shipment." In other words, if Russia were to say "No, I don't accept this certification," the USDA doesn't want to be held accountable for that.

Every time we ship to Russia or to Europe I have to apply for this USDA permit. The permit really only has to cover meat and dairy products. Vegetables, nobody cares about that. It's mainly your meat and your dairy that these customs people care about. That was another challenge because we had to figure out how to set this up with the USDA. That went all the way back to when we were first shipping on the Progress. So there've been a lot of issues with import/export and certification.

We also receive special import permits to allow us to import space food from Japan, Europe, and Canada, because we can't meet the normal commercial import requirements. It would just be too expensive. We basically have waivers, special permits, that allow us to import with the understanding that the food will never go anywhere other than space. We can't sell it; we can't let it out to the general public.

Because these permits are only good for a year, we're in the process of renewing the one for European foods. We have an Italian crewmember who's going to fly next year. A couple of companies in Italy have made some special food for him.[47] In order for it to get there they have to import it into the US, and we have to stow it in our containers. We're working on that permit. We just received a shipment of food for Chris Hadfield from Canada. A couple of the products they're sending are meat products. Because they're meat, they require an import permit. We just got that permit in place so Canada could ship to us.

Japan, they're not going to ship food over here anymore. They can just launch it on HTV. They're handling that internally now. That helps us. Of course, ATV doesn't really help the Europeans because ATV launches from South America. I joke to my boss that my job description needs to include import/export.

My biggest challenge since working at NASA has been to try to educate folks about what the limitations are as to what we can do based on the fact that we don't have refrigeration in orbit. A lot of people don't really think about that, about how certain foods just aren't going to work because you don't have refrigeration. It's been a challenge over the years and interesting to see how the whole philosophy about food has evolved. When I first came to work here, I would say for 95 percent of the crewmembers who went on the Shuttle, food was way down the list as far as importance. "Oh, it's a camping trip, I'll find something I want." No big deal. That's really the way it should have been, because it's a short duration flight. There were very few crewmembers who flew on the Shuttle who were that concerned, "Oh, I'm going to be without something for two weeks." It just wasn't that much of a psychological issue on a short duration flight.

We started talking to those first crewmembers who were going to go to Mir about food. "Oh, no big deal, don't worry about it, just look at when I flew on Shuttle. It'll be fine." We had all that data, so that's what we did. We looked at the preferences based on what they had picked on their Shuttle flights.

When they started coming back from Mir they said, "I should have paid a little more attention." When you're there for a longer period of time, all of a sudden it takes on more importance, because it's one of the few creature comforts they have. It was different from the Shuttle. On the Shuttle it was more like they were grazing, maybe they would do one meal a day. But on Mir and then on the Station, they settled into a more normal routine where they were eating together more often. It became more of a social thing, and food took on a higher importance.

Quickly the word got out to the other astronauts who were going to fly long duration missions: "You need to really pay attention to what you're taking because when you're up there for a long time it takes on more importance."

So there was a process whereby first the astronaut corps got educated about the psychology of food and then management got educated. All the managers over in the Station Program asked, "How can food be important? I worked Shuttle. It was never important during Shuttle. How can it be important that we get certain food there at a certain time?" They just couldn't relate to that.

Over time, when they got enough complaints from crewmembers on orbit because they didn't have the kind of coffee they wanted or other issues, they began to realize how important the psychological aspect of food was for long duration spaceflight. Now that is tracked very closely by the ISS program management because they know how important it is to these crewmembers and to their whole frame of mind when they're on orbit. It has changed dramatically as far as the relative importance over my career.

6

Astronauts

The competition to become an astronaut has always been fierce. Since 1959, NASA has chosen fewer than four hundred people to fly space missions—three hundred fifty to be exact as of 2020. The most recent selection in 2017 had more than eighteen thousand applicants and, out of those, the space agency selected twelve. Theirs and future classes will train on missions to the International Space Station (ISS), Moon, and Mars. Astronauts of the twenty-first century will continue to explore space, beyond low-Earth orbit, where the first women astronauts initially made history.

Dr. Kathryn D. Sullivan is an explorer at heart. She was one of six women in the first class of Shuttle astronauts and the first woman from the United States to walk in space. At the time of her selection she was completing her dissertation at Dalhousie University in Nova Scotia. Sullivan, like the other five women, excelled in science and was accustomed to being one of a handful of women in her field. As a graduate student, she had participated in several oceanographic expeditions where she spent time at sea. A female scientist onboard a ship was about as common as a female astronaut in the 1970s; she helped to blaze a trail for women at sea just as she later would for young girls aspiring to fly in space. Sullivan went on to fly a total of three Shuttle missions, including one that deployed the Hubble Space Telescope. After fifteen years, she left NASA to become the chief scientist for the National Oceanic and Atmospheric Administration (NOAA). In 2020, she became the first woman to reach the Challenger Deep, the deepest known point in the ocean.

Joan E. Higginbotham represents an elite group of space explorers, having been one of only three African American women to fly onboard the Space Shuttle. She began her career at the Kennedy Space Center (KSC) in 1987, just as NASA began ramping up to support STS-26, NASA's

Return to Flight mission following the *Challenger* accident. Higginbotham worked on the Orbiters for nine years, eventually becoming the lead project engineer for *Columbia*, the world's first reusable spacecraft. During her time at KSC, she supported more than fifty Space Shuttle launches while earning two master's degrees, one in management and another in space systems.

Eileen M. Collins was the first female Shuttle pilot and first woman to command a Space Shuttle mission. She came from the Air Force, which opened flight training to women in the 1970s. Collins and three other women participated in training at Vance Air Force Base in Oklahoma, but—as the women learned—the military remained a man's world. Military traditions were hard to change. In 1979, Collins became the first woman at Vance to receive her wings. After graduation, she served as an instructor pilot on the T-38, C-141, and T-41, and taught mathematics at the US Air Force Academy. In 1990, she became the second woman pilot to graduate from the Air Force Test Pilot School at Edwards Air Force Base.

In 2007, Pamela A. Melroy became the second—and last—woman to command a Space Shuttle mission. She had previously flown two ISS assembly flights as pilot of the Orbiter. Melroy came to the space agency from the Air Force in 1994, where she was a test pilot and a combat veteran of Operation Just Cause and Operation Desert Shield/Desert Storm. She holds a baccalaureate degree in physics and astronomy from Wellesley College and a master's degree in Earth and planetary sciences from the Massachusetts Institute of Technology.

The first six women astronauts pose for a photo. From left to right: Kathryn Sullivan, Shannon Lucid, Anna Fisher, and Judy Resnik stand, while classmates Sally Ride and Rhea Seddon kneel.

Kathryn D. "Kathy" Sullivan [2]

My brother, the avid pilot, was all over this Shuttle astronaut selection application and process. He followed every single piece of it. He was a highly rated pilot. He had thousands of flight hours already, an engineering degree from UC San Diego, and he had already applied. He applied both as a mission specialist and as a pilot to increase his chances. [3]

He started pestering me about applying. "They say they want women and minorities to apply, and how many twenty-six-year-old female PhDs can there be in the world? You should give this a try."

My head at that time was still just on the oceanography side of things, so I blew him off. "I'm working in fourteen thousand feet of water. It's already hard enough to understand the bottom of the ocean from a surface ship, and now you want me to go two hundred miles above that! This is not what you do to understand the bottom of the ocean."

We jousted and teased about the application a lot while I was at home in California, then I went back to Nova Scotia, dismissing it. Within a week or so I saw one of NASA's small ads about the recruitment in one of the US science publications that the library received. When I read that, a different gear clicked. I recognized a strong parallel between the mission specialist role as they described it and the oceanographic expeditions.

I fired off the application probably somewhere in January or February of 1977 and promptly fell back into trying to complete my fieldwork. I have another cruise to do; I'm doing all the data analysis. I'm in the converging phase of pulling my PhD dissertation together, which is all-absorbing, and putting out feelers for postdocs and the other avenues that one pursues to have some gainful employment when you finish your degree. [4]

That all goes along until about September, October, when I get a phone call from Bill Ryan down at Columbia University's Lamont-Doherty Geological Observatory. Bill was one of the sea floor geology authors whose work I had read in detail, and I had applied to have a postdoc with him doing deep-sea marine geology.

He calls me up and says, "Are you going to take my postdoc? I'm getting to where I need an answer. You're my top pick."

I said, "Oh. Oh, yes. Well, probably?"

"Have you got something else?"

"Well, no, I don't have another postdoc, but there is this one thing; I just haven't heard anything. I put my hat in the ring for this NASA thing, and that's the only outlier, but I haven't heard anything from them."

Turned out Bill Ryan had applied to the 1967 astronaut selection; from those applicants the group that became known as the XS-11 was chosen.[5] So he had a little sense of the process, probably more sense of it, frankly, at that point, than I had. He didn't make the cut, had gone back to New York and built a fabulous career. In retrospect, he was just as glad that his path had taken the turn that it had. It gave him a certain sympathy for my circumstance, and, importantly, it really gave him an empathy for the importance of playing this all the way through.

We had a good conversation. He told me his story. He was one of the people who also put the marker clearly on the table, and he said, "You do understand that the numbers are astronomical, and the probabilities are vanishingly small. In all likelihood you're going to come take my postdoc."

I said that was fine with me. I thought it was a fabulous opportunity, but could he wait? I hadn't heard anything at all from NASA. Could he give me some time?

He said, "Absolutely. Come this direction because you know the other answer. Don't walk away from it."

I rummaged back through whatever papers I had and found a Houston phone number and called somebody up. I don't recall who it was. The first thing they said to me, I think, was, "Oh, haven't we told you 'no' yet?"

I said, "You haven't told me anything yet." Their heads are absorbed with the people they're bringing down for interviews. My name didn't ring a bell, so I think they presumed I was someone they had already turned down calling back to pester.

"Oh, let me go look," and they go off and rummage through files and come back and confirm that, "Oh, well, actually, yes, you're going to be invited for an interview." They don't look at too many Earth scientists even today. They've got enough test pilots to make up full blocks, twenty folks at a time. They've got enough MDs to make up full groups of MDs. I was back in the tail of the pack in what probably was called the "cats and dogs" group, the test pilots whose schedules didn't fit any of the other windows, the one mathematician or oddball space scientist, and this girl geologist from Nova Scotia.

I called Ryan back and told him that, and we again commiserated about the numbers and the likelihood, and I said that was fine. They said they would have the selection completed by the end of the year. Could he wait that long to let me see how it plays out? He said sure, he could do that. The interviewee packet arrived not long after that phone call, and off I went down to Houston. I'd been to Houston once before. I'd had an aunt that lived there when I was a teenager.

The week starts with an evening pre-brief at the Kings Inn, the hotel just outside the center's front gate. I walked into the Kings Inn. There were clusters of people. I can't tell you the whole composition of that interview group anymore. Fred Gregory, who became a classmate, was one of the very memorable ones.[6] There clearly were some clusters of folks who knew each other from science communities or knew each other through military circles, and I felt very much an outlier. I didn't know anybody in this room. It just looked like a room of people that belonged there and seemed to feel they belonged there. I felt very much out of place and thought, "It beats me; I've got no idea what any of this is."

I remember thinking, "Kath, enjoy this week a lot, because these people really seem to have some sense of what the hell is going on here, and you really don't. So have a really good week, because this may be the end of the road."

After that, it's kind of a blur. It's one appointment after another. Drinking from a fire hose; deliberately inscrutable in some respects, the whole "good cop, bad cop" game with the psychological interviews. One or two folks deliberately planting a sense that everything about you through this whole week is part of the interview. The implication was that if you're strolling aimlessly across the campus versus marching purposefully, someone will note that and write it down.

It's just all those things to try to figure out how do you react if you can't tell what matters, but you suspect everything matters, and you suspect big consequences hang in the balance. How do you behave? Do you go bonkers? Do you settle down? That was just all a little surreal. I can remember bits of the interviews with Terry McGuire or Dr. Joe Atkinson.[7]

What I really remember about the interview week is that Fred Gregory and I had a ball. We fell into getting out to the gym and playing racquetball, and that was quite fun. Stayed out playing racquetball so late one night that they secured the front gates to the space center, and we had to slide under the front gate on NASA Road One. We both looked at each other and said, "This either makes it or breaks it," and just slid under the fence. We made our way back over to the Kings Inn.

I got home from the interview feeling pretty settled on a couple of very simple things. I knew I could do the mission specialist job and was confident I could do it well. My sense of it at that point was it's going to be a pretty cool thing to get to do this. I would love to get to do this, but I've only seen nineteen other people. There are who knows how many other groups besides ours, and I have no way of knowing where I rate. I have no way of knowing how astutely, objectively, or politically the folks that make the final call are going to do it. What I hope they're going to do is take the best ten, twenty, or thirty people they can find, and I hope I'm in that group.

That was it. I know I can do it. I know I would like to do it. I know I have no way of telling where I am on the ladder, and no realistic way of thinking I can control the final judgment. So, great, I'm going to go back and finish up my PhD and do that well. When they figure it out, I'll either turn right or turn left and either one will be pretty cool. So that's what I did. I have it in my head that it was around the second or third week of January 1978 that we finally got the calls. I've not gone back and checked the dates.

Halifax, Nova Scotia, is one hour ahead of Eastern time, and of course they're trying to get the word out by phone to everyone before they drop a press release so that—as it turns out—you can brace yourself for the media onslaught, not that I realized that at the time.[8] The phone in our little apartment, which I shared with about four other grad school girls, rang at something crazy early, you know, six thirty or seven in the morning.

It was George Abbey.[9] I know now that that means I was getting the "happy call," but I had no idea who was going to call me back, if ever. It was Mr. Abbey in his renowned laconic way, saying hello and asking if I remembered my time in Houston. Would I still be interested in coming down to work for NASA? You would have thought he was asking if you were still interested in working at the grocery store. It was sort of, "Just wondered if you were still interested." You know, your life is going to turn upside down in a minute, and it's just this completely low-key question.

I said, "Yes. Yes, I would. Thank you."

Of course, the whole rest of the day, the phone at the lab and the phone at the house were just ringing off the hook. It was fun. It was just an exhilarating, very crazy, very wild, very fun quasi-disorienting day. It was like riding the top of a tsunami and just sort of going through the world, "Look at this! This is amazing."

In Halifax it settled down fairly quickly after that. Number one, it's mainly a US story. Number two, at the time Canadian media and Atlantic Province media wasn't anything as voracious as they might be today. It wasn't a twenty-four-hour news cycle. It was a big wave, and then the media wave left me. There was a nice little buzz, but I also still had a PhD to finish up, so I had things to do.

Somewhere in the February time frame we went down to Texas. We were all invited down to Houston for the public introduction of our class. My biggest challenge and probably insecurity going down for the starting interview was realizing how we were going to be plopped up onto a very visible public stage. Based on the little dose of media that I had gotten, I realized that it's going to be a big public event. We're going to be introduced to the world. There are five other gals. There are probably people who actually have a wardrobe and are going to be there.

I'm twenty-seven years old, living on four thousand dollars a year, not much of a wardrobe, and not much of a bank account—so the wardrobe picture is not going to change radically overnight. Happily, Carolyn Huntoon, who was one of Chris Kraft's key folks at that time, and Ivy Hooks, I think both those two ladies, bless their hearts, had the insight and foresight to recognize the introduction was not about appearance or clothing.[10] At that time, three of the six women were working full-time. Judy Resnik was out in the working world and Anna Fisher was in the working world; Shannon Lucid was doing biomedical research.[11]

Two of us were still grad students, Sally Ride and I.[12] We were both finishing up PhDs, and I really appreciated however much Huntoon and Hooks weighed in and tried to recalibrate everyone's focus.

I don't remember the particulars like each being called out on the stage. I think that's what happened, but I don't have any real vivid memory of standing backstage, hearing my name, and walking out on stage like actors at the Oscars. That didn't register. There we are out onstage, and there's this click, click, quote, quote, and about two minutes later they say, "Thank you. This is the end of the formal event, and any of the astronaut candidates are available for the rest of the day for media interviews."

We'd been told that was the deal. You walk out, you appear as a group, and then there are media interviews. Well, you know, the six of us gals plus Fred, Ron McNair, and Guy Bluford plus Ellison Onizuka were odd people.[13] There had never been critters who looked like us admitted into the astronaut corps. I remember the conversations before we all went out there, probably some of them with Carolyn telling us gals in particular, "There's a lot of media interest. There might be some for you guys. It's going to be a long day." So I braced for that.

We eventually came to refer to our class as "ten interesting people and twenty-five standard white guys." This event was one of the things that started that, because the twenty-five standard white guys were done about 4.3 minutes after the formal event and had the whole rest of the day free. The other ten of us, we were there till Lord only remembers. I don't remember what time—way late.

I really give credit to Carolyn Huntoon on this one, because none of the six of us gals had ever been through anything like this. Carolyn was a fabulous counselor for us. We're all very accomplished. In a certain sense there's a lot of similarity between the six of us: career-oriented, technology-oriented, smart, capable, confident, composed.

But we're also different. I mean, style, manner, personality, personal history—married, single, dating, whatever; that stuff's all different. And that's going to be a lot of what the press is trying to scratch into. The press was going to be basically starting to compose a mental model for themselves and for the US public about "who are women astronauts."

So we had a quick little—not a lengthy conversation—but just a bit of time to chat before all the interviews started, in which Carolyn helped us appreciate the issues we might be grappling with. If you are the first gal who a reporter asks about—pick anything—dating, or will you wear makeup on the Space Shuttle, any of those silly questions that the six of us would probably get—how would we respond? We'd either blow them off because we thought, "Who cares?" or we'd answer it not imagining what they would make of it. They will want to stretch that to be an answer for all time for all women, the complete image and declaration of what women astronauts are.

If one of you is more comfortable than another talking about very personal details, they will want to pry for the same level of details from everybody. However any one of you answers something, it's going to create a set of feedback loops that will affect the others. You might just want to take a moment and think and talk about the degree to which you want to each do and go your own way, or the degree that some agreements in principle between you might serve you well for today and serve well in the long term, because these are going to be the first data points.

That was really helpful. I think by and large most of us, we'd been accomplished, but we had been able to lead pretty private lives. I think most of us closely agreed that in taking a public role our professional performance and the dimensions of that are fair game to talk about or critique in public. But the color of my living room or who I like to hang out with, those kinds of things are off-limits. I'm planning on still having some private life.

I don't recall that we made an explicit agreement to do this, but we each went off for our first interview, and as we were about to be shepherded from interview one to interview two, we each, as I recall it, independently—or maybe somebody thought about it and it rippled through us—said, "Potty break," and bailed into the ladies' restroom. Most all the reporters were male, and all the Public Affairs Office staff was male, so we could have completely private meetings in the women's restroom.

We'd jump in there and say, "Okay, who did you have? Who did you have? Who did you have? What are they after? What do they want? How far did you go? What do you think?" We just synched up after almost every one of those interviews. We'd only just met each other, but we learned a little more about our differences, our similarities, and our boundaries. This was the first testing of whether we were inclined to share a common interest versus pursue an individual interest. That's my main memory of the introductions day, bailing into the ladies' room to have these little tag-up meetings.

There continued to be lots of media requests for all of us. It never, for me, came to being chased to the house. It was the first time in my life I had a personal private phone of my own. I went ahead and got an unlisted number just as a bit of protection against that kind of thing.

The media struck me as being to some degree pretty predictable. I don't think this was all that consciously formulated; it sprung fully formed into my consciousness. When I met everybody and looked at our six women, my own assessment on what's the media going to do was that they're going to follow the stereotypes they normally do.

They've got a cute blonde. They've got a cute kind of curly-haired, flirtatious, single gal. They've got two married gals, and of the two of them, Anna was slighter of build, younger. Rhea Seddon, Anna, and Judy are going to be their lightning rods and their icons.[14] They're very telegenic, photogenic. Sally might get in there; she fits the physiological stereotype, you know, slender.

Shannon and I are taller, stockier. Shannon's married. If you were going to line the six of us up, put our six photos up, and pick cover girl shots, I'm not a cover girl type. I've been on covers, but I'm not an archetypal cover girl look or face; neither is Shannon. The other four are closer to what I reckoned the typical lens would be on. Who do you want to put on the cover that will signal the things that you're trying to emphasize to sell a magazine?

I pretty quickly said, "I'm not real likely to be one that's going to have everybody chasing me, because these four women outwardly look more obviously like the stories they want to chase." Apart from the marriage milestones, you might get a different answer from someone like Rhea, or Judy, if she was still around.[15] You know, Rhea shows up: blonde surgeon driving a Corvette. Bam, there you go. Okay, let's guess who will get the attention.

It's about fitting the mental models that the writers or the reporters have, and I've never really fit very well into people's standard bins. I formed that assessment. I reckon I'm not real likely to sit in anybody's standard bin. Too tomboy, too smart, too strong, too all those things; not the archetypal little girl.

We, the six of us, through especially that first year or two, would periodically get together, often with Carolyn, or often as not catalyzed by her, to touch base back on the bigger issues—the dimensions of women in the astronaut corps at NASA long-term that were emerging day by day, from who we were to what we were doing, but that oftentimes you don't see that so much, young and in it at the beginning yourself. Shannon probably had more sense of it in some respect than any of us did, because she was further on in her career and had more experience with seeing how these things all evolve. The meetings were more episodic than regular, especially in our first year. We probably got together four or five times, maybe six times that first year.

It was not always issue-generated as much as it was just building the professional acquaintance of staying in touch with each other, being a little bit aware of what's going on at NASA. Every now and then Carolyn would have something to raise, because she was working at the senior management level. So every now and then she'd alert us to something. I don't recall particulars because they were not really momentous, big, "you six must make a decision" kind of things. It was a place for us to, to some degree, vent and share experiences with each other.

Lots of the guys, not all of them—some of them had scientific back-grounds—but lots of the guys came out of the military squadron tradition. They'd go off drinking and have happy hour together. They'd change in the locker room at the airport together. They'd change in the locker room at the gym together. All the rest of the astronaut corps was male, so they had locker room time to see Gordon Fullerton and see Dick Truly in a very unstructured moment, maybe do one of those coffeepot or locker room exchanges where a lot of information exchange happens in an unstructured way.[16]

I think the main thing Carolyn was trying to do was foster some of that networking for us, because we otherwise weren't going to really have any network. Most everybody you were working with at the center was male. It's going to be challenging enough to have a little cohort of six in a population that just became something like sixty. I think there were two dozen astronauts when we arrived, and we were thirty-five, so I think we rounded the number up to something on the order of sixty.

I've never actually asked Carolyn if her main intention was to catalyze some networking among us, but it seemed to me that she was trying to do just that because of her own perspective on the importance of it. With our respect for her and her seniority, if she suggested it, we would convene. We might well not otherwise convene ourselves.

We didn't want to become "the girl astronauts," distinct and separate from the guys. None of us had ever followed that model. I don't think any of us in our prior professional life had taken the path of being the girl X, the girl Y, joining all-female groups. All of us had been interested in places that were not highly female, and just wanted to succeed in the environment, at the tasks, and at all the other dimensions of the challenge.

I think Carolyn was just trying to give us that initial cohesion. I think she probably in some ways realized more than we did how important unity was on some level. It won't matter on everything, but it will be highly important on some other things, and only if you get a little bit of that initial sense of relationship going can you take on some of these systemic issues that might come up.

There were one or two little things, and I think that bit of cohesion and that sense of there is something that affects all of us here helped. One that I encountered was the first time I was assigned to work in what's called the Cape Crusader role.[17] Cape Crusaders are support crewmembers that look after the prime crew's interests on vehicle integration, on testing, and preparation so that they can focus on the training. On launch day, you can tell them, "Strap in. It's ready to go. So say we, your colleagues and buddies, who have been keeping an eye on everything and checking it out." Pretty fun job.

I was the first gal in our class to be assigned this task. Dan Brandenstein, I think, and me and Loren Shriver were new kids assigned to be Cape Crusaders.[18] At the time, a lease had been taken on a three-bedroom condo down on the beach, and that's where the Cape Crusaders stay, so you could leave some clothes and toiletries and just fly back and forth.

There were rarely more than two or three guys there at a time. Each guy would take his own bedroom, drive out, and work on site. It was more pleasant than staying in the crew quarters, which is miles away from everywhere. If a fourth or fifth guy came, they'd hoof it out to the crew quarters or stay at a hotel. So it was first come, first served, take a bedroom at the condo.

I get assigned, and Don Williams, I think, was going to be our group lead. He's another classmate.[19] I went around to see Don and said, "I need the key to the condo." I had something coming up, and I needed to be down at the Cape.

Don starts very nervously back-pedaling, "Well, actually, you know, I've been thinking about this," and the essence of what he was thinking about was he was really nervous about the prospect of a three-bedroom condo with a woman in it. "What will people say? I don't know. Ooh, I don't know if we want to do this."

My manner is not to punch somebody in the nose right away, so I heard him out. I replied, "I think what people will say and how long it will take to get anybody past it is completely a function of how we handle it and what we say we're about. I'm about getting down to the Cape and doing my work. That's the thing we need to do, so I think how we handle it is entirely in our hands. I think we just need to saddle up and go do it."

"Well, let me think about that."

I passed the word among the other gals, and they rolled in on him. We just did a little good cop, bad cop thing. I was going to have to work with this guy. They rolled in and asked, "What are you going to tell your wife when I'm assigned to a crew with you? If we can't handle this, if we don't understand what this is and how we're going to handle it, how will we handle flights? I don't think the media gets a vote. There aren't enough women to make up a crew, so it won't be an all-female crew. I don't think your wife gets a vote on flight assignments. You're crazy. You've just got to get over this."

My strategy was, I'm going to go get a key from one of the guys who's already been there. I'm the first guy in our new group that's got something to do at the Cape. I'll go down. I'll get settled in the condo. When one of the other guys shows up, I'll tell him, "You've got to go sort that out with Don because he has decided that we just really can't have mixed gender in the condo, and I'm already here, so you'd better just go. Not my problem."

It was a bit of a pincer strategy, and it took him about a day to realize that this was not a tenable position. He got over it. So we helped each other out there.

When Sally was getting ready for STS-7 and they had bench check coming up—where you go look at all your personal equipment—it was another time that I think she recognized we needed some agreement on an issue. They had a personal hygiene kit that contained the basic toiletries; for the women, if we wanted, they designed another kit that had makeup and other female items in it. This was the first time a female astronaut came over to do a bench review on all these pack-out items. She patrolled the hallways to grab one of us and be sure there was a second view, a second voice, a second awareness, not have it all pivot on just one viewpoint. I ended up being the gal who was around, so I went over with her.

We would do things like that. "I'm going to get asked *xyz*. It's going to cut this way. They're going to take this as everybody is in agreement. I need to talk to you guys about it."

Later, I drew a spacesuit assignment, which tackled the issue of, "How are we going to solve the urination problem for women in a pressure suit?" We were not launching in pressure suits at that time. We were launching in shirt-sleeve clothes. The EVA (extravehicular activity) suit proved to be an engineering challenge as no women had ever been in the corps before and therefore outfitted in the suit.[20] The men peed into a cuff. When we arrived, they realized, "Oh that's right. Women are in the group that does spacewalks. The cuff isn't going to work for them." Of course, they realized it might be a good thing to include one of us in the discussions.

Most all of the calls that I would make were really straightforward, and there was no question in my mind that there was a sound answer here; there was not really much personal variability around it. They imagined they were going to create a form-fitting device for women (since a form-fitting device worked so well for men, except when it leaked, which was often). The other easy option was diapers, which they (all male engineers) couldn't imagine asking a macho male astro to wear. The form-fit idea was a disaster, of course, and we all ended up using diapers in the end.

I think we all were pretty alert and not playing games with those cases where there might be some variation that actually mattered to one of these other folks. If we thought there was, we'd say, (a) let me find out, and (b) if there is, let me make the input in a way that sets up that full range as sensible, normal, and what the system needs to provide. Because if I come in and say, "Exactly. That one's perfectly fine forever," and another woman comes along two flights later needing something different, the system is going to say, "I told you we shouldn't let women in, because we had an answer and it worked, and then every one of them comes along and wants to make it different."

The guys tweak stuff all the time, but that's the norm and it's not commented upon. When it's a new group, a new cohort, a new identity, all those little things can become cited as points of difference and cited as arguments as proof that this isn't working. "They don't get it. They can't just take a standard and live with it." Is it accurate? No. Is it equitable? No. Is it evenhanded? No. But it is part of what happens when you're shifting the makeup of the group. We did a pretty good job, on the whole, I think, of watching out for that and watching out for each other on it.

Our arrival on the scene was a huge encouragement and motivation to women selected later. It ought to not be the case that we only take inspiration or learn from a human being who looks just like us, but the fact of the matter is that it's a quicker path. If I see someone just like me doing something, it's a quicker intuitive path to recognize that, "Oh, I could probably try that." So that sort of sounds subtle, but in some respects it's a pervasive impact. The door has already been cracked open.

Who knows? If I had watched Jane Smith walk out to the launch pad in a Mercury flight or an Apollo flight, maybe it would have registered differently to me that it could be one of the things I could do. I think that probably matters a lot in a systemic fashion, in motivation, and inspiration, the ability to imagine that you could go do that. It probably adjusted a number of young gals' academic paths or sense of how to go about that.

I'd give us six good marks, personally, professionally, and technically, how we carried it off. We got the assignment to carry a banner, and in different respects different bits of it fell to each of us along the way. Sally got the nod to carry it first up into orbit. I think any of the six of us would have succeeded at that and done fine at it, but, having said that, the ball came to Sally, and I think Sally played that one very, very well. Nothing came out of STS-7 that steepened the hill for any of the other five of us coming behind, and I don't think any of the other five of us did anything other than help flatten the hill for gals who came behind us.

The other big thing that's happening in the time frame that the six of us are there is that the astronaut corps itself, all-male beforehand, is starting to change in other ways besides gender. When our class arrived, the astronaut corps really had to triangulate around civilian scientists, women, and minorities. All of those were provocative forces in varying degrees.

If you read Mike Mullane's book, you get the sense from him that it was the unwashed, unproven, you know, soft-skinned scientists, never tested by harsh conditions, that unnerved him the most or that threatened his sense of pride in the caliber of experience.[21] The gender part of that was clearly also a factor. All the guys we walked into, every woman in their lives before we arrived was wife, girlfriend, daughter, or secretary, period; occasionally a nurse. But that's it. That's all you ever were, and so whether we wanted it or not, it fell to us to change their views about women.

I think if something had really gone badly, if one of us had really, really muffed an assignment, their personality clashed with someone, or lacked the personal strength of character to stand up to the combination of scrutiny, joshing, and public exposure, if any of that had fractured on even one of us, I don't know that it would have reversed the process, but certainly it would have probably created a deep, long detour, and just left more debris for the next wave of folks to have to pick through before they could really regain the path and start making the kind of progress that we would want.

What about an Eileen Collins? Eileen and everybody of those subsequent classes is, if I'm right, to some degree a beneficiary of the bits we were able to figure out, the foresight we were able to muster, the way we went about it, and, most importantly, the way we behaved and the way we performed. One key thing that we had nothing to do with was that society at large was also shifting, and so eventually, by virtue of nothing that we did, women gained access to the test pilot school lines in both the Air Force and the Navy. If that had not happened, I think it would have been a lot longer road for the agency to reach a point of saying that that was not a pivotal "yea or nay" credential to have. As long as that stayed a "yea or nay," that door either needed to open or the pilot astronaut door wasn't going to open.

But we didn't have anything to do with that. Eileen and the gals who made their way into that realm and succeeded there—what they did in test pilot schools is a direct parallel to what I've been talking about. Show up and help that world learn who you are as a pilot and the fact that you're female, where does that matter? Where does that not matter? Start to pull those pieces apart and be able to make more insightful judgments.

Eileen's track record will get her anywhere she wants to go, and she's proven that over and over again. The interesting question that can't be answered is this: Suppose that in the time frame that the world set about to select us, women already were in some numbers coming through the test pilot ranks. I'd have to check the calendars. I think there probably had been one, or a very, very small number.

The interesting question might be: What role did a number of years of being able to watch us perform and behave, what role did that play in helping the selection process make sense of someone like Eileen? And, in particular, in the little bit of time you have to look at someone in an interview, you're judging Eileen and yet you're judging her for command, and that puts a whole different lens with a lot of both technical and cultural factors to it. You're judging them, their composure, and their aptitude for command.

Commanding something on an easy day is fine; command means you're still keeping it together when everything is falling apart. You're still leading the keeping of it together and mustering everybody's energies. That's a very important judgment to make and to make well, and I guess I'd like to think that elements of how we handled ourselves in the first couple of waves of flight helped paint that picture and create some of the sense by which Eileen and company were evaluated in a meaningful way and with some insight. I'd like to think that maybe some of what we did helped create a better picture for that. But those gals walked in that door on their own right—I don't want to take anything away from that.

Astronaut Kathryn Sullivan uses binoculars for a magnified view of Earth through the Space Shuttle's forward cabin windows.

I felt treated pretty equally within the astronaut corps. I had been one of the first gals into geology field camps. I had been one of the first gals on the geology expeditions given the vehicles and driving. I had been one of the first gals at sea, so I'd been a lot of places where you're dropping into an environment that's not used to having people like you around. In these sorts of operational environments, there's a habit and a pattern of a certain kind of joshing and a group dynamic and teasing, or hazing, that can be pretty harsh. I think that's much of what I observed and what was happening when we arrived.

I felt like everyone around our class table genuinely believed and respected the fact that we had all passed the same screening. I didn't feel like any of those guys felt that we'd been let in free or had been let in easy. They didn't know us. They didn't understand our backgrounds. They had never spent time where we had come from. But I felt like they reckoned, "You got in the same door I got in, so you get a fair shot to start out and show what you can do."

I thought we were given that by our classmates. In terms of the space center and NASA in general, it beats me what everybody's individual opinion actually was, and who knows what the gossip at the bars or in the boys' locker room was, but we didn't walk in as a buck private, you know. We walked in carrying the title "astronaut." That would be like walking in the Navy wearing the rank of admiral. So you might well encounter someone who is flustered by seeing you carrying that title and having that role, but in the first instance, probably in a flash they're going to be processing an important decision: "I've never seen anyone who looks like you who carries the title 'astronaut.' Everyone I've seen that looks like you, I can dismiss them. I can flirt with them; I can ignore them, whatever I want. They're minor people, maybe—minor people in my technical life—so I'm not sure I know how to treat you. There's a way I treat people who look like you; and there's a way I treat people called 'astronaut.' They have never merged before, so which way am I going to treat you? I do know that any time I've encountered someone called 'admiral' or 'astronaut' and not treated them the way the title implies, I got my head handed to me."

I think dynamics like that probably got us a grace period from almost anybody, whatever their personal opinion actually was, because you just don't ever treat admirals that way. Even if the next thought in their head was, "I really doubt you're going to pull this off," you were going to get to move first and get a little window to show what you can do.

I think the six of us stepped up in those windows well, in personal composure, behavior, and in technical performance, and, over time, at different speeds, with different people. Probably still have some skeptics out there that still to this day think the whole thing was a bad idea. By and large, I think we stepped through those doors and through those windows in ways that were positive, solid demonstrations of equal opportunity.

Was there any competition between the women astronauts in terms of who was going to fly first? It was not really something that there could be an overt competition on. In a certain sense, everything you're doing and how you're doing it is somehow, probably, going into this decision process, whatever the heck the decision process was. That's also deliberately inscrutable. Again, you could watch all the doings, create your mental model, speculate about it, score keep the whole thing, and you might be right, or you might be totally out to lunch. All that struck me as sort of a pointless way to spend a lot of your time and energy.

We would all get caught up in it now and then and worry and obsess about it, because it was just deliberately a quiet black box. I think this is very conscious, and maybe it's another part of the test, I don't know. Now and then the hand would come out of the box with flight assignments, and then it would go back down. What the hell was in that box and how it was being done, I, to this day, have not a clue. So, yes, I would have loved to have gone first. I was confident in my abilities.

There clearly was some kind of a horse race or beauty contest. The agency was eventually going to have to pick one of us. I certainly never believed there would be two of us on a first flight. You might get away with putting one on, but don't ever put two on the first mission.

It's idiotic, but I've seen this too often. It's how the world works. Let's test with one, because if it's two, Lord only knows what would happen. It's going to be one, and they're going to get around to it eventually. They're going to have to get around to it. We're all in line behind the guys that were here before we got selected, so none of us are going before all of them are lined up. It's going to have to be sooner than later within the Shuttle sequence, because they'll just get torn apart if it's not. If they've flown our whole class and the six of us are at the end, they're dead, so that's not going to happen.

Somebody is sitting around watching what we do, how we do it, how we dress, or how we'll look on the covers of magazines. Beats me. Somebody is looking at all of that, deciding who goes first. My bet, just my own mental model was, it's NASA, and there will be a major league effort to try to draw the public interest out. I've got to bet there's some big factor that, all other things being equal, they will pick a cute one who gets on lots of covers. This is all guys making this choice. Somehow that's in there. So that probably means Anna, Judy, Rhea, and Sally have a qualitatively better chance of getting the nod first rather than me or Shannon.

I would have loved to go first. The decision was not unexpected. As I went through my calculus formula, on just my own assessment of bearing, style, acumen, and assignments that had been had, I said, "Sally's probably in the lead." I didn't think any of us would do a bad job, but if my scoring sheet was kind of about right, I had reckoned Sally was probably in the lead. It partly made perfect sense, you know; partly wounded my pride because I'd like to get that nod.

Until the STS-7 launch, I was in the conference room with everybody watching every launch. But in 1983 I was interested in getting my scuba qualification finished because we got all the scuba quals done but we hadn't had an open-water check yet. If you wanted to actually get a scuba card, you had to make your own arrangements to go do an open-water dive with a different instructor and have them sign you off.

I decided I really wanted to do that, had that all set up, and received an invitation to give the commencement address at UC San Diego on the day before or after the STS-7 launch. I decided spur of the moment when that came in, "You know, what the hell? I'm going to accept this. I'm going to line up with this guy at Scripps Research Institute. I don't get to go on this flight, but I don't need to be sitting in the conference room watching the launch. I want to go do something else."

I watched the launch. My brother came down to go out on the ship and do the dive quals with me. We woke up early and watched the launch. He turned it off. He said, "Okay, so you should have been there."

I said, "Yes, I could be there. It would be fine."

He said, "Okay. Let's go dive." So we went off, and I got my open-water dive checked that day.

The end of STS-7 also had an interesting story, though, because they were meant to land at the Cape, and there was this immense crowd of VIPs who were going to be the very first special people to get to meet the now-famous first American woman to fly in space. But the STS-7 crew landed at Edwards Air Force Base in California.

As soon as it became clear they were going to send the vehicle to Edwards, NASA recognized there was a big problem. There's two thousand, or however many, VIPs in Florida expecting to meet someone really neat and interesting, and she won't be there. Someone went patrolling the hallways in Building 4 and grabbed me.[22] "We need a pilot, and we need another one of the women, and we need to go placate these people."

They grabbed P. J. Weitz, and they grabbed me.[23] They said, "Go to the Cape. You're the replacement. You're the designated hitters. Faint substitutes, to be sure, but too bad. You're it."

We flew down to the Cape and walked into this immense sea of people. It was a really interesting moment. Two thoughts instantly went through my head. One was, "I am really happy that Sally's in California and gets the six or eight hours to just digest what she's just done, absorb it, and let it be hers, because it ain't much going to be hers. It's quickly going to become everybody else's." I was just instantly really happy for her that she had this little hiatus to make her own initial sense of the flight—to enjoy it and bask in the moment.

The next thought was, "And if this is what you get for going first, she can have it." I told her that.

Joan E. Higginbotham[24]

It wasn't necessarily my preplanned career path—or my intent—to go work for NASA. There are some people that always wanted to be an astronaut, or they always wanted to work for NASA, and I wasn't one of those. I was going to work for IBM. I had interned with them for several summers during my college years. They were just a really great company, and they thought I was a good employee.

Unfortunately, about the time I was graduating, IBM had a hiring freeze on engineers. What they did offer me was a position in sales and my choice of ten different cities, with the understanding that when the hiring freeze for engineers lifted, they would move me over to the engineering department. That sounded great, but I'm not a salesperson, so I didn't say no because I didn't have any other job offers at the time. In the interim, NASA, who normally interviewed at our campus—and I didn't know this at the time—was not going to come on campus that year. They asked for the résumés of graduating seniors majoring in mechanical and electrical engineering.

This was in 1987. It was in preparation to gear back up for Return to Flight, because after *Challenger* there was a reduction in workforce.[25] NASA was beefing up the engineering staff. One of the gentlemen in Florida received my résumé. He called my dorm room at Southern Illinois University one evening, introduced himself to me, and said that he was hiring two positions in his department. Was I interested in working for NASA?

At first I thought it was a joke, because I didn't know that résumés were sent to NASA. I never applied to or inquired about NASA and really didn't know a whole lot about NASA, except for *Challenger*. Obviously, that wasn't a plus. After he convinced me he was on the up-and-up, he told me about the two jobs that he had in his department, and I thought about them for a while.

I asked if I could get back to him within a week. It sounded intriguing, but, in all honesty, I really didn't know anything about NASA. It meant leaving friends and family and everything I knew. It sounded exciting.

Who doesn't want to go work on the Shuttle? I expressed those concerns to him. The outcome of that was a trip down to Florida, which is rare, to interview. They don't normally do that. I flew to Florida and toured the Kennedy Space Center and interviewed. I think I pretty much made the decision right then that I would accept the position, although I didn't share that thought with them. I figured I'd give it a shot. Five years or so later if I didn't like it, I could always go back to private industry. I ended up staying twenty years and two months with the agency.

When I worked at KSC I served in different positions. Once I was Jay Honeycutt's "Bubba," as we affectionately called the role.[26] Bubbas were the up-and-coming engineers who would job shadow Jay for four to six months or so. During my tenure as Bubba, Jay mentioned that astronaut selection was coming around and that he thought I'd make a great astronaut. I thought we were just making small talk. He chuckled, and I chuckled. This little conversation went on for several months. He would ask me, "Did you apply?"

"No."

A week later he asked, "Did you apply?"

"Okay, I get it, you're going to keep asking me until I apply." I applied just so that the next time he asked, "Did you apply?" I could say, "Yes, Jay, I applied."

I applied in 1994. I was at home one day, and I got a call from somebody at the Astronaut Office asking me if I wanted to come to Houston and interview. I'm thinking, "Yeah right," but they were serious, and so I went down. I wasn't necessarily expecting anything to come of the interview but thought it would be cool to be selected. It's an extremely competitive process. The year I applied they had nearly six thousand people apply, and ultimately selected fifteen. I wasn't one of them. They only take a dozen or so every time.

They normally do it every two years. I think when they selected the 1995 class, it should have been the 1994 class. They had skipped a year. They knew Space Station was coming up, and they were really trying to enlarge the corps so they had enough resources for Space Station. NASA literally turned around a year later and conducted another selection, and I was selected in the 1996 class.

That was a large class. They'll never do that again, because it probably strained every resource they had. There were forty-four of us. They called us the Sardines because there were so many of us. We had thirty-five Americans and nine internationals in our class. It was the largest astronaut class ever.

One of my first assignments after astronaut candidate training was Cap-Com (capsule communicator).[27] Initially there were four original Cap-Coms for the Space Station. When the Space Station was first manned, we were in Mission Control twenty-four seven. I must have worked the first five or six expeditions, probably even longer than that. I worked them all, because we were the only people who were trained as Space Station CapComs. This wasn't like a Space Shuttle mission, and it had been some years since Skylab (the first American orbiting laboratory). We were navigating new territory again.

Initially, the thought was that we needed twenty-four-hour coverage. Since there were only four of us, it was very taxing. It took away a lot of time for us to do flight training, get our T-38 hours, do simulations, and all the other things.[28] It was really grueling at first. It was one of those jobs that would burn you out quickly because it's all-consuming. Eventually

Astronaut Joan Higginbotham poses in front of a T-38 aircraft.

we went to a different model where we had to train other people because we were seriously overwhelmed at the expense of our other training.

There wasn't a lot of communication at the time. We were using the Russian assets, the satellites. We'd get a pass maybe every hour-and-a-half or fifteen minutes if the communication was good. Of course, the Russians would speak to the crew first, because it was their asset. We would coordinate with the Russian Glavni to discuss what needed to be relayed to the crew during the next com pass.[29] The Glavni would speak first, and then we'd get some time to speak to the crew. It's funny how many times we never got to talk to the crew. The Glavni would talk about mission-related items, then they would talk about non-mission-related things like the weather. Mind you, we haven't had our chance to speak with the crew. We were very happy when we got our communication assets and could talk any time we wanted without having to coordinate with our Russian counterparts.

Being a CapCom helped tremendously once I was assigned to a mission because I understood the mission from both perspectives, crew and CapCom. It helped me better determine what's really relevant in the short amount of time that CapComs need to relay information to the crew, especially if something's going wrong. As a CapCom, we need to have the ability to determine what is crucial to relay to the crew and what information can wait to be relayed. It also helped knowing how Mission Control operated, because a crewmember awaiting information from the ground may think, "How come it took twenty minutes to get us an answer?" Once you've been CapCom in Mission Control you're more cognizant of that. For example, the flight controller is talking to another flight controller to gather information for the crew. Then they need to give the information to the flight director and CapCom, who then relays it to the crew. Being a CapCom helped me understand the dynamics of the mission.

So I think I had a better feel for what was going on in the room. As a matter of fact, when we were up on orbit, something happened, and I said, "Oh I can tell you what they're doing right now. They're doing this and this and this. We should have an answer relatively soon, depending on who's on console." I also learned the capabilities of the flight controllers who were on the console—their strengths and weaknesses. I had the ability to gauge what information is going to come up, and how it's going to come up. It's a really good position to have. I think it gives you a good perspective for being a crewmember and for working in Mission Control.

I was in the car when, I think it was Rommel, head of the Astronaut Office at the time, called to inform me that I was going to be assigned to STS-117.[30] I said, "That's pretty cool. Who else is on there?" That's important too. It so happened that one of my really, really, really good friends was on there so I was excited. I called him, and we chatted. Of course, we couldn't tell anybody. I don't think they were making the announcement until the following week, so we could only tell our families. I went out and celebrated that Friday and no one but my boyfriend knew why.

We started training, and then the *Columbia* accident happened.[31] I had three classmates on board, but I knew the entire crew. It was a really difficult time. I woke up that morning, I remember, and I think I'd had a bad day in the robotics lab the day before or something, so I was having a little bit of a pity party. The first thing that was on the news was that they had lost communication with the Shuttle. I'm thinking to myself, it must be a slow news day, because as you're coming in there's always a period of time where you could possibly lose com with the Shuttle. The communication instructors tell you this.

This was the first I was hearing of this. Then the news anchor went on to another story, but then came back to the Shuttle story. They said, "They've lost communication with the Shuttle, and they've also lost tracking." You never stop tracking the Shuttle. Even when the Shuttle goes into the zone of exclusion, you know where the Shuttle is. It's still being tracked.

That's when the story caught my attention, and I started paying attention to what was going on. Within five to seven minutes, it was reported that someone had seen debris coming in over Texas and I was thinking, "This is so not good."

I just remember being glued to the TV. One of my girlfriends, who's also an astronaut, called to ask what was happening because she was at a five K race and had begun hearing rumblings. She called me and asked, "What's going on?"

I told her, "They lost com with the Shuttle." She said something. I said, "I think they're gone."

Several of my friends started calling me to see how I was doing. Someone called me to see if I had eaten, and I really hadn't. Probably about two or three o'clock that day we got a call. NASA wanted all of the astronauts to come in to the office. Right where the big NASA Johnson Space Center sign is on NASA Parkway, the public had already started this big memorial with stuffed bears and flowers. I went to my STS-117 crew office. One of my crewmates, Rick Mastracchio, was in.[32] We just looked at each other in disbelief. Then all of the astronauts and support staff assembled in the conference room up on the sixth floor. I remember we were waiting for the meeting to begin, and I just started bawling. I just lost it. The management team was talking about what had happened and what they knew so far. They asked for volunteers to work various aspects of recovery and family support.

That was a really difficult day. Everyone went off to do various things. For a while management kept telling us we were going to continue crew training. For a short period of time we actually continued training. We were supposed to fly in September of that year, 2003, and this was February, so we were on a pretty accelerated pace to get ready for September. Then we were told that we probably weren't going to fly for a while, so training slowed down a little bit.

I was assigned to work at the Cape reassembling *Columbia*, since I had worked on that vehicle and had a lot of knowledge. We were working in the big warehouse putting the Shuttle debris pieces back together like it was a jigsaw puzzle.

Was it hard for me, having previously worked on *Columbia*? It's funny because someone—a reporter or someone—had asked me about *Columbia*. "So how did you feel about losing a national treasure?"

I responded, "I didn't think about it just now, till you asked me that. It's a Shuttle. Yes, it's sad that we lost a Shuttle, but my mind was mostly focused on the crew." There were just so many things that we had to get through on a personal level. The families went through so much. So who's caring about a stinking Shuttle at the time? I worked on the vehicle, and I was sad because it was lost forever, but that didn't cross my mind for quite some time.

I remained assigned to STS-117 for three years. Then I got a call saying, "Come up to the Astronaut Chief's Office." I thought I was in trouble. Nobody wants to get called to the "principal's office." But the visit was just to inform me that Mark Polansky and I had been taken off 117 and put onto STS-116.[33] There were several crew reassignments. Our flight date was December 2006—I think we probably started training a year to a year-and-a-half out.

It's funny; you get used to your crewmates. You bond. I was on a really good crew—STS-117—so when I was told that I was moving, I was a little ambivalent. Then I was told that Mark was moving with me, which was a good thing, because he's a great guy, a really good friend of mine. After getting married in 2012, and getting "bonus children" I can tell you from experience that it was like being in a blended family. I moved into the new crew office thinking, "Are you going to like me and accept me into the crew 'family'?" The other good thing was Robert Curbeam was already on the crew. He's just a great friend of mine and Mark. Billy Oefelein was the pilot. He's a really great guy. Then Nick Patrick came in. We all got along. Christer Fuglesang was already on the flight too.[34]

It was important to forge new relationships in the newly formed crew. The management in the Astronaut Office recognized that and suggested the crew participate in a National Outdoor Leadership School trip to Utah. I thought, "Really, we're going to be camping in the Canyonlands National Park for how many days? Nine days? Great." My idea of roughing it is a three-star hotel.

We hiked twelve miles that first day. We rappelled the next day up a five-hundred-foot cliff, and I hung in there. The instructors were really impressed with me, telling me I was a trooper. Like I said, it's not necessarily my favorite thing to do, but you do what you have to do. They gave me a lot of credit for just muscling through even though that was not necessarily my cup of tea.

When my in-flight assignment changed from STS-117 to 116, that was being part of the whole blended crew. Everyone had pre-existing crew roles. I was on the flight deck coming and going on 117, but they had already had their crew assignments on 116. When Mark, as the commander, asked everybody what they wanted to do, I said, "I'm not trying to kick anybody off, but I'd like to be on the flight deck at least one way. I'm not going to say put me up there both ways. If I'm not up there either way I'm good, but I would like to be on the flight deck one way." I got to fly on the flight deck coming in.

It was great because we all knew that roles were going to change a little bit and people were flexible. The basic roles didn't change. I was still going to be the robotic operator. The EVA crew didn't change. It was the secondary roles that changed a little bit.

My primary role was to be the robotic operator, and that was really tricky because we had a lot of robotic operations.[35] We were taking up a truss, so we had to get the truss out and install it.[36] We also had two or three EVAs that were going to require robotic operations as well. I practiced the procedures in the lab by myself first. Then I practiced at the NBL (Neutral Buoyancy Laboratory) with the crew.[37] There was another robotic arm simulator at the dome, and VR (virtual reality) lab. There were plenty of opportunities to practice.

My secondary role was to be the loadmaster—that's the person who takes care of all the stowage, equipment, and other items that we brought up in the Shuttle. We were taking up about three thousand pounds of equipment. That was something that was really good for me because I'm just a little bit meticulous. You need somebody who's meticulous and detail-oriented, because you don't want to lose stuff or leave it in the wrong place. They knew that was going to be my job, because I am just a little—they had another name for it, but we'll just call it meticulous.

Then we had tons of other tasks. I was the com guru. When something doesn't go right, "How come we don't have communications? What's wrong?" It is better than being the IT (Information Technology) guru, because the computers always crap out. I'd rather have been the com guru.

Launch—launch attempt number one, the weather is iffy. Despite that, I'm excited, because this is the first time I'm getting suited for real. Let me go back. We have our last meal. The press is taking photos as we're eating. We start getting dressed and go out to the pad. It's really surreal; all the training and studying culminates in this moment. I'd never been at the launch pad, obviously, when we've been that close to launching.

The vehicle is alive, spitting and spewing gasses. We're looking up at the Shuttle thinking, "This is so cool!" We get in the elevator and go to the 195-foot level. The commander and Suni Williams get strapped in first simultaneously.[38] Mark is on the flight deck, and Suni on the middeck. I think I was third or fourth to get in, so the first thing I did while waiting to be strapped in is go to the bathroom. I just could not "go" in the diaper; that doesn't work for me psychologically. Then I called my family and said, "Hey, we're good, we're launching." They were heading to the launch site.

I got strapped in and waited. The launch team is discussing the weather the entire time; low clouds at RTLS (Return to Launch Site).[39] The crew is thinking, "Seriously? Come on people. Let's launch. How often does the Shuttle have to return to the launch site right after liftoff?" But in reality, we know the rules are in place for a reason, and we respect the decision of the launch team. We get to T-minus nine and still had an issue. They weren't "go" for weather. They decided to count down to T-minus five and polled the weather guys. Weather was no-go. "We're going to just have to call it for the day."

So we get out and go back to the crew quarters and get undressed. Major bummer! The launch team was trying to decide whether or not they were going to try to launch the next night, because there wasn't going to be much change in the weather.

They opted not to go the next night, but tried on the ninth of December, with similar weather conditions. That morning I'm thinking to myself, "This is dress rehearsal number two. We're going to get dressed and go sit out there for two-and-a-half hours, and we're going to come back." The decision to get us suited for a launch attempt came a little later than usual.

Out to the launch pad we go again. I remember the tech was putting on my gloves, and we were a little bit behind the timeline. Everyone's feeling a little bit pressured. As he's putting on my gloves, he jammed my finger. Before I entered the vehicle, I remember calling my family, and nobody was answering their phones.

I called my boyfriend at the time. I said, "Where's my family?"

He says, "Oh, I think I saw them. Let me go look."

I said, "Well, I don't want to hold up the phone. I'll call you back." I let somebody else use the phone and then called him back.

He said, "I don't see them anywhere."

I began panicking and asked, "What do you mean you don't see them?" I called my brother and said, "Where are you?"

He says, "We're on our way." I think he was still at the condo.

I screamed, "Leave now!" and hung up the phone. I called my girlfriend who was the crew support person taking care of my family. "Don't let those buses leave until my family gets there. They're on their way."

She says, "I won't let them leave."

As I'm getting strapped in, I'm thinking, "Where's my family? I wonder if they're at the launch? I will be so mad if they don't make it there in time!" Then I think, "I should not be thinking about this right now. There are more important things to think about, such as the mission." I decided not to ask anyone if my family had made it to the launch site because I would be so torqued off if they didn't make it there in time to see me launch.

We're counting down again and get down to T-minus nine. I think they gave us a go for weather. We count down to T-minus five. We're holding our breath. Working in the space program for so long, I know the count can get down to zero and the launch still may not occur if the main engines shut down, so I refuse to get excited.

The count continues down to zero. Then, liftoff! Suni is on my right and Christer is on my left, so we all joined hands and raised them in celebration. It was so cool! It was just like we were in a simulation, except there was real motion!

I could hear the harmonic of the vehicle. I could feel it going faster and faster. I heard this little hum. You really didn't feel a whole bunch until we got to roughly launch plus six minutes into the flight, where the Gs kick in and my breathing became labored. It felt like there was an elephant on my chest.[40]

Then they called MECO (main engine cutoff).[41] Christer had to get up really quickly because he had to get up and photograph the external tank falling away. We had practiced this a million times. He takes off his helmet, gives it to me, and I stuff it in his equipment bag. He takes off his gloves and grabs the camera out of the locker.

Suni and I unbuckle and float to the ceiling; we're giggling like school-girls. Then the fun is over. It's time to prep this vehicle, to get it ready for operations. It was a pretty cool day, long in the making.

When we rendezvoused and docked with the Space Station, I was supporting the commander, ensuring we were following the checklist and making sure we were properly configured for our maneuver burns. As we're coming in for the last couple hundred feet, my job was to ensure that we were coming in at a steady rate. It was just like we were in the sim-ulator, at least for me, because I wasn't the one flying. The commander, however, was a little bit on edge.

We saw the Space Station the day after we launched. It was this bright white star, far off in the distance. As we're getting closer I'm looking at the thing saying, "Holy moly, this thing is huge!" It was just looming over us. We're coming in for our rendezvous with the Space Station, and we're doing really well. I'm calling off ten feet, five feet, and so forth. It was perfect, a perfect docking. As soon as we got the capture call, our other crewmates began the actual latching of the CBMs (common berthing mechanism).[42] The docking went very smoothly.

Our crew had many responsibilities, including installation of the P5 truss. The day began with the guys going out for the EVA. Suni and I were in the lab working the robotic arm. Both robotic arms were being used. The commander and Nick were operating the Shuttle arm. They had taken the truss out of the Shuttle payload bay and positioned it on the port side of the Shuttle near the nosecone.[43] Suni went in, and we grappled the truss and positioned it off the port side of the Space Station. Everything was going smoothly. It was the first time Suni was at the controls—first time for either one of us—but she was actually doing the flying that day and I was backing her up. It was really interesting because I could tell she was a little stressed.

During training, we were told there was a possibility that before we got the truss fully installed, the arm could possibly stall, and we'd have to perform this contingency procedure. Just like clockwork, the arm stalled. We ran the contingency procedure, and it worked like a charm. The EVA crew helped us tremendously. They were at a great vantage point to tell us how to maneuver to get to the truss capture latch indicators, which let us know the truss was fully installed.

The next day when it was my turn at the controls, I asked Suni, "How's the clearance over there?"

She said, "Joanie, we have really good clearance in this camera here. We have a secondary cue here." I understood the way she was feeling the first time she was at the controls the first day. It's a little daunting. I was thinking, "Okay, so I have what, a thirty-five-million-dollar arm, and I have a crewmate, and I can bash them both into the side of the Space Station. That wouldn't be a good day, but no pressure."

I mentioned I was the loadmaster. We moved a little stowage and equipment every day. There was a transfer book with details of what item got moved in what order. It was similar to a game of Tetris. There was some stuff that couldn't be transferred because it was needed for EVA and other tasks. My goal was to get ahead of the curve, because everyone who had been a loadmaster told me that the last day you're throwing stuff through the hatch before they close it. I was determined not to be doing that. The first day of transfer went really well. I looked ahead and called the ground to ask if we could transfer some items early. I would verify with the ground to make sure they were okay with us moving stuff early, and they usually were.

Like I said, the loadmaster needs to be detail-oriented; you need to look at your notes and just keep really, really copious notes about what you're doing. If you do anything differently than the checklist, it's good to let the ground know. They are your friends. That way, if you don't remember what you did, they can remind you that, "Oh yes, you still have four things over there that you took out of the kit that you need to put up."

Due to the solar array getting stuck, we had unplanned robotic arm operations that took me away from my transfer duties. We had the backup loadmaster and others running transfer during that time. It got a little disorganized and the commander said, "No one touches the book but Joanie. This is her book."

Landing—again, we were late leaving and late coming home. The weather was crappy everywhere. We had stayed up an extra day, and we had to come home that day. The weather was cruddy at the Cape. It wasn't any better in California. At one point, they were considering taking us to White Sands.[44] We were glad about that, because that meant no poking and probing by the flight docs. That's the only place they don't send the medical team. We were excited, "Ooh! We can go get margaritas." That was also the only place we didn't have clothing, so that would have been interesting.

The landing team decided to bring us home, so we suited up. The team still wasn't sure where we were landing: the Cape or California, or maybe White Sands. As we were getting close to the deorbit burn, the weather wasn't looking good anywhere. They said, "We're going to wave off this rev and hold off on your fluid loading," which is the most important thing, because you don't want to have to do that twice.[45] Everyone took a nap. We set the alarm so we'd wake up on time. The team literally waited until the last second to tell us we were go for deorbit burn. Because of that, we actually had a lot of cross-track to make up, so as we were coming in, we were pulling a couple Gs that you normally don't pull.

Part of my job was to get some footage of the flight deck with the lipstick camera. It's called a lipstick camera because it's about the size of a tube of lipstick. It does not have a viewfinder, so the only way you know what you're filming is via the flip-top video camera that's connected to the lipstick camera via a wire. As I was scanning the flight deck, I would look on the monitor to see what I was recording. I had the recorder Velcroed on the back of the pilot's seat, which works well in zero gravity. However, as we were coming back in for entry interface, gravity took over. The monitor fell off the seat to the ground, and it yanked the cable out of the camera. I was thinking, "I can't mess up now. My crew will never let me live it down."

I tried to pick up the recorder and plug the cable back in. Unfortunately for me, the bulky launch entry suit and seat belts hindered me from bending down and physically picking up the recorder. I used my feet to lift it. I lifted it high enough to reach it and plugged it back in. The reason I knew I could not get the recorder with my hands was because I stupidly tried to bend my head down and pick it up, and that was not a smart thing to do. When I tried to get down there, I couldn't reach. Then my head was so heavy that I couldn't lift my head up. I just remember all of a sudden laughing to myself. I said, "I went to space and got a big head." I had to take both my hands and lift my head back up. That's when I decided to use my feet to pick up the camera. Engineering ingenuity at its finest.

When we landed, I was trying to get off the vehicle; I'm up on the flight deck and the ASPs (astronaut support person) came in to help me off.[46] I went to stand up and immediately fell back down in the seat. I thought, "Wow!" I felt so heavy, like an elephant.

He says, "Take off your parachute." I took off the parachute, stood up, and fell right back down. He said, "Just sit there for a while."

So I sat there, got my wits about me, stood up, and climbed slowly down the ladder. I turned around to exit out of the hatch, which is actually at a lower level, so I had to get on my knees and crawl. I crawled out the hatch and several flight docs were standing out there. I looked up at them and said, "I can't get up. You've got to help me up." They helped me up to my feet with all that gear on. I walked assisted to the crew transport vehicle to get changed and slowly tried to get my land legs back.

After this mission, I went to talk to Steve Lindsey, who was the chief of the office at the time.[47] I wanted to know what the plan was for me. The class behind us had not really started flying, and then there was another class of unflown astronauts. We all knew the Shuttle Program was coming to an end. I'd been there twenty years, so I'm trying to figure out my future: go into management at NASA or pursue another career. I was ready to go back to space, "Give me like a week to recuperate, I'm ready to go back."

Steve provided little insight. I had two or three conversations with him. It wasn't so much that I wanted to know if I was going to fly again, rather I was trying to gather facts so I could figure out what to do with my life. I was sensitive to the fact that there was a long line of people who still needed to fly. It'd be nice if they flew before the program ended.

Long story short, Steve couldn't provide any useful information, so I just went fact-finding. I wasn't necessarily trying to leave NASA. It so happened that I was talking with the folks at Marathon Oil. We were in talks, nothing concrete. Then I was assigned to STS-126.[48] Probably about a week or two later I got an offer from Marathon. Now I was faced with the daunting task of deciding if I wanted to fly again or take this position. It was a really, really, really tough decision, because flying is an incredible experience. Not a lot of people get that opportunity, and NASA is a great place to work. My first flight was so incredibly phenomenal; I didn't think the second flight was going to live up to that. So I said, "If I'm going to leave, I'm going to leave on a high note," and I did.

There were challenges at NASA. I don't know if there was a major challenge. Learning how to fly a T-38 was hard, but then it just clicked. It was the same for robotics. For me the first couple of months, it was just like Greek and then all of a sudden it clicked. It just made perfect sense.

When I first began working for NASA, I was very young. They used to call us "fresh outs," meaning fresh out of college. At that time NASA had a whole bunch of fresh outs, then you had guys that were fortyish, then you had the seasoned guys that were older. There was this huge age gap. It took some time to gain the respect of some of the guys who had been around for years and years. Some people were just really very accommodating and nice, and with other people I just had to say, "Look, I'm very capable, so get over whatever issues you have." I remember I had to take one of my counterparts' bosses to task one day, but I would probably have that in any environment, just being the newbie.

Gender and race could have had an impact on my career at NASA. There were not a lot of African Americans working on the Shuttle at the Cape. Matter of fact, I was the only African American female engineer working on the Shuttle. They had one other African American woman that was working on the Shuttle software, and they had contractors. For a long time, I was the only one working on the Shuttle side. I think I was so enamored with the job and what I was doing that it didn't faze me; probably fazed a few other people, though. I was fine with it. My bosses were the greatest people in the world. That was never an issue with them.

There were one or two times where some remarks were made in the Astronaut Office. One of them was rather disparaging. I was a little bit surprised and upset, because it came from a colleague. I was like, "Really, that's your mindset?"—and don't ask me why I thought this, but you would think that at that level people are above that. But obviously they aren't.

I was very blessed in going to NASA in that I had really good bosses who, like I said, were not color-conscious or gender-conscious, and saw me for who I was. I've been lucky that my bosses have been incredible people.

Eileen M. Collins[49]

In 1989, I applied as both a pilot and a mission specialist. As an active duty military officer, you have to apply to your service first. Because I was a student in the Air Force Test Pilot School at the time of the application, they selected me as a mission specialist. The Air Force would not select me as a pilot until I had graduated from the school, and I was six months away from graduation.

During the astronaut interview, I was asked, "Would you rather be a pilot?" I explained to them how I had applied for pilot, and they understood how the military services screen applicants, but they still asked me, "Would you rather be a pilot?" I said I would love to be a pilot or mission specialist, either one.

They asked me at least five times in that interview if I would rather be a pilot. I stuck with my original answer, because that's how I really felt. Looking back, now that the interview and my career is past history, they made the right decision, because I was better qualified as a pilot. I had been flying airplanes my entire professional life. I was close to graduating from the Air Force Test Pilot School. I was just better qualified as a pilot.

It turned out that the Shuttle Program had never selected a woman as a pilot. Because of the timing, I became the first woman pilot. The call that you get from the board once you've been selected happened on January 16, 1990. January 16 is the anniversary of the class of 1978 getting the call, and several board members were from the class of 1978, so they wanted our anniversary to be the same as theirs.[50]

I was out flying an A-37 aircraft, doing spins, when the call came in. I was pretty wrung out. I was solo, and I had somebody chasing me, so it was a pretty busy flight. I came back and I walked into the squadron room, and one of those yellow government phone slips was posted on the bulletin board. Nobody told me about it. I just happened to look. "Major Collins, call Duane Ross at NASA."[51]

My heart started beating faster. I was the only one in the room other than the secretary. She was an enlisted gal, a wonderful person, and her name was Denise. I called, and Teresa Gomez answered the phone, she was one of the assistants to Duane Ross.[52] "Oh, Major Collins, let me get you Duane Ross."

Duane Ross gets on the phone. "Hello, Major Collins, let me get you Don Puddy," who was the chair of the selection panel.[53]

He gets on the phone and says, "Oh, Major Collins, let me get you John Young."[54] I had been handed off through all these people, and my heart is beating.

John Young gets on the phone and says, "How would you like to come to Johnson Space Center?"

"I would love to do that."

John Young went on, "You wouldn't believe all the great programs we have here. We fly T-38s. We have the simulators. We have the Weightless Environment Training Facility."[55] He's going on and on. He must have gone on for five minutes. Then he asked, "Well, do you have any questions?"

I asked, "Am I going to be a pilot or a mission specialist?"

"Pilot. You're going to be the first woman pilot."

I knew that was the right thing for me. I didn't get that big euphoric happy feeling that everyone thinks they might get. I felt a huge sense of relief. I think the gal that was in the room with me knew what was going on, but I went back in the squadron room and I didn't tell anybody. They told me to keep it quiet until the announcement went out, and I went home and told my husband. I was very happy. For me it was a dream come true, and who knows what the future is going to hold? But I was happy to come to Houston and spent sixteen years.

There was a huge difference between my reception at the Air Force and the Johnson Space Center. A lot of it had to do with the timing, and a lot of it had to do with the history of the organization and the culture of the organization I was going into. I was in the first class of women pilots at Vance Air Force Base in Enid, Oklahoma. The class actually started in September 1978. Back then every class had forty students; we were four women in a class of forty. There were over five hundred pilots on the base, other instructors, other classes, but we were the first class of women at this base. A footnote here—there were several classes of women who had already gone through pilot training at Williams Air Force Base in Phoenix, Arizona.

Going into Vance Air Force Base in 1978, a place that had been training pilots since 1944, there's a culture that didn't really understand women. There are many stories. I'll tell you one of them.

I wore a green flight suit walking into the commissary. They'd never seen a woman in a flight suit, and we were very self-conscious walking around on the base. I'm checking out, and I knew that there was some resistance to women coming into pilot training. Most of the men were very accepting of us, but there were some men that just were old school and didn't want the women there. I don't fault them for that. It takes a little bit to change people.

As I checked out, the lady behind the cash register said to me, "The wives don't want you here." I was surprised, because I thought some of the men don't want us here, but the wives don't want us here either? That bothered me.

I asked, "Why?"

She said, "They don't want you going cross-country with their husbands."

I said, "Thank you for telling me that." I decided I was going to get to know the wives. It was important for me to let people know that I was there to be a pilot, and I wanted to be the best pilot I could be. I wasn't trying to steal anybody's husband. I wasn't looking for a boyfriend. People wonder what your motives are. I wanted to fly.

I did get to know the wives. At our first class party, the other three girls didn't go, so I was there by myself. I walked in to the party. All the guys were on one side of the room talking about the T-37, the boldface and procedures, what the tests are like, what flying is like, and what the instructors are like, all the stuff I needed to know.[56] All the wives were on the other side of the room talking about the curtains in their house, and the friends that they had—it seemed like they all knew each other.

I didn't really fit in with that group of wives. I needed to get over there with the guys. I remember standing there, and I went over to the wives. I spent the whole party with them, trying to get to know them. I was so mad at my female classmates that didn't go to the party.

By the time we all graduated, I think it was understood that we were there to learn how to fly. Of the four of us, one of them "washed out" (did not graduate), one "washed back" (delayed) to another class, and the other failed all of her check rides. I did okay; the point here is that no one was giving the women special treatment! I wasn't the top graduate in my class. It was hard for the women. I'm not faulting any of them. The classes at Williams, there were some mistakes made there with the first class of women. The Air Force higher-ups said, "Every one of them," there were ten, "will graduate, period. Make them graduate and make them graduate on time."

A couple of these girls struggled. If they were guys, they probably would have washed out. The Air Force pushed all ten of them through, and that was a mistake, because it doesn't help the women to have a second standard. Lesson learned by the time my class went through, although we were at a different base, we were set to the same standard as the men.

I think it was very fair, and that was good, and I think by the time we graduated, we were accepted. But it was a very difficult year. I say we were accepted. It wasn't really that important to me to be accepted into the culture. It was important for me to graduate and become a pilot, but for many of the women it was difficult.

I came to the NASA Johnson Space Center in 1990 and fit right in. I cannot even think of one story where there was any discrimination, rude comments, a feeling that because I'm a woman people would set a different standard or expect something different. I credit that to the women who were flight controllers, engineers, and mission specialists before me. They came into Johnson Space Center and did a fabulous job, so my experience was smooth. We, the following groups to include women, were able to focus on the mission.

I was very pleased. The other comment in that area is the history of NASA—it's a relatively young organization compared to the military. There wasn't quite as much of the cultural attitude of "This is the way it is here." NASA was established in 1958. That does seem like there is some history there, but it wasn't quite as ingrained as it was in the Air Force. NASA was very nice, very easy. I had no complaints.

You may have heard stories from the other women, but let me say after being in the Air Force, I was hardened to it. I think a lot of it bounced off of me. There might have been a little comment here or there that another woman might have taken as insulting. I would try to bring some humor into a situation if I thought somebody was making a derogatory comment.

I just made sure that people knew that I was there because I wanted the mission to succeed. I found if you focus on a safe mission, successful mission, then the team that you're working with will come to respect you without you having to say anything about it directly.

By the way, the first women astronauts did parachute training in the summer/fall timeframe of 1978 at Vance Air Force Base. They were in the local papers. They were in the local news. I heard about them, read about them, but I didn't cross paths with them. That was when the reality hit me that I could apply to this program someday.

I knew back then that I wanted to be a Shuttle pilot. I thought we would have had many women Shuttle pilots before me, because I was only twenty-one years old at the time. I'm sorry I didn't get a chance to meet them at that time. I eventually got to meet them. I don't know if you really have to meet somebody to be inspired by them. Just the fact that I knew they were out there doing what they were doing was all I needed.

My first flight was a Shuttle-Mir mission, which rendezvoused with the Space Station Mir.[57] It meant a lot to be selected as the first pilot for a Shuttle-Mir mission and the first woman pilot of a Space Shuttle. That doesn't do it justice. Obviously, getting called into the boss's office and told you're assigned to a flight, and this is the flight, is pretty exciting. Jim Wetherbee, the commander—I was very fortunate to fly with him.[58] He had a great reputation, very knowledgeable on Shuttle systems and Shuttle flying. He was an experienced commander.

Mike Foale, Bernard Harris, Janice Voss, and Vladimir Titov, who was one of the first two Russians to fly on the Shuttle, made up the remaining crew.[59] Titov came to the United States in 1993 to train, and he ended up assigned to our flight. He had very little English initially, but he picked up the language very fast for an adult.

The six of us trained, and we had a very long delay. We were assigned in September 1993. Within a couple of weeks, the flight was delayed. I don't remember all the dates, but I think we were scheduled to fly in May 1994, so we had only seven months to train for this flight when we were assigned to it. Management told us that it was probably going to be delayed, but to go ahead and start training. We did. The flight was delayed nine months to January 1995. It was very hard for a first-time astronaut to see their flight get delayed nine months. We hopped over five other missions.

The mission stayed scheduled for January 1995. We ended up with a little delay, and we launched in February 1995. I just kept focusing on my mission and learned as much as I could about the Shuttle. We all knew that what we knew could save our lives someday, so we wanted to know the systems, to be able to run the procedures correctly, quickly, and understand why we're doing each step.

I memorized several time-critical procedures that were ascent-related, for example, main propulsion system helium leak, APU (auxiliary power unit) shutdown, and fuel cell shutdown.[60] You don't run the procedures by memory, but if you have them memorized, it's much easier to read the checklist and work quickly and efficiently through it. That's what I did for the long delay period, and it worked.

I felt that my simulator ability was very rusty, because I had been down at Kennedy Space Center for over a year, maybe fifteen months, as an astronaut support person, so I had been out of the simulator for a long period of time. I was slow. I thought, "This is not me. I'm going to get back up to speed." That's what I did for the next sixteen months. That's what all pilots do.

We were the first Americans to see Mir. The community and the team developed the rendezvous procedures with Mir, which was huge. How we were going to do not only all the programmed burns from a distance, but how we were going to hand-fly it once we got within one thousand feet. Jim Wetherbee did the hand-flying, but I backed him up and I also trained for that.

We started using these laptop computers; 386s is what we called them. The software was very primitive, but it was enough to tell us what our closure rate was and the distance. We were able to feed the radar data, which also had range and range rate, into that laptop computer to give us a prediction of where we were going to be in the future if we did a certain burn. It was very helpful in saving us gas. That's what I remember as one of the most substantial things we did on the mission: develop the Mir rendezvous procedures.

Obviously, we'd been doing rendezvous with the Shuttle and satellites for many years. But this is the first time we were one hundred tons rendezvousing with another one-hundred-ton spacecraft. We were concerned with how the navigation system was going to work, how the flying qualities of the Shuttle were going to act when we were close to the Mir Space Station, which had solar arrays, and how the communication system was going to work with the Russian system.

Another very interesting story is how we negotiated with the Russians. Sometime in the summer of 1994, our whole crew went over to Russia and did quite a bit of training. Most of it was getting to know the Russians, working with them, developing trust for each other. I was the right-hand person to Jim Wetherbee. We went into the Mission Control in Star City. We met Viktor Blagov, a Russian flight director, and we started speaking to him through an interpreter about how we were going to do this rendezvous.[61]

At that point in time, we were only approved to go to one thousand feet of Mir, and we believed that wasn't close enough to get the kind of data we needed. We thought certainly we could go in closer than one thousand feet. How about three hundred feet? Viktor Blagov was a very good negotiator. We had to justify to him why it was safe for us to go in to three hundred feet. We had some engineers with us. Obviously, we weren't making any decisions there, but it was important for us to build a relationship where he could trust what we were saying.

They eventually agreed to let us go in to three hundred feet, not in that meeting but at some point in the future, at which point we thought, "Why should we stop at three hundred feet? Why don't we go to one hundred feet?" We're not in Russia anymore; we're back here and we worked the system. Jim Wetherbee was really the leader in all this, so I credit him with the idea. I was more of an observer as a pilot, supporting him. We got both the Russians and the American leadership to agree to one hundred feet. Then we thought, "Why should we stop at one hundred feet? Why don't we just go to thirty feet?" Very close to the mission, we were approved to go to thirty feet. That was close enough for us.

On launch day, right after main engine cutoff, we had three jets fail. I forget the type of failure of the other two, but the third jet was R1 upper.[62] The master alarms are going off, and that's where your training kicks in. We looked out the window after everyone had gotten out of their seat, and we could see R1 upper leaking. If you looked out the back window, right at the payload bay, that jet was leaking straight up away from the payload bay, in the same direction that Mir was going to be on our rendezvous.

It's spewing all this oxidizer. We thought, "they're never going to let us do that rendezvous to thirty feet, because it's going to contaminate Mir's solar arrays," which would affect their power.

Mike Foale got on the radio and said, "It looks like a geyser."

I thought, "Oh, they're never going to let us rendezvous." Sure enough, the rendezvous was temporarily canceled. It was originally scheduled for flight day four.

We eventually did it on flight day four, because over the next couple of days we ran these procedures called a staged repress. We turned the Shuttle toward the sun. Maybe the sun could bake it out. We did several little tricks that the guys at Mission Control in Houston came up with, and we finally got the leak down to almost nothing, so they let us do the rendezvous.

Jim Wetherbee had written on a piece of paper, handwritten while in quarantine before the flight, the speech he was going to make while we were at thirty feet. It was something like we are coming together as our nations are coming together. Mike Foale and I are like, "Stop reading that message and fly!"

Mike Foale is yelling, "We're getting too close! We're at twenty-nine feet!" Mike Foale had previously calculated that two masses are going to attract each other if there's no other force working. He said, "If we don't be careful, we're going to dock." That was a joke by the way. It was all in fun. We wouldn't have docked because we didn't have the attitude or hardware. We didn't even have a docking mechanism on our Space Shuttle that we were flying. But there were a lot of jokes around Newton's law of universal gravitation. All that worked out, and we had a good time with it.

Obviously, we knew what we were doing was a pretty big deal, not just the technical part of it but the part about two nations that were enemies during the Cold War now working together. I don't think the significance of it really comes to your mind until many years have gone by and you think back.

My second mission, my commander was Charlie Precourt, who used to fly F-15s in Germany.[63] The commander of Mir was Vasili Tsibliev, and Charlie and Vasili, in one of their conversations, realized that they were stationed on opposite sides of Germany back in the 1980s.[64] Potential enemies, and now they're flying in space together. As the years go by, we look back at that. I think it's a huge transition in our relations with the former Soviet Union, now Russia, seeing all these changes take place. I think we worked very well together.

At the worker bee level, astronauts, flight directors, engineers, we did very well, considering the potential conflicts. We really learned to work together. I think NASA and the Russian Space Agency have a lot to be proud of, that they really got with the program and followed the nations' direction for us to get together and work in space. It's been good for us.

It was the spring of 1998, late February, early March when I was named the first female commander. Jim Wetherbee was the chief of Flight Crew Operations Directorate. He called me up and told me, "You are going to be assigned to STS-93.[65] You're going to be the first woman commander. We have total confidence in you." The stuff that they always say. George Abbey was the center director at the time. He said, "Let's go up and talk to Mr. Abbey. He'd like to say a few words to you before you go up to Washington." It was all really positive.

I had to prepare a couple of remarks, which I did, and headed up there. Bill Clinton was the president, and Hillary Clinton was the first lady at the time. My husband went with me, and I have a cousin that lived there, so I invited my cousin, Mary Kay Morin, because they said, "Invite any relatives." She was excited to meet the president. We go into the Oval Office, and there were several people in there: NASA Administrator Dan Goldin, and a few other managers from NASA.[66] We met President Clinton, and he shook hands with all of us, and then he started talking about Stephen Hawking.[67]

Oh, by the way, before I went into the Oval Office, I met Sally Ride in the hallway. That was the first time I had met her. I didn't even know she was going to be there. Wow, Sally Ride! We had talked on the phone several times before that, but we had never met. I never asked her why she was there. I thought, "Did she come up here just for me, or is she up here on other business?" It was great to meet her. She was very supportive, and she talked about all the media stuff, the "Hey, if you ever want any help with that, give me a call. You're not going to have any problems; you'll get through it" kind of advice.

Fast-forward to when we met the president. We crossed into the hall-way between the Oval Office and the Roosevelt Room, and I was fine up to this point. But I looked into the Roosevelt Room, and I saw Sam Donaldson in there and all these cameras in the back, all these people, and the room was completely packed.[68] I think for the second time in my life I had this panicked feeling, which I never get. I recovered from it rather quickly. I just said, "Eileen, just be the woman commander. That's a different person. Just go into that role. You'll be out of it before you know it. Just do your speech."

We walked in, the president talked, the first lady talked. There were a couple of other speeches. I went up and gave my little spiel. Eileen Hawley in JSC Public Affairs helped me a little bit with the speech.[69] She didn't write it for me, but she said, "You need to have a theme, something that people can grab on to, relate to." We came up with "it was a dream." I started my speech with it was just a dream that I had, and now dreams do come true. We thought it would be important to inspire young people to not just be an astronaut someday but to study math and science and set their goals high. Set them above what you think you can do. I was hoping that was the message I would get across. That day I thought I just want to give this speech and get out of here and get back to the Johnson Space Center and start training for a mission that I was really, really excited about.

At the time at least, I didn't consider myself a smooth operator with the media. I did feel like my talents were flying, even running procedures and analyzing failures, and trying to figure out how to get a system working again, or, if you have a lot of failures what order do you work them in. This logic, a big puzzle, that's what I like doing. I didn't really care for the media stuff that much.

I got back into training. Steve Hawley, Cady Coleman, Michel Tognini, and Jeff Ashby, the five of us, we really nurtured a great team.[70] Fabulous training team led by Lisa Reed.[71] Fabulous flight control team, all of Mission Control. The payload officer was David Brady.[72] The flight director was Bryan Austin.[73] We still keep in touch. We really developed a great team.

The Chandra X-ray Observatory was AXAF, the Advanced X-ray Astrophysics Facility.[74] It was named after Subrahmanyan Chandrasekhar, an Indian astrophysicist who had been up at University of Chicago.[75] We got to learn a lot about him, and we met the teams that built the telescope. It was just a dream mission.

It wasn't the rendezvous International Space Station build, which was the focus at the time, but it was such an important mission to the science community and to NASA that we put all of our heart and soul into it.

There were a lot of delays. NASA had had the problem with the Hubble Space Telescope, which we didn't discover till it was deployed.[76] We were not going to allow that to happen to Chandra, so there were a lot of little delays at the end to fix things, do more testing.

We did a lot of work with the Chandra Control Center, which is up near Boston. We traveled up there. We traveled all over to the factories. Chandra was built by TRW, so we made several trips to southern California. It was a really great experience.

That mission having the first woman commander was a side note, a footnote. I don't know if it really made that much difference to me, because I wanted our mission to be safe and successful. This is important to say. In our day-to-day operations, we were a team. Who cares what you look like, what race you are, what age you are, whether you're a man or woman? It was all about safely deploying the observatory and successfully deploying it.

As we got close to the launch, there was a lot of, "Time for you to go talk to the media." We did the standard crew press conference. We did the standard interviews with commander and pilot. I think we had a bigger turnout because people were curious as to what is the woman commander like. The media folks that normally cover space knew me, people like Todd Halvorson.[77] But there was a whole new group of media people that came in that just didn't know that much about space, and they wanted to know if I was Danica Patrick, or am I like James Bond, or what is she like.[78] I think they learned that I'm just a normal person.

I really learned, when it came to talking with the media, that I couldn't shy away from them, because I was a spokesperson for the space program, and that is a responsibility that astronauts have. Shying away from that or walking away from the media was walking away from my responsibility. I worked at speaking with the media. How did I do that? I learned their names. I learned who they worked for. I was never intimidated by them. I saw them as people. The American people, people around the world wanted to learn about the space program. Yes, they asked silly questions. I never belittled them for asking silly questions. I would tell myself that they are projecting themselves into what the American people want to know about a woman commander, the Space Shuttle, the Chandra Observatory, what it's like to be in space. I respected all the questions they asked, and I thought that was very important.

I didn't run away from them or shy away from them, which is what I did earlier in my career. When I was in the Air Force, I shied away from the media because I thought it really irritated the guys I worked with. It irritated them for several reasons. They thought it was a distraction to them. They thought the women got special attention. Some of them were jealous of the women for getting their names in the paper. They all had different reasons.

The reason we need to be out there is not for personal reasons, but we need to be out there promoting the mission, why it is so important for a country like the United States, a very strong leader in the world, to continue being the leader in space exploration, particularly human exploration. It's very important that we continue to do that. As a Shuttle commander that was a sounding board for me to get that message out.

The other thing that changed me while I was assigned to the STS-93 flight was a trip to Kennedy Space Center, where I was walking through one of the work areas with my crew, and a woman came up to me and said, "Thank you for doing what you're doing. Because of you, we are more respected by the guys down here." Pow! I never saw it that way. I still remember her saying that, and how important it was for me to hear that, because that was another experience I had that made me realize that I had a responsibility to get out there and talk about human spaceflight.

I'm not perfect at it. I've made my share of mistakes, but I was very grateful that we had the media coverage that was there. I would deflect it from myself. The "non-space" media asked silly questions. I tried to respectfully answer the question but then tag onto that silly question something about the Chandra Observatory or something about some of the secondary experiments that we were doing on the flight that were very interesting.

I'm trying to remember—what were some of the questions they asked? There was a reporter that asked Jeff Ashby an embarrassing question in the crew press conference. It was something about, "You have two women on the mission. How are you going to get them out of the bathroom so you can use the bathroom?" I don't remember exactly what it was. Jeff made a joke about it. We laughed at it.

When the press conference was over, this woman reporter came up to me, and said, "Why do you put up with that? That is very demeaning."

I was surprised. I said, "You've just got to laugh at some of this stuff once in a while because there's a lot of stress in this job. I just want to keep my energy focused where it needs to be." But out of respect for her, I had to try to understand her position. Maybe what had transpired was bad for the image of women. Maybe, possibly, I didn't really see it that way. I tried to learn from that.

I talked to Jeff about it. We laughed about it. We both agreed that we shouldn't make a joke like that in a crew press conference. I've always laughed at Jeff's jokes. I had to realize that maybe there'd be people that were offended by that.

When it came to the media for STS-114, it was totally different.[79] STS-93 was a lot of what's it like to be a woman in space, what's it like to be a woman commander, all the "woman-commander" questions, which were very low-intensity silly questions. For STS-114, there was none of that. It was all about safety, because we were the first flight after the *Columbia* accident. It was all very serious, no kidding: "Are we going to fly the Shuttle again ever?" and "How are we going to fix these problems?"

I don't remember when I was named Return to Flight commander. I

Following the safe return of STS-114, Commander Eileen Collins waves to the crowd at Ellington Field.

was assigned, I think, sometime in 2001—the crew was supposed to fly in 2003. Four Shuttle crewmembers, three Space Station crewmembers going up, and three Space Station crewmembers coming down. The accident happened in February of 2003, and NASA Headquarters told us to continue training. But I couldn't, because my Space Station crew was sent to Russia for a Soyuz launch, and there were four Shuttle crewmembers left, and two of them were CACOs, casualty assistance calls officers.[80]

My crew, the original four Shuttle astronauts, stayed on. There really wasn't a reason to break us up, but the mission changed from a crew exchange to a Shuttle test flight. If there was any talk about changing the four original crewmembers—I'm sure there was, there should have been—I was not involved in that. I would think that they should have discussed it. Now we've got this Return to Flight mission, are the four that we're going to fly the right people?

At that point, we were assigned Andy Thomas, Wendy Lawrence, and Charlie Camarda.[81] I actually worked with Kent Rominger in getting those crewmembers assigned to the flight. It was his decision. He had pretty much selected the available crewmembers who had the training to do the robotics and the spacewalks and the other tasks that needed to be done. We really picked our crewmembers based on the skills they had and their availability.

On STS-114, we had the rendezvous pitch maneuver (RPM).[82] I could talk about the maneuver for quite some time, but there are two things I'll focus on. This is a tribute to the change in the culture at NASA after the accident. Part of the blame of the accident went to the NASA culture.

An engineer in the Rendezvous section came up with the maneuver. It was not my idea. I was in a meeting in Building 1 when I first heard about it. I listened, and I was very interested in it. There was somebody in the meeting that said, "You can't do that because there's a flight rule that says the crew has to maintain visual contact with the Station inside of X number of miles." This pitch-around was supposed to happen at six hundred feet. You're way inside of this flight rule limit.

I remember thinking, "That is such a good idea. Why didn't anybody think of that before? In hindsight, it makes so much sense. Why didn't I think of it?" I was a little bit brutal on myself. I asked myself, "Why don't you think out of the box like this guy does? People need to start thinking this way. It's important that we are more creative in the way we think."

We started talking about the idea. Although I was skeptical at first, I liked it, and it was developed with crew input. It wasn't just my crew, but there were people in the Astronaut Office that were assigned to rendezvous, and they did a lot of the development work. I thought it was very well done.

They all deserve awards, I think. We as a crew were involved as much as we could be.

We looked at the procedures. They brought procedures to me that were in draft form, so we helped make inputs. We flew in the dome simulator. We made a couple tweaks here and there. The engineers would come over. We had flight controllers there. We had the instructors there. It was a classic test pilot–type creativity and development, and the final execution was so successful that they did it on every single flight after that to the Space Station.

The second thing I want to say about it is that the actual maneuver itself, the actual pitch-around portion, is flown with the autopilot connected. We were pretty much just monitoring during the actual pitch-around. The difficult part about that maneuver was flying to the start point of the pitch-around. There are six degrees of freedom, which are your X-Y-Z location, which are your up, down, your right, left, and your fore, aft. Then you had to hit a very specific spot in space six hundred feet directly below the Space Station, not out of plane at all. You have to be exactly zero in the Y plane. You also had to have a very specific closure rate in the X direction, a specific closure rate in the fore-aft, and no movement in the Y plane. Not only in location, but no movement. So, six degrees of freedom. We had to hit all six of those, and it was difficult to do that. It was all hand-flown. Over the year or so that we practiced this, we had to figure it out, with the help of all the flight controllers, of course. We give them credit because they dedicated every day, five days a week, to figuring out how to do this thing.

It was very difficult for the pilots to hit those parameters, but we learned how to do it, and it went outstanding on the flight. I flew it, but Jim Kelly was running the checklist.[83] Charlie Camarda and Wendy Lawrence were running the laptop computer. Wendy had the handheld laser pointer. We had this laser in the payload bay that was feeding continuous data into the laptop computer, and we could look at that. It would tell us, "If you make this input in this direction then you're going to go this way." It has little predictors that tell you where you're going to be in a minute, two minutes, three minutes. All of that made the maneuver very fuel-efficient.

The point I'm trying to make here is that the difficult task for the pilot was flying the rendezvous to the six-hundred-foot point and meeting all six of those parameters. When you engaged the autopilot you could go hands-off, and it would fly the 360. During the 360, all we did was monitor. When you come out of the 360, you see the Space Station. You would then take manual control and you would fly a quarter fly-around up to the V-bar. The V-bar is the plane that parallels the surface of Earth. From the point that you came out of the rendezvous pitch-around maneuver, everything was hand-flown. But the flip itself was not hand-flown.

Is that important? Because people have asked me that question over and over again, I felt that I should describe it. In the future someone's probably going to be interested. I was extremely happy with the way all of it went. I wouldn't have changed any of it, including that first RPM maneuver. The Space Station astronauts photographed our heat shield during this six-minute maneuver. The photos showed two protruding gap fillers between the tiles. We actually performed an emergency spacewalk on flight day nine, where Steve Robinson removed these gap fillers in an effort to prevent any damage to the Orbiter on entry day.[84]

There were so many memorable events from that mission; it's hard to pick just one. I would have to go with after wheel stop. We landed at Edwards Air Force Base just before sunrise. It was a nighttime landing. After wheel stop, we ran a few checklists. The Kennedy Space Center technicians approached the Orbiter after they gave the all-clear. I looked out the window and I saw them down there, and they were emotional. They waved at us. You could see the emotion in their faces. They must have just felt great, getting the Orbiter flying again.

Obviously, we felt wonderful about it also, but it's a very intense time, the landing. You're just so focused on what you're doing, you don't really think about the big picture. But once we were stopped and had much of the checklist completed, when I saw them, it hit me. We did it. We're back flying again. So maybe that's the most memorable.

Was landing the Return to Flight mission the proudest moment of my life? My children being born were the proudest moments. But I'm a woman, so I have a little different point of view.

We lost foam on our flight. I wasn't too happy about that. I am not faulting anybody for that in particular, because I personally agreed to not fixing the PAL ramp. The PAL ramp is the protuberance air loads, which is a ramp that runs down the starboard side of the fuel tank.

Go back six months after the accident. On one of our trips to the

Lockheed Martin Michoud Assembly Facility near New Orleans, we're crawling around on the tank. I think it was external tank 120. It was horizontal. We were looking at where the foam fell off of *Columbia*. It turns out that some of the geometry is difficult, so the technicians hand-spray the foam. If it's a smooth area, it's automatically sprayed. The automatically sprayed areas typically didn't separate on launch. But the parts that were hand-sprayed, which is the difficult geometry, are more likely to fall off.

I remember looking at the PAL ramp, and I asked the technician, "What are we doing with this?" He explained. I think I asked, "Why don't we just take it off? Why do we need to fly with this PAL ramp?"

He explained that we have to fly with the PAL ramp because of the way the airflow comes around the tank. The PAL ramp protects these high-pressure fuel lines. If you don't have the PAL ramp to protect them, that high pressure could put the fuel lines into a flutter mode, and they could break apart, and you'd have a big boom.

Okay, I understand that. We talked about it a little longer. I talked about it with NASA management, and the decision was made not to remove the PAL ramp. I agreed to fly with the ramp. So, we launched with the PAL ramp, and foam fell off of it. It was seen on the video camera. The foam was about the same size as what had fallen off of *Columbia*. It went right underneath our right wing. It did not hit us.

While we were up in space we were informed of this. As soon as they knew, they told us about it. We did the Sunday morning shows, *Face the Nation* and *Meet the Press* and a few others. Every single one of them asked us about foam. I think the only reason they wanted to interview us was because the foam fell off. We candidly and honestly answered the questions without being disrespectful, but we pretty much said, "This needs to be fixed."

The Shuttle was grounded for another year. We flew in July 2005, and the Shuttle didn't fly again till July 2006.[85] It was determined that we could launch the Shuttle without the PAL ramp. It was removed, and it never flew again. As far as I know, we never had any large foam losses after that. I wasn't very proud of the fact that we lost foam on launch, so I can't honestly say it was the proudest moment of my life, because we had made some mistakes.

I think it's important to say this. One of the things I learned in the

Return to Flight period and in my training is even the most intelligent people make mistakes. We developed a four-step process for dealing with mistakes. The first thing you do is you admit it. You acknowledge that you made a mistake. You don't have to be blatant about it. You just have to face it. The second thing is you fix the mistake that you made. The third thing is you need to put a procedure or a policy or something in place to make sure it doesn't happen again to somebody else. The fourth thing is you move on, you don't dwell on it, and you don't let it get in the way of your future decisions or your ability to do the mission. You just move on. The reason I still remember it is that four-step process even applies to your daily life and your family and anything that you do.

I have learned in the post-accident period how to be a really, really good listener. Yes, passively, but also actively going out and seeking out ideas and what people think.

The other thing I learned is not to be an intimidating type of person where people are afraid to speak up around you because you're going to squash their idea. That type of thinking helps people be more creative, and they're more willing to come out with these ideas. All of that, although I knew it before, I truly learned in the post-accident period.

I think the way I approached command for STS-114 was by changing my leadership style from STS-93, where I thought the commander makes the decision and we all follow. In 114 it was more of a team effort; I didn't feel like I had to prove that a woman can be a commander. There is a lot of that that goes on, I think, whether you're the first black person doing something or the first woman doing something. You feel like you have to prove yourself. There was a little more hands-on directing type of leadership style, for me, in STS-93. In 114 it was more collaborative. Some people might see that as weakness, but it's not, because in the end if it's your decision, make the decision, and you've got to live with it. You've got to be confident, and you have to let all the people who work with you know that this is the right decision, I am confident, and this is what we're doing.

But that comes after you let everybody have their say, with respect, and not just squashing somebody's idea because, "Oh, it wasn't my idea," and just not listening to them. You'll see that with some bosses who are insecure. You will see that attitude where "I don't want to hear what anybody has to say. I don't have time for that." That's not good. You don't want that kind of leadership style in a risky business like the space program.

I'm getting off the subject here; this is a whole separate talk. I share it with groups that are interested in learning the lessons from the *Columbia* accident. It's important, I think, that we continue to pass on those thoughts.

There's one more thing I would like to say. Speaking for myself, it's important to say that the women mission specialists that came to NASA before us did a fabulous job. They made it easier for us to come in here and be pilots and be commanders, because they were so well respected. They are intelligent, dedicated, and really believed in the mission. I think it's important that I say that.

In the same light, I think that all the women who worked here at Johnson Space Center, back in the 1960s, made it easier for the future generations of women and the Shuttle astronauts when they came in. This is opposed to what I said when I first came here about going into the Air Force—there were no women at that base at all except for a couple of secretaries, and there weren't that many women secretaries. There were just a couple. There weren't that many women there at all.

Whereas at NASA, it was nice and gradual in bringing women into the center. No matter what those gals did, whether they were a secretary, a photo lab tech, or working in the clinic, whatever they did, they all contributed to women being gradually and fully accepted into the program here. Big thanks to them.

Pamela A. Melroy[86]

As a military pilot, I had a lot of issues with credibility. A lot of it has to do with my physical appearance, which I'm very proud of. I ask kids all the time, "Do I look like an astronaut?"[87] The little three- or four-year-olds say, "No!" I like to remind them it's not what's on the outside that matters; it's what's on the inside. On the inside, I'm an astronaut. I always knew I was going to be an astronaut, from when I was a little girl.

I accepted and bought into the fact that people looked at me and they were like, "Oh my God, she looks like my kid sister!" You're going to have some credibility issues with that. That was okay with me because from my perspective, if I tried to behave in a more masculine way, not only was I not going to be true to myself, but it would never ever get across the real point, which is you actually don't have to be a man to fly an airplane.

That was very much a part of who I was even when I showed up at NASA. I spent all those years as a pilot and instructor in the Air Force, building my credibility, starting all the way at the bottom. I truly expected the same experience would happen at NASA. It did not. That really surprised me.

The one thing that was clear to me from the beginning was that the smarter people in the office—and believe me, astronauts are nothing if not smart—figured out pretty darn quickly that because I was a pilot, someday I was going to be a commander and potentially their boss. It really changes the dynamic, in my opinion.

It's an interesting situation when you come in with that known quantity. I would say, especially because Eileen Collins came in ahead of us, that really helped a lot too, and then the fact that Sue Kilrain and I came in together.[88] So the office went from one woman pilot to three women pilots overnight.

We'd go out to go fly, and the maintenance guys would make some comment about a broken jet or coming in late the other night. You're looking at them going, "I have no idea what you're talking about." We began to realize they just mixed the three of us up all the time. It was like there was *a* woman pilot.

Sue used to joke about it. She would say, "I didn't have the brain that day. I gave it to one of the other two bodies." It was so funny, it really was. Of course, it's a little different now. There were more women in the office later, after 1994. A mission specialist was just as likely to be talking on the radio as a pilot, because you trade those duties off. At Edwards Air Force Base that was a huge deal. You didn't hear women's voices on the radio at all. So if you made a goofed-up radio call, everyone was teasing you about it in the officer's club that night.

It wasn't that bad here in that regard; it was just a little funny that people mixed us up. I never got a sense of that credibility gap from the moment I walked in the door at all. It was just a very, very different place from the military in that regard. I think that reflects the number of people in the office, the fact that Eileen went ahead of us, and that she bore the brunt of being the first one. I think people expected there to be a lot more women pilots after Sue and me. So people were like, this is just how it's going to be. I really never felt an issue with that. It was great, because that liberated me to evolve my leadership style.

One of my first experiences in the Air Force, from a leadership training perspective, was being shown the movie *Twelve O'Clock High*. This is military leadership: you come in, you hammer them, you're tough, and they respect you. When people start performing, then you can back off and be a little bit of the nice guy. I'm rolling on the floor laughing, going "Uh yeah, might work for him. Wouldn't work for me, at all." If you tried that kind of tough behavior as a woman, it just doesn't fly. It really doesn't. You end up losing your credibility and just becoming someone who is not rational.

It's so interesting. This behavior that was acceptable in men is irrational when it comes from a woman. It became a topic of interest for me, especially in the military, trying to understand why there were differences between the way men and women led; not difference inherent in me, in the person, but inherent in the society in the way that you had to be successful. It took me a long time to really understand that leadership has a cultural and a social dimension. It's how people expect you to behave in certain ways.

There's a great military book by John Keegan, a famous military historian, called *The Mask of Command*. He talks about the different leadership and behavioral styles of leaders, all the way from Alexander the Great to

George S. Patton, and talks about how individual styles would never have worked if you mixed the cultures. That person would have disappeared. You would never have heard of them because they would not have been this charismatic, amazing leader. It wouldn't fit the needs culturally. So that was a great experience for me to understand that, because I thought, "Oh, my gosh. One of these days I could be a charismatic leader even though I'm a woman." I'd come to the conclusion long before that because I wasn't six-foot-four and didn't have a deep voice and didn't look like everybody's vision of a great military leader, I could never be charismatic.

I can't honestly say that I ever have been charismatic, but I can also say there's no doubt that in certain situations that social dimension allows you to be more natural and to evolve your personal leadership style. So for me, what I was able to do was to become more of a Pam leader, not a woman leader, or to copy a male leadership style. I was able to be more individual and authentic. That's actually a really important part of leadership, to be who you are. I think it's really just trying to be true to yourself.

I will tell you that I have a lot of traditional male personality traits. I am extremely comfortable pursuing loud and dangerous things. I didn't have any trouble mixing and blending with men, because I was truly interested in a lot of the same things. I think it's more about just being true to myself, every piece of me, not just the piece that liked flying and is proud of being a warrior in the military. I take those things really seriously. If you're talking about flying and those kinds of things, it's very easy to blend with men.

If I said, "Hey guys, you're going sightseeing, but there's a really cute shoe shop right around the corner, I'll see you later. I'm heading off over there"—I would never hesitate or pretend that I wouldn't worry about whether somebody would say, "What's up with that?" The parts that are traditionally feminine, I'm not trying to hide them or camouflage them or pretend to be a different person. It's really about being as true to myself as I can be.

People ask me a lot about women's leadership styles. Trust me. There are dimensions to that. The social dimension is very important. If you can get beyond it, you can really evolve to a much higher level. For me NASA was a place that I could do that, which was really cool, because I didn't have to worry about that stuff.

I have really tried to pull stories out of things that people said to me or times that I felt uncomfortable, and I just haven't been able to do it. That was a terrific experience for me in the Astronaut Office. I know it's substantially different than the earlier women, and I want to acknowledge that. But it was great for me.

My first flight was tough because I wasn't the last person in my class to be assigned, but I was the last one to fly because we were waiting for the Russian Service Module to be launched.[89] It was so exciting to be assigned, but we were assigned for almost three years. The rest of the crew loved it. They're having a ball. They know that being assigned to train is the best place to be in the Astronaut Office. It's really the best it gets in a lot of ways. So they were all having a blast, but I just wanted to fly my first flight. It was hard watching all my classmates go fly, waiting for my turn. I was very anxious, but what an amazing crew.

There's a couple of observations that I would make about astronauts. One of them is that they're very, very competitive people. They would not be where they were if they weren't competitive. They're perfectionists. They like the hardest thing out there. Real challenging for a commander, because everybody wants to do the hardest thing on board and nobody's worried about who's running the computers and stowage. You have to have those little jobs that are less fun, and someone has to do them.

You've got all these really smart people. You're starting from scratch. You don't have to break someone's culture because it's a new crew. You're not trying to change things. You are trying to create a new culture, so what you're really limited by is the willingness of the people to merge into a team, whether they really value that or not. I believe that. The STS-92 crew was very diverse from a personality perspective, but everyone wanted to merge.[90] Everyone wanted it to be close and fun, to know each other's families.

Brian Duffy, who commanded STS-92, is a fantastic leader and taught me so much, but his leadership style is very different from mine.[91] Jeff Ashby, my second commander, was my classmate, and we really had a lot of things in common. The way we approach problems, the way we think, the way we solve problems, all made us a very harmonious team in a lot of ways. The way the information flowed from him to me was he would tell me about his thought process or relate experiences that he was considering when he was making a decision. It was about the information. He was transferring experience and information to me.

Brian, on the other hand, rarely did that. Brian instead is someone who would rather have everyone in the room think it was their own idea. So he never tells anybody what to do—or very, very rarely. Instead, he creates an environment where questions get asked. He knows how to make things happen without making it appear that he had a hand in it. It really is such an evolved form of leadership. A good leader is someone who everybody says, "That's a good leader; she's a good leader. Look at what she just did, that was a good leader." Brian is the kind of leader where everyone says, "Aren't we a great team? Didn't we do such a fantastic job on that?"

It's only if you know all the mechanics that are going into it behind it. He did amazing things by just making a phone call or just asking a question in the right place that directs everyone's attention to the core problem and gets everyone else and all the levers moving. He is a master at organizational dynamics. He would call one person and ask a question, and stuff that really needed to happen—that you could have spent hours or days trying to work up a chain—it would just happen naturally. Everybody thought it was their idea. It was amazing to watch. He's really an extraordinary leader in that regard. As you can tell, he's obviously somebody who has absolutely no ego whatsoever and is about getting things done. I think for him, the joy was really in the social relationship and the fact that we did work together so well, and we had so much fun doing it.

I was the baby on STS-92. Everybody else had flown in space. We waited so long to fly that I had worked myself into a complete frenzy because I was so afraid that I was going to do something wrong. I was going to let my crew down. I was going to let our nation's space program down. I just needed to go fly.

I was very happy to land on that flight. We diverted to Edwards Air Force Base in California, and we waved off a couple of days. We were very rested when we came in to land. Edwards was where I had been a pilot. It was the best flying there of my whole life. When we rolled out on the runway there, and I looked at this place that meant so much to me, and I realized I'd made it all the way through the flight without screwing anything up—too badly, anyway—you couldn't touch me for like a year. You couldn't make me mad. You couldn't upset me. You couldn't make me sad. I was as happy as I had ever been. It was a really great experience.

One of the things that the first flight teaches you is that you can do it. What you also learn, when you get into your second training flow, is all the things that could have gone wrong. When I got ready to fly STS-112 with Jeff Ashby, that was a different crew.[92] Jeff had flown twice in space as a pilot. I'd flown once. We also had Dave Wolf, but Dave's last flight had been to Mir.[93] He hadn't flown as a Shuttle crewmember for like ten years. We also had three rookies: Sandy Magnus, Piers Sellers, and Fyodor Yurchikhin.[94]

Sandy and Piers have never been close. They weren't close when they were classmates. Fyodor, really, I think he felt like he had a lot to prove as a first-time flier, so it wasn't the same experience at all. Jeff used to say, "Mom and Dad and mean Uncle Dave take the kids to space." That's what he used to joke, because Dave was like "suck it up" all the time. Jeff and I really had such a close relationship. I think that always happens with pilot and commander. Your whole training flow is together, but we really had such a compatible sense of how things ought to be done. I think that was such a great experience for me because he just transferred huge amounts of knowledge to me. I think that was very important.

It was a really stressful flight for me because I was the IV (intravehicular activity person). We suited the crew up, and then I talked the spacewalks.[95] That was actually a fantastic experience for me. I loved being right there with them all the time. I was as close as I could be to right out there, but it was so much work.

I didn't send any email. I felt like I never got any rest for about the first five or six days on the flight. It was much harder than STS-92 in that regard. I had so much more responsibility. I really felt it acutely because of the rookies. I vaulted from the protected baby—lots of people available to show me how to do things and to help keep me from failing—to feeling like failure was around the corner every minute and that it rested on me to make sure that it didn't happen in the things we were doing.

Sandy and I are really close. That was another great friendship. Dave and I actually got really close on that flight as well, so there were friend-ships, and it was fun. We had our crew jokes, and we had a great time. It was probably very much like an average crew, where there's a lot of great feelings and a great camaraderie, and that's great, and that's about it.

That was fine, actually. I didn't feel the loss because I felt so crushingly overwhelmed with all this work to do. That was really a grind. It was just tough. I can't honestly say I had a great time on orbit the way I had on STS-92. There were moments. There were flashes. One of the things that I did enjoy the most—other than being IV—was being in charge of photo/TV. I had so much fun with that. I really enjoyed it.

One of my favorite pictures from that flight was when I was setting up the crew for their crew photo, and Dave Wolf grabbed the camera and took a picture of me. I had a video camera going, and I had four cameras set up for different lenses and different shots, and I was holding a camera. He had just taken another one away from me. Over in the photo/TV area they had that picture for a long time. I called it my paparazzi shot because I'm just surrounded by cameras. In fact, most of the pictures that you see of me from that flight I had a camera with me. I was carrying it, it was around my neck, the earphones, the holster, it was somewhere on me. I just absolutely adore photography, and so for me that was a great part of the mission. But it was tough; it was really hard work.

During STS-112, an International Space Station assembly mission, Pilot Pamela Melroy surrounds herself with cameras.

After that experience, I felt very ready to command and eager to command. Rommel—Kent Rominger, the chief of the Astronaut Office—asked me if I would take over as the lead Cape Crusader to train the new Cape Crusaders. I finished up my postflight in January 2003 with the STS-112 crew. On February 1, I was in Florida on the runway ready to unstrap the STS-107 crew to get my refresher.[96] Of course, they did not come home so that was really hard. It changed a lot of things.

The guys in Florida were tasked to do the reconstruction. They called the chief of the Astronaut Office and said, "During *Challenger* we had a separate room for all the crew module stuff. We're going to do that again. We need an astronaut to come help us with that." Who else is going to do it but the lead Cape Crusader? That's the liaison to Kennedy Space Center. That's how I ended up dropping into that. It was completely by accident. It's not like somebody handpicked me. I didn't even have a safety or mishap investigation background. I would not have been the person picked to do that, but it just so happened to be my job assignment at the time.

When I was working on the investigation, we had people on the team from JSC Engineering, Safety, MOD (Mission Operations Directorate), some of the flight docs, Life Sciences. We consulted with Biodynamic Research Corporation in San Antonio, a world-leading bioengineering investigation company, and with a friend who'd worked on the *Columbia* accident investigation, who was both a medical doctor and a PhD engineer out at Stanford University, Dr. Greg Kovacs.[97] We had a lot of help from the materials science people in Florida.

I had an experience that not a lot of astronauts have, which is that I interfaced in a very deep way, and in fact supervised the activities of a lot of people from different parts of the center. It exposed me to a lot of aspects of project management, but also the different cultures, the different styles, the different capabilities, and the different approaches of all the organizations at JSC and at NASA.

That was a great experience for me because it really opened my eyes up to the world beyond the Astronaut Office. The world was not just about this small group of people flying in space. It is true that I saw them when I worked in Florida. People do work on technical teams with folks, but they come and go. It was nearly three years, I think, that we ended up working on the report—and the work that we were doing was very hard and took a huge amount of time. It really gave me a huge sense of confidence going into the mission as a commander.

I really felt like I had both the experience in space and on the ground, because being a commander is not just about landing the Space Shuttle. A lot of people think it is. But it really isn't. Flying is an important piece of it, and you have to train. The commander is responsible to work with the lead flight director. You participate in the IPT (integrated product team) that brings together everything.[98] There's stowing so the stowage issues are addressed there, plus any issues with the installation of payloads or other things.

It's a much broader community. I had relationships with those broader communities and gained a pretty big perspective on that, so I felt very confident going into that. Again, it's the most fantastic leadership position in the world in that you get to start your own culture. That's good and bad. If you screw it up, that's bad, but really you don't have to worry. Most leaders, most positions that you come into, you're not starting with a new team. You're taking over as the head of the team. It's very rare that you're in the position where you can start something. As a commander you get the opportunity to do that.

One of the things Brian taught me was that I needed to let the crew be themselves and that I shouldn't dictate the style to them. I needed to know who they were, without me telling them what to do or what my standards were. That didn't go as well initially. We, the crew of STS-120, went off to National Outdoor Leadership School.[99] I think certainly a couple of the crewmembers were uncomfortable because they really wanted me to be in charge and tell everybody what to do and to set my expectations. It's interesting training because you rotate the leadership on a daily basis. Two people co-lead every day. When people are making leadership decisions, they really expose their thinking to everyone. That was a great experience. I think it was hard for some of the folks on the crew because they just wanted to know who the boss was, but I wanted to know who they were without me and to let them be themselves.

What I found out about the crew was that they were comedians, not everyone, but that was a huge part of the culture of that crew. They had all the movie lines going. When you're looking at guys like Dan Tani and Doug "Wheels" Wheelock, those guys can just do stand-up comedy forever.[100] They're really funny, funny people. That was clearly a critical part of the crew dynamic. It was also a special crew in that like the 92 crew, they were all very committed, very deeply committed, to bonding each and every one to each other. Yet we didn't have so much time that we became out of touch with everyone else. In a lot of ways, I really look back at that as the perfect crew. I don't mind saying so myself. It really wasn't about me; it was about them. We had just enough time to bond together.

One of the places that I did put my foot down was in the way we inter-acted with the people that we worked with in Mission Control and the training teams. We would never be a high-maintenance or a high-overhead crew. We would be respectful of the roles that everybody else played and didn't carry any special requests or demands too far; we weren't difficult to work with, and we were cooperative. I think that paid off. I have had people tell me that we came across that way, so I think that was very successful.

One of the toughest parts about being a commander is knowing that when you're in space you can't possibly be there for all the critical moments. You can't be in every place at every time. Building this culture is incredibly important; it's like when you're raising children and you know that they're going to have to make their own decisions at some point. What is their point of reference? They go back to what they were taught about what's right and what's wrong. You have to work a lot up front at setting expectations for where the boundaries are. "We're not doing it that way. Do anything you want, but when you hit that wall I'm going to let you know." That's what happens in training. You set that culture, and you set the expectations.

"We're going to always follow this rule; we're never going to deviate from it. That's just not a rule that's open to any flexibility." Those kinds of things. Yet try to keep it as open as possible, especially for the humor. The only bad part about that for me was that they leaned on me to be the person who said, "Okay we've had enough fun, we need to get to work." We could kid around the entire time through a tough scenario in a simulator, and as long as we were getting everything done it didn't bother me. But there always comes a time when you have to stop kidding around. You've got to talk about something, for example. It's not like you have to stop entirely, or that anything bad has happened. We just need to take a few minutes to talk about this, or, we've got to go somewhere else now.

That's always tough, because you're like the line in the sand—"Oh, Mom!" You can enjoy it, but you also know that you're the safe boundary for everybody, because they know they'll stop when it needs to be done. Then they don't need to worry about it, they just have fun. You can tell them when it's time to go home. That was the only hard part about that.

We had five spacewalks planned. This was a monster mission—delivering Node 2 and moving P6, holy cow![101] It was great. We were all so excited when we got up there. I enjoyed the experience so much more than STS-112, because as the commander your job is really not to have too much on your schedule. It's to be with everyone and to make sure that you're keeping track of the timeline. You pitch a hand in where things are falling behind.

When the missions changed and the dates slipped and it was clear that Peggy Whitson and I were going to be on orbit together, we were like, "Hey, that's pretty cool."[102] We had the opportunity to talk quite a bit before she went in space, and we shared a lot about each other's crews, and talked about some of these culture issues.

We were together on STS-112 because she was on the Station. We didn't spend a lot of time together because Station crewmembers have their own stuff going on, but we'd spent enough time. I knew Peggy well enough. When you fly in space with somebody, even if you don't spend a lot of time with them, there are a lot of things you get to know about them.

For me, I was just really looking forward to going in space, but the funniest thing happened about a week before launch. Somebody asked me in a preflight press conference, "So how's this going to be—two women commanders."

I said, "I'm really looking forward to that first moment of shaking hands through the hatch with Peggy."

Well they picked up on it, and it was in the paper. George Zamka, my pilot, he was photo/TV.[103] It dawned on him he's going to have to get that picture. He came to me and he said, "We've got to get that picture; you've been talking about it. What happens if we don't get the picture?"

"We'll call the photo/TV guys up, and we'll ask them for help in setting up this shot." The photo/TV guys have a Hall of Fame and a Hall of Shame. The Hall of Shame is much bigger. We called them up and said, "Hey, we really need to get this picture. We'd like some advice. Can you show us some previous pictures like this from other missions so that we can get a sense for framing and positioning?"

They showed us a shot from STS-71.[104] It was the first Mir docking mission. Hoot Gibson was the commander.[105] I said, "What—you're joking me. We haven't done it right since then? The last time was on Mir?"

They said, "Well it's really dark in there."

"Yes, I know. But I filmed that video on STS-112."

They said, "Yes, but the video camera is much better with low light than the camera is." We spent a lot of time talking about this. We had a big light. We had to get it there. We didn't have a cord that was long enough. Everybody's freaking out as we're trying to get *the* shot.

Well, George got *the* shot. I think of him every time I look at it, but I

*Expedition 16 Commander Peggy Whitson greets STS-120 Commander Pamela Melroy after
Discovery docked with the ISS.*

didn't know how much it meant to him until after we landed. We were
doing crew PRs.[106] It's always a great moment because you find out what
everybody else thinks about the mission. Somebody asked him was there
something you were really proud of. He said, "I'm really proud of that
picture." He should be, because it was hard to get. It really is a wonderful
photo. So that was a little funny thing about us being together. We didn't
spend a lot of time thinking about it. Obviously other people did. That
was our one bow to that, to make sure we got that shot.

Peggy and Dan and I had trained to deploy the P6 solar array, which
we had moved over between EVAs 2 and 3.[107] At the same time there was
a separate drama going on. I think it was during EVA 2. They'd noticed
some vibrations on the Station that they were not really sure where they
were coming from. They got to the point where they started to worry about
the solar alpha rotary joint (SARJ), one of the two.[108] They thought there
might potentially be some vibration coming from it.

Dan Tani went out there and pulled back all the thermal blankets. Sure
enough, there was some kind of contamination over the little teeth that
grind around each other. It was not a pretty picture. He took a little tape

and dabbed it on the contamination, which was obviously a very precious thing to bring back to try to figure out what was going on.

In the end, that turned out to be a very big deal for the Station, resulting in about a year-and-a-half's worth of maintenance efforts and multiple EVAs to fix it. Huge deal. On the ground they're going, "Well, let's jettison EVA 5. Let's just plan for them to go out there and really pull all the blankets back, unbolt some things, and get a really good look at this thing." First big hiccup. This is all going on there on the ground—they're talking about us getting ready to look at the SARJ. Scott Parazynski and Wheels were on their way back in from the EVA.[109] Dan and Peggy and I—I wish we had video of us, I didn't think it was going to be very interesting—were unfurling the P6 solar array.[110] We had a lot of problems because the angle was the highest beta angle we'd ever flown to the Space Station. Typically, the really high beta angles mean that the sun is coming in at a very high angle. That can be a problem for the solar arrays collecting power.[111]

They just opened up the tolerance on that a little bit and let us fly at a higher beta angle.[112] The camera angles were tough with that high angle. I was getting a lot of glare from the sun. The sun was shining directly into our camera views. In fact, I had to call an abort at one time, because I was watching the big picture. I was in charge of the cameras and everything looking right. Peggy was on the PCS (portable computer system) laptop with the commands. She was looking very carefully for when the fully deployed command was there. Then she would send the stop command. Dan was counting bays. It's a backup way of knowing how far it's gone, because they didn't have a special marker. You had to literally count them as they came out to know where you were in the deployment.[113]

I called an abort: "It's too much glare. I can't see anything." So we let the sun angle change a little bit. Then we picked back up, and we got the first half of the solar array out. Then we went to the second half. We got it about two-thirds of the way out. There was a big glare again. I was just at the point of going, "I can't stand this!" and then it came out of the glare. That was when I saw the rip.

I knew right away that something was wrong. It looked really bad, so I called, "Abort! Abort, abort, abort, abort, abort!" Peggy's hitting the keyboard. She's not even looking. She's going as fast as she can. Then we're looking at it. There was a lot of glare. We can't really tell.

You couldn't zoom in too much for a clear view. So we zoomed—

rushed—from the Space Station to the Shuttle side, sat in the commander's seat where you could look out at it, got the binoculars, and oh my gosh, oh my gosh. At that moment, the whole crew was like we can't even imagine what we're going to do about this. There was no way to get out there. The whole thing is electrified. Spacewalkers are told not to even get within X number of feet of it because you can't touch it. It's 120 volts DC. There's no way to get out there. There's nothing to hold on to.

So we had to shift our focus then, because we're being uplinked the procedures to do the SARJ contingency EVA. We went into overdrive getting ready. We set up all the tools, worked all that stuff, and got a whole day's work done when they called us and said, "Maybe we're not going to do the SARJ contingency EVA. We really think something's got to be done about the solar array."

We talked to everybody. They were so good to us on the ground. We talked to the chief of the Astronaut Office. We talked to our Space Station flight director, Derek Hassmann, several times, and Rick LaBrode was our Shuttle side flight director.[114] That was, I think, a great moment when all those relationships on the ground really played out perfectly the way they needed to. It paid off that we'd worked so hard on them.

Usually the Station crew helps you do things like get ready for a spacewalk. They know where everything's stored, so it's really great to have somebody there who's helping you to suit the crew, but then they go on their merry way. They've got their own assignments. They've got lots and lots of work to do on the Space Station, as we well know. They've got their own experiments and things going on. It was a moment when we realized that was just not going to happen. We had to all come together as a single crew.

The ground just basically tossed both of our flight plans out. Everybody played a part. The Russian crewmembers were helping set up. They had to go find Kapton tape.[115] Used every piece of Kapton tape on the Shuttle and about half of what the Station had because we had to wrap every tool and everything that was metal on Scott's suit.

They called up the plan and said, "We're going to use the Shuttle boom and attach it to the Station arm."[116] I thought, "Wait a second. That's going to kill the cameras." Those sensors can't go unpowered more than an hour by the flight rule. "Oh my gosh, what are we going to do for the inspection?" We have to do an inspection of the Orbiter after we undock.

"We're throwing the whole book out on this one."

Then they're telling me, "We think he's going to be about forty-five minutes away from the airlock." I knew from being an IV that thirty minutes is the inflexible rule. You start to realize all these things are going out the door.

I certainly understood that undocking was going to cause the solar arrays to probably rip even further if we didn't do something, because the Shuttle thrusters as they fired would cause that to happen. There was also a snarl. We were very afraid that one of the vertical cables that is part of the structural integrity that keeps it stiff was snagged. The concern was this fully electrified cable—if it snapped, there was a problem. If we just let it hang free, to unwind without the tension of the solar array anchored at the top, they were pretty sure it was going to damage the mechanism permanently. But you also had this electrified cable that could just float away, and God knows what was going to happen to it then. There were so many things that were bad about this.

It was obviously the right decision. We didn't have any issues with it, but there was so much at stake. For me, my biggest concern was Scott's safety. I also knew from working in Mission Control how little you can actually see there. You think when you're on the Shuttle, "Oh, they can see all the same video we can, they hear." They can't. Your situational awareness is not the same. There's a time lag. There are communication dropouts. So, for me, I think the realization was that if anybody was going to call it off and say we've gone too far, we were going to hurt somebody, we weren't going to be successful, and we needed to stop what we were doing before something bad happened, I really felt it was going to be me, that that was really the ultimate, "Oh Mom, oh crap."

One of the best moments on the whole flight for me was when we had one big massive net meeting together with the EVA and robotics folks. We all crammed ourselves into the airlock looking at the screen. They were talking us through the whole scenario: what they were going to do with the arm, what the concerns were with the boom and the spacewalk. I had my crew notebook open. As they're talking through the scenarios, I had this list of things that needed to get done as we're going through this. Then the meeting ended, and the crew started to talk.

I floated back up into the upper corner of the airlock, and I watched. They were talking about everything on my list. Everything I could think of, and a few things I hadn't thought of. They're all firing at each other, "We're going to go do this. We need to go do that. I'll send an email down and make sure we get this file." It was all happening. It was that perfect moment when you realize all that stuff you invested in back on the ground to set the culture and the expectations was complete. They didn't need me.

That's what Brian taught me. That's the ultimate moment as a commander, when all the right things happen without you being there because it means you set the culture right. All the work that you put in up front is happening. It's all happening perfectly. I don't want to take credit for that because it was the crew. At the same time, the fact that the up-front investment had paid off made me feel fantastic, just fantastic. It was really one of the best moments. I just watched them do their thing and was incredibly proud and amazed and thrilled.

The only thing I really still was worried about was Scott's safety, so I had a long talk with the doctors. I'm thinking about my poor spacewalker out there at the end of an arm, which, if it gets electrocuted and shuts down, how are we bringing him back? I sat down with Scott privately, too, and I told him that I was nervous enough about the spacewalk that I wanted to be IV for the actual activities. I know that was hard for Paolo Nespoli.[117] He was our IV. It was hard for me to take that away from him. I just felt like if Scott's life was potentially in my hands that I needed to be right there with him. It needed to be my voice that he was hearing on the radio. I spent a lot of time up front setting up my little nest. I had video cameras running, a little row of tapes all ready to go, and two different lenses out there. One really tight focus, one for more of a wide shot. Binoculars. Then George was running the helmet views.

Watching Scott go further and further away, I said, "Dr. Parazynski," something like, "I hope you're ready for your surgery."

He said, "The patient is prepped. Let's go."

He got out there. We saw the rips and the snarls. The wire was very frayed. He was supposed to cut it. But because of that problem with the rewind, Wheels had to be at the base of the solar array. He had a clamp, a pair of pliers, that he used to hold on to the bottom of it. It had a high-tension reel at the bottom. He was holding it, and when Scott cut it he then would release it just a couple of inches at a time. It would very gently and slowly reel back in. That solved that part of it.

That's in fact what they did. That worked out successfully. That was a big relief when that was done. It was a perfect example of what I mean by everybody in the crew had to get involved. Peggy and George had turned Node 2 into a machine shop. They had constructed the cuff links that Scott would use to stitch it up with wire.

Scott got out there, and he went to shove the first cuff link in. We're all tense watching his hands shove through the hole. Wheels was sitting at the base of the solar array shivering because he was so cold.[118] He could have been so distracted. He sees this billow start going back up toward Scott.

He sees that it's poofing out. "Scott, look out!" Scott looks down, sees this big thing coming at him. He holds the "hockey stick" out as far as he can—that was his "stay away from me, don't touch me" stick. It was a piece of plastic shaped like an L that was wrapped in Kapton tape. He leans all the way back. Of course, I'm looking straight on, so I didn't see this until he called it. Then I'm looking through the helmet camera.

I thought, "Oh my gosh, he's going to do this nine more times." By that point, it had taken us some time to get him out there, to figure out what's going on, and to really understand the rip and all that other stuff.

The ground is starting to get worried about the duration of the EVA, because we still have to unfurl the solar array. They said, "Okay, it's time to get going."

Scott had to pour it on. With all this, he and Wheels had to figure out in real-time how to communicate because every time he put a cuff link in that happened. Wheels had to tell him when things were getting crazy and to settle down and lean back and look out. They made it through.

Then we pulled him back away from it and finally, one bay at a time, opened it up. I was so relieved when it went all the way out. Everybody was congratulating each other, "Oh my gosh, wow!" Scott is saying, "Great work on the arm!" Dan and Stephanie Wilson are going, "Great work by the spacewalkers.[119] It's awesome."

But Mom had to say, "Guys, we will celebrate when Scott is back inside. Let's keep going. We've got to get him back to safety," because he was still forty-five minutes away, and they still had a big robotic maneuver to do. Actually, a lot of people commented to me about that statement, but that was my job on the crew, to be the party pooper, so I was. I was so worried about Scott.

Somebody asked me recently, "So when did you feel good and know that it was over?"

I said, "When the airlock door opened, and I could throw my arms around both of them." Spacewalking is very dangerous, and it ain't over until it's over, and even the repress and depress of the airlock has its dangers. For me, declaring success on the solar array repair didn't happen until I could be with them inside.

After we'd undocked and we were doing inspection, I was sitting up on the flight deck of the Shuttle. I was floating, looking out the window over the Earth and listening to the sound of my crew doing their job. Inspection is pretty busy, and there's a lot of moving the arm and getting the cameras ready.[120] By the way, the cameras worked just fine. Yay! That was a great outcome, that they didn't, in fact, fail. Thinking to myself, "Is this okay if I don't do this again? Yes. I've got all these memories. They're pretty much burned in there now. I think it's okay if I don't do this again." I could not top that flight.

Trainers

Training is a critical component of space exploration. Mission preparation and simulations easily exceed the time spacefarers spend in space, for good reason. Crews and flight controller teams train hard, so when NASA is finally go for launch, the on-orbit tasks—rendezvous, docking, landing, or walking in space—are rote. As they say at NASA, "You fly as you train." Astronauts and flight controllers constantly work different scenarios on the ground in case something fails or goes wrong in space. Astronauts complete a series of training courses before a flight, selected depending on the mission objectives, that eventually culminate in integrated simulations with Mission Control. Instructors like those featured in this chapter monitor their progress and are vital to mission success.

Instructor Cindy Koester explains the regenerative environmental control and life support system to Expedition 23 Astronaut Tracy Caldwell Dyson.

Anne L. Accola had recently graduated from Colorado State University when she received an offer to work at the Manned Spacecraft Center (MSC) in June 1967. Assigned to the Mission Planning and Analysis Division, she worked on dispersion analyses for two missions, Apollo 9 and 11.[1] Dissatisfied with her assignment, which was dull and separated from the thrill and excitement of space missions, Accola decided to pursue a graduate degree in computer science and took a leave of absence from the center. After completing two semesters at the University of Wisconsin she returned to Houston in the summer of 1970. With the Apollo Program coming to an end and cuts to NASA's budget, morale was low, and then Personnel announced a reduction in force (RIF).[2] Having failed to accrue enough time in service, Accola resigned under notice of a RIF. She went back to the University of Wisconsin to complete her master's degree in computer science, but a crippling economic recession curtailed hiring for recent graduates. Just four months after resigning, she returned to the center and went on to become part of the Simulation and Training Branch.

Lisa M. Reed joined the Training Division at JSC in 1987, as both the center and the agency were recovering from the *Challenger* accident.[3] During the two-and-a-half-year hiatus from spaceflight missions, offices across the center reviewed procedures and made improvements to their processes. Reed, who had a background in computer-based training, helped to design and implement new training for both astronauts and flight controllers using the PC as a training tool. Eventually she became a technical instructor for the Space Shuttle. Reed continued to focus on Shuttle and also International Space Station training through 2002, when she accepted a position with a technology consulting firm in Colorado. During the Shuttle-Mir Program—a joint effort between the United States and Russia and the first phase of the International Space Station (ISS)—she became an instructor on the Orbiter docking system. Following the loss of *Columbia* and the STS-107 crew in 2003, she served as an investigator for the *Columbia* Accident Investigation Board.

Anne L. Accola[4]

When I finished my master's degree, I went home to my parents, who lived in Denver. I sent a Christmas card to my old section at MSC, so they knew I didn't have a job.[5] Unbelievably, NASA was hiring again in 1971. This was only four months after I had been RIFed. They offered me my old job, and I resisted for a while because I hadn't really liked the work. But I thought, "Well, it's better than sitting around," and there wasn't any other prospect. The economy was not improving.

But first I talked to Personnel and asked, "You aren't going to have another RIF, are you? Because I don't want to drive all that way down to Houston and then have to turn around and go back to Colorado."

He said, "Oh, no, no."

I went back into the Mission Planning and Analysis Division. I think I'd been there just a couple of weeks when they announced a RIF. It was guaranteed that I would have to go out because I still had not accrued enough time, but they were going to work on it.

Well, it turned out that the Flight Control Division had a training program for new hires, and for some reason those people would be protected and couldn't be bumped out in a RIF. Gene Kranz wanted to hire four people.[6] Personnel's solution was to transfer me there to keep me being RIFed a second time, especially since they'd promised me there wasn't going to be another RIF.

But Kranz didn't want any women, and it was quite a while before I got transferred. Ultimately, he ended up taking me, but it was because I would have had the *Federal Register* for the entire Houston area blocked, so he couldn't have hired anybody that he wanted until somebody hired me, and he had the only hiring authority at JSC. I think this went on for a month before he finally took me.[7]

That was the most unhappy, chaotic four years of my NASA career. When I went into the Flight Control Division, Gene Kranz was the head of it. He had a couple of male deputies, Mel Brooks and Jones "Joe" Roach.[8] I remember Joe got the job of having me come in earlier, even before I started there, and he just talked to me. "We haven't had a female here before." I remember him asking, "How do you want to be treated?"

I thought that was a strange question. It seemed like I'd want to be treated like everybody else, if they were treating them fairly and with respect. I was assigned to the Simulation and Training Branch, because obviously that way I couldn't be on a flight control team. Heaven forbid a woman would do that.

Gene, when I went into his office—he met with the four of us—gave us his tough and competent, discipline and morale talk; you know, here's your marching orders.[9] Everybody laughs about it, but it really is very impressive, and it inspires you to do good things. It makes you feel like you're part of a team.

Everybody knew I was coming by then. What I know about the difficulty of getting that transfer was all thirdhand. It comes from guys that I knew talking to other people and finding out what the status of things was. The head of the Training Branch, Dick Hoover, didn't know which section he wanted to put me in.[10] He wanted to think about it before he made a decision, so he parked me in a chair in the secretary's office outside his office. So I looked like a secretary without a typewriter, but I did get started on the training program. I think it was two weeks they had me sitting there without an assignment.

I did finally get assigned to a section. What I did was go through the actual training program, which consisted of a lot of workbooks. Also, they had videos, like lectures, people talking on various aspects. Then I had to take all the lessons they taught the flight controllers on the trainers. I also had a familiarization in the command module simulator and the lunar module (LM) simulator.[11]

I was actually fortunate that this was 1971—it was May when I was transferred—and the feminist movement was just getting started. Betty Friedan's book *The Feminine Mystique* had been out for a little while, and polyester pantsuits were in.[12] I was very lucky that I was able to switch my wardrobe over to pantsuits, which I then wore every day because I never was certain if I would have to climb up a ladder and crawl into something. I remember I went somewhere where I didn't have to do that, and one of the crusty old contractors who was running the place said, "Young lady, you'd better invest in some pants if you're going to stay around here." It was funny.

The reaction within the division was varied. Some guys just treated me like a curiosity. They would come by to take a look at me. Others were friendly, just as nice as could be. A lot of men had daughters, and they were thinking how they would want men to be around their daughters. Then there were a couple that went out of their way to be a problem. Some of it is just male behavior when they think there are no women around listening; you know, they forget. It took them a while to realize I was on the other side of the wall. Generally, it was pretty good.

The secretary whose office I was parked in for a while was a feisty woman. She told me that I'd better succeed; if I didn't, Kranz would have an excuse never to hire another woman. She implied—I don't know if it's true—that there had been others who had worked there for a while, and they just couldn't take it and ended up in the restroom in tears and left. I don't know whether there's anything to that or not, but I was not to cry and go to the restroom in tears.

There had been a woman that worked for TRW, the contractor, and on some of the Apollo flights she had worked on a team that got called in to support in the Control Center at times. Some have thought of her as the first female to work in the Control Center, but she wasn't actually assigned to a team; she didn't work shifts regularly. She became a very strident feminist and very outspoken in a public way. She quit and became a lawyer.[13] So I was the first civil servant woman in the division. I'm glad that I not just survived but succeeded in the Flight Control Division as the first woman.

I've seen things Kranz has said that suggested I was not on a flight control team, so I don't know if he defines a flight controller as somebody who had to sit in front of a console seeing telemetry from a vehicle. We were considered to be part of the team when we were working during the flight in the Control Center, but we didn't have any real-time telemetry to look at in the position I was doing. But I was the first woman to work there in some aspect, so that's my claim on history.

The first assignment I think I had, they were coming up on Apollo 15.[14] That was the first flight that had the lunar roving vehicle (LRV) for the crew to drive.[15] It had to be navigated, and it had heading and distance indicators. The crew was told to go at this heading for this distance and then change heading to go to that heading at that distance. They had a readout of range and bearing back to the lunar module so somebody could corroborate they were where they were supposed to be.[16] If the astronauts said what the range and bearing were, then they could pinpoint them and say, "Okay, they're in the right place."

The person that was in the Control Center doing the navigation needed to be trained in simulations. In that era, they were done two different ways. Our branch actually had math models of all the vehicle systems that could put out telemetry that looked like it came from the vehicle, and they would put in malfunctions. They would actually go through a crew timeline. When they were using those math models, they also had what they called astronaut simulators, and they had training people who pretended to be astronauts doing the procedures. For the lunar surface EVAs (extravehicular activities), we did some training with astro sims.[17] The astronauts also had simulators that they practiced in on their own, and they could connect the simulators to the Control Center so that the flight controllers were training with the actual astronauts in the simulator.

To train this navigation person for Apollo 15, they just thought they would write the information on a piece of paper, stick it under a camera, and project it onto one of the channels in the Control Center so that these guys would be talking about where they were going and then we'd figure out, "well, they should be here," and write down numbers on a piece of paper. Obviously, that was not a very satisfying way to operate.

Somebody up at NASA Headquarters decided they needed to have a better system for training. They assigned me to it, because the solution was to get a computer, write a program, and give them more representative information. Since I had just arrived with this master's in computer science, I got the assignment. The computer, unfortunately, didn't exist. Hewlett-Packard was making them but hadn't produced any yet. So, we had to wait, and I got—I've forgotten now—it was either number five or number seven off the assembly line.

I wrote the program, and then I had to wait for the computer to come. It could hold one hundred instructions, and my program took too many. I had to work on it and work on it until I got it down to exactly one hundred. You typed in the instructions on a keyboard. You saved them on a magnetic strip about nine inches long, about one-and-a-half inches wide. After you saved it, you'd just put the strip in to read it. The only output method was a printer that used paper tape about the width of cash register tape.

When we were going to have a simulation, I would load this computer up on a cart, put it in the elevator, take it down the office side of Building 30 over to the Control Center side and up. Of course, the command and service module (CSM) people were still there and the lunar module people were doing their part of the simulation. While we were training the astronauts out on the EVA, they all manned their regular consoles.

It was just three of us. It was me and my computer and these two guys who were pretending to be astronauts on the surface of the Moon. We were in a separate room, on another floor by ourselves, away from everybody else. One of them was a contractor in the Training Branch, Hiram Baxter. The other one was a PhD geologist from Brown University; his name was Jim Head.[18] He had the real geology knowledge, Hiram was doing all the Lunar Rover stuff, and I had the computer setup.

At the beginning of the EVA, they would give the crew the heading and distance numbers to start with. The crew would periodically report the range and bearing so that the Rover navigation guy could plot them. I had set the program up so it was based on one-minute intervals. I could just punch in five times right in a row, so then I was five minutes ahead and I didn't have to literally do it every minute. When they would call and ask for range and bearing, I would read it out and tell the astro sims, and they would read it to the flight controllers, so that's how we trained them for those flights.

We'd try to make everything seem really real, but everybody knew that it was Anne and her computer, because I had a computer and nobody else did. One time the thing just wasn't responding. I was hitting the keyboard, and nothing was happening. I kept trying and trying, and I couldn't figure it out, and so I thought, "Well, I'll reload it."

So I reloaded the program, tried it again. I tried to catch up; I couldn't. It just wasn't working. Meanwhile they were asking the guys next to me, the astro sims, for the information, and I said, "I don't know what's the matter." We were trying to figure out what to do. I unplugged it. I plugged it in again. I did just about everything.

An hour had gone by. They kept asking, and then the flight director, Pete Frank, came down to ask what the problem was.[19] I decided, "I'm just going to load everything up and take it back to the office. I'm wiped out." Then I noticed it was out of paper. It didn't have any way of telling me it was out of paper, and it had never happened before. We all had a big laugh about that at the party afterward. That was one of the funnier things that happened.

The people that worked on the CSM and the LM were on three teams and worked three shifts. The lunar surface group was split into two teams, two shifts, and there were four on each team. They were the planning team and what you would call the execution team, or the EVA team. One group was working in the Lunar Surface EVA room in the Control Center when the astronauts were going out and doing the EVA, and then the other team was there when they were getting settled back into the LM, sleeping, and then getting ready to go out again. The planning team actually worked about fifteen or sixteen hours. The execution team had a shorter shift, but they were on during the actual EVA.

So, because of shortages, my sidekick from the training team who was pretending to be an astronaut went and worked in that room on the LRV systems for Apollo 16.[20] Then for Apollo 17, when they had another hole to fill, I was the only one that had the background to fill it. So I did the navigation for the planning team, which amounted to, when they landed, trying to figure out where they were, so the geologists could decide what they wanted to do in terms of sending them out on a traverse.[21]

This was a really low-tech job. There was a photo-based large map of the landing area on an extremely large table at my end of the room, and I had some tape about a quarter-inch wide. The geologists would just rough out the direction where they wanted them to go and where the stops were. I would take that tape and turn it into straight-line segments and use a protractor and a ruler to figure out what heading that line segment on the map was headed and what distance it was according to scale. That was my job, taking the outlines from the geologists and doing these little segments and then figuring out how to refine where they landed. And it was pretty much just the four of us in there overnight.

The geologists, of course, were in the Control Center while the crew was doing the EVA, sending up stuff about certain kinds of rocks or getting all the information, but then they would go home and sleep after they'd given me enough information to lay out the traverse. I didn't get a lot of sleep because I wanted to know what was happening on the EVA, and the only way to do that was to stay awake when the other guy was actually doing the navigation.

There was one interesting thing, though, about the landing that never did seem right. Something was off. Either they weren't quite where everybody thought they were, or the whole thing had gotten skewed a couple of degrees, like you needed to rotate the map around a couple of degrees. We worked and worked to figure out how that could be and couldn't figure it out.

One morning Gene Kranz came through and asked how things were going, and I told him about that. He went out and had the capsule communicator call up to the crew and ask them to describe it again, which I was sort of pleased with, that he took me seriously and passed it on.[22] But Jack Schmitt got a little annoyed.[23] He said, "Well, we know where we are. We already told you all this stuff," but then he did it again, which didn't change the location. That was the end of that as far as I was concerned, but it never solved the mystery.

At the same time that we were doing the Apollo flights, we were getting ready for Skylab.[24] We had these trainers that were just mockups, so they looked like the panels in the actual spacecraft, but they weren't flight-type hardware. They didn't have much of anything behind them. But they were good enough to train the flight controllers so they could see the types of actions the crew would have to take; they would get a feel for where things were located, what they were asking them to do.

They had the trainers for Apollo, and we had to get them ready for Skylab. There were actually four people who ran those trainers and built the new ones. I don't know actually where they were built, but they were running into problems. With the workload of Apollo training, they weren't getting the Skylab ones done. We were told that we were going to have to do the wiring diagrams for these things to help the people downstairs because if they did them all they'd never get them done in time. I didn't know what a wiring diagram was. I hadn't seen one before, so I had a lot of tutoring. I kept looking at books on electricity trying to get a handle on it. I was assigned to do a familiarization of the orbital workshop, the multiple docking adapter, and the airlock.[25] Others would be doing more detailed lessons on other things.

We did the wiring diagrams, then they were going to actually do the wiring, but they got behind on that. I found myself down there with these big boards and these long patch cables, figuring out, "Okay, this is this switch, and I need it to do this." That was another novel experience for me. I got that done and then wrote the familiarization for the actual class and had to check it all out and record it.

We were also preparing for Skylab simulations. I was assigned experiments, and the set that I was assigned included basically all subjects but the astronomy. It included Earth resources, biomedical, and then something that was called corollary, which was just a name for some that didn't have a better name. So I was to be the person in the Simulation Control Area (SCA) training those three flight controllers on those sets of experiments.[26]

There were a couple of contractors who actually were assigned the more specific experiments. They would come up with the faults and put them in, and they would operate the models as if there actually were experiments there. I was just there to monitor the flight controllers, see how they did, watch if the malfunctions went in and looked okay. I don't remember how long that training went on, but it was almost all with the first Skylab crew, as I recall.

The plan for Skylab was to have four MCC (Mission Control Center) teams for the first manned mission and then expand to five for the second and third. The training people would be assigned positions all across the teams and positions, and we would do on-the-job training (OJT) for the first mission, since nobody was going to run simulations to train us how to do their console jobs. There were a lot of aspects of their jobs that we didn't see in training because we weren't there looking at it. We would be scattered throughout the teams, and the teams would be rejiggered to have five teams so the schedule wouldn't be quite so bad, although it was bad enough with five.

For that, I was assigned to the biomedical experiments data. We got those assignments far enough in advance that we could actually be included in meetings, and the investigators could give us some background and help us get ready to go on the console. Well, the Skylab launch had problems, and the first mission was delayed. Skylab 2 didn't go according to plan, and they didn't get the biomedical experiments done as soon as they should have because they were fiddling with that parasol outside.[27]

My title was Med Data. From the Flight Control Division, we had three people. One was Biomed, and he was in the Control Room, the main room you're used to seeing on TV. There were two of us in the Science Support Room—one looking at the experiment systems, and one responsible for the data. There was also a medical tech who would look at the EKGs, which we only got when they were over a tracking site.[28] And there was somebody else from the Science Directorate in there.

The plan had been for this big information system to be the backbone of everybody's data analysis, so all this telemetry data would get played back from the sites and get put in there. You were supposed to be able to save the data to a computer. We were now up to a computer terminal information system, sort of. It just wasn't robust enough, and it was completely overloaded, so nobody was getting any data. In the case of the biomedical experiments, after the first week they had basically not gotten any data, and this was supposed to be the first-look data. The experimenters would see enough results to continue the experiment and then they could go where it was all being archived and request runs to be made; that's where they would do their more detailed analysis, but it would be later.

Jack Knight was the Biomed that I was going to be working for on this team when we went on console the next mission.[29] He had surveyed the whole problem and figured out that we were asking for a lot more data than was really needed, so we weren't getting any. He talked to the investigators that understood exactly what they needed so we could narrow it down to smaller chunks of time, smaller pieces of data; we got smarter about asking for which data to be played back.

He pulled me off my on-the-job training, and we worked overtime for a week or so trying to get the new protocol straightened out and we caught up on the data extraction and reports so everybody else would understand how to do it. Then I got back to my OJT, but everybody was falling behind again, so I ended up going back and helping to pull the data. So I got gypped on my OJT, but I think I learned everything. That was sort of fun.

The thing I hated the most about it was the hours, because we'd usually be five days on and then two, sometimes three, days off, and then we'd come back. We were rotating through all the shifts, so one set of five days we'd work during the day—that was considered the execute shift, which was when the crew was up working. The next time we came on console, we'd work five days on—really, the swing shift—and that was a little bit of execution, mostly winding down and starting to plan. The next time we came on for five days, we'd work the graveyard shift, and that was catching up on what hadn't gotten done during the other two shifts so that they'd have a clean slate to start with. The crew's activity and procedures, if any new procedures were needed, were planned overnight.

At the time, I don't think I had aspirations to move into the flight controller position. I didn't think it was realistic. I didn't think they would let somebody from training, especially me, cross over into the MOCR (Mission Operations Control Room).[30] I didn't really think about it, because it was a lot of fun just doing the training. I was finally up close and seeing it all; it was interesting and busy.

Then it suddenly wasn't, because once Skylab was over, Shuttle was a ways down the pike. ASTP (Apollo-Soyuz Test Project) was the next program up, and Hiram Baxter and I wrote the requirements for the experiments that were going to be on that flight, for the training models.[31] That amounted to a few weeks' project. ASTP took so few people compared to the number that had been involved in Skylab that they actually picked who was going to work on ASTP, and that group went off and did their thing. The rest of us were left to get ready for Shuttle, which was pretty slow because the program just wasn't coming along. We were going to be doing the same thing, training flight controllers.

There were some changes at that time. There was a reorganization at a couple of levels. They rearranged the two directorates. They had a Mission Support Directorate, which was going to build the simulator and operate the simulator, so they had divisions for the simulator and mission support. They also had a division for flight software, to get that developed, and they had MPAD, the old Mission Planning and Analysis Division. They were all in the Support Directorate. The Operations Directorate then included the Flight Control Division, the Training Division, the astronauts, and the Aircraft Ops.

The change that took place in the training world was that the Training Division in the Ops Directorate was going to actually do all of the training. The group that had formerly worked with the astronauts in the simulators got split. Some of them came over to training, but most were left behind to develop the simulator and operate it. We had to look at it from the point of view of training the astronauts in the simulators that were not developed yet, as well as training the flight controllers for their jobs in the Control Center.

In this early period we were just trying to understand what the systems on the Shuttle were going to be like, start thinking about writing up models for trainers, and learning everything we could and drawing diagrams to teach ourselves. I was assigned the environmental control system and something else, I think, which was a very odd assignment, considering my background.[32] But it didn't last for too long, because we reorganized within our branch, possibly even within the division.

They were creating a new section. It was basically going to be covering the data processing system and navigation. My branch chief said I could stay where I was with this environmental control assignment and keep the section head that I had, or I could move into this new section that was being created. They were bringing Bob Holkan in to be the section head.[33] Everybody else was just assigned but I had my choice, so I thought, "Well, I'd like to try something different," and this was more down my alley than environmental control.

I went into that section, and we were all just sitting around reading and learning everything we could about the flight software and the data processing system, which wasn't a huge amount at the time. The flight software was just growing. The requirements were way too big, so the software wouldn't fit the memory. The computers weren't all that big to start with. They had three cathode ray tube (CRT) displays, which the crew could call up different displays on.

The program had to go through a big software scrub in order to be able to fit and meet schedule. The two crewmen who were involved in designing the displays and the flight software were Bob Crippen and Dick Truly.[34] As a result of this scrub, they were busy in meetings to redesign all this stuff, and they needed somebody to write it down. They asked Bob Holkan for someone to do it, and he asked if I'd do it. He said he didn't know what it entailed—it didn't sound like much; I may not like it. If I didn't want to do it, I could opt out, but they needed somebody so I agreed to do it.

It turned out that they were in these meetings deciding what got scrubbed, and they would design the display for that particular system for the crew to look at. They would come in early from the meeting the night before and rough it out, and then they would go off to that day's meeting. By now we had advanced from blackboards to whiteboards, and so they were doing it in their office on a big whiteboard.

My job was to sit there at a table and read it all and read what other material I could so that I could lay it out and write a description of everything that was on the display. This was so the software developers would know what they were supposed to be displaying and what controls the crew had over the system. So I'd do that, and when I finished, I'd go back to my office. The next morning I'd show up and do it again. That went on for a couple of weeks until we got that package done.

That was for the Approach and Landing Tests, which is where they took the *Enterprise* up on the back of the 747 and flew around with it and then landed.[35] Then during the later flights they separated, and the Orbiter glided into landing. The Approach and Landing Test displays and software got scrubbed first, and then they had me do them for the orbital flight tests, so I did the same thing.[36]

Somewhere along in there, they decided they needed a flight software handbook, and because I had done the work on the displays, I already knew a little more about the software than others, so I got that assignment. That started out with Dick Truly pulling out the T-38 pilot's handbook and showing me how that was laid out and what it was like.[37] He really wanted it to be user-friendly, not some dry, hard-to-find-what-you're-looking-for manual. There were a few people that were assigned to help me with some sections, but I had overall responsibility for the book and I did most of it. We made it user-friendly. I used the approach of the CRT displays package. For the keyboard, we had a foldout with a picture of the keyboard with arrows pointing to a description of what each key did.

It turned out to be a best seller for flight software. IBM was supposed to be developing a handbook for the flight software, so I was supposed to be doing just a pared-down version for the crew. Well, it was mostly for the crew, because I think it was called the *Shuttle Flight Operations Manual*. It was in two or three volumes, and I did the volume on the flight software. I don't know how many we printed and reprinted. Everybody got one, and then IBM plagiarized most of what I had done in their handbook. Everybody knew it. NASA thought it was sort of ridiculous. IBM added some material, but not a lot; at least not a lot that was really useful. If you wanted to find something, everybody went to my book instead of the official one.

In the meantime, some other people were working on trainers for the data processing system and lessons, and we were gearing up for the simulator. Requirements had been written for it. Because the flight software on Shuttle is run in four computers at the same time, continually being synched up and voting against each other, and then there's a fifth computer that's a backup if those four fail, there was really no way to try to simulate the flight software. It was far too complicated. There was no way to run that software on other computers and have it work right and look right to the crew and the flight controllers.

So, for that reason, we had to have the real flight computers in the simulator. We couldn't model them to train the flight controllers separately. The crew had training sessions that were standalone. I've jumped into the future now, but the flight controllers, all of their simulation training was with crews in the simulator since we didn't have that standalone MCC capability for training anymore. Also, nobody would pay for another set of computers and flight software just for the Control Center training. We had to get early versions of the flight software for the simulator.

As we were getting ready, there were two sets of instructors named for the Shuttle Mission Simulator and one for the Simulation Control Area over in the Mission Control Center.[38] It would be the first time—in my experience—that the crew instructors were from our group. They were at the simulator. The people in the Simulation Control Area were watching the flight controllers, observing their performance and how the simulator was working, and telling people in the other building when it was okay to put in malfunctions, because we didn't have this other whole batch of people pretending to be astronauts in the simulator.

I was named the lead instructor on one of those two Shuttle Mission Simulator teams. Jerry Mill was the other, and Denny Holt was named the simulation supervisor on the team that would work in the SCA in the Control Center, but not until later.[39] The plan was we would go over, check out the simulator, learn how to work it and to be instructors. Then we'd start working with the two assigned crews, John Young and Bob Crippen and Joe Engle and Dick Truly.[40] Some months after that we would start doing integrated simulations with the Control Center.

Well, the simulator was a basket case. It absolutely did not work. You'd turn on a switch, and nothing would happen. Or you'd turn on a switch and instead of turning on what it should have turned on, it turned on something else. Or you could open the circuit breaker, and the switch would still work. It was absolutely unpredictable, so we would write discrepancy reports.[41] We had these procedures where we were going to check that every switch worked right, that every system worked right, that our displays were reflecting the models of the electrical power system or the thermal control system, that they were reflecting what was happening.

It was really a mess. We struggled through that. We wrote all those problems up. They worked on them.

The crew insisted they had to come over and get in the simulator; they were losing training time. They came over even though we knew they shouldn't; we needed to get the simulator ready. After a while they realized they were not getting any positive value out of it and they weren't really helping us, so we got them out and worked on it. We actually just tried to document everything and then shut down for a while. I've forgotten how long, but we just walked away and told them, "Fix it."

When we came back, it was better, but we always carried a backlog of things that weren't right with it. Eventually the crew came back and we trained. I think I did that for about a year. The Shuttle itself kept slipping, so we were always a year from flight. Then, three months later, we were still a year or more from flight, so we tried integrated simulations after the simulator was good enough to run with the crew, mostly just to find out how bad the telemetry was. We made it look mostly normal for the crew, but it was not what they were seeing in the Control Center. We had terrible problems, so we did just enough to figure out what all the problems were and then shut integrated sims down for a while so they could fix them.

In the Approach and Landing Tests—I want to go back to that—the crew was in the simulator that was developed for it, and it worked relatively well. It was very simple because there were no systems, really. They hadn't developed those models, so it was mostly just getting off the 747 and flying.

I worked in the Simulation Control Area for the computer navigation systems for the Approach and Landing Test simulations, and we did a number of simulations. That was a very small group. I think there were only about five of us in the Simulation Control Area. Of course, there were just the four astronauts and a handful of instructors at the simulator. The flight control team was only about twelve people. I think everybody that worked operations and training there probably didn't total forty. Everybody knew everybody.

I mentioned that I became a lead instructor, which meant that there were other people on the team. I was responsible for a couple of things, I think, at the time, but there were four, maybe five other people on the team who were responsible for individual systems, training the crew, and then monitoring the simulator and putting in the malfunctions. I was the overall instructor, watching them, saying, "Oh, no, you can't put that malfunction in, because they haven't figured out this previous one, and it's going to complicate things." And then telling the people that were running the simulator to get it started or stopped and turn it around. I had the most interaction with the crew in the simulator. About a year into this, I think, we were ready to go into more training at the simulator and more integrated simulations. Other crews were coming along, and the simulator was stable enough that we could start getting them some time.

Congressman Ronnie Flippo talks with Anne Accola.

They expanded and created a second SCA team, and I was the sim sup (simulation supervisor) on that SCA team. They added more simulator teams of instructors; we ended up with five before we finished. I went over to start running simulations from the Control Center, which was interesting, because the SMS, the Shuttle Mission Simulator, still wasn't totally reliable. Sometimes it would just crash. Then you'd have to start over again, try to get it restored. If you were running launches, that wasn't bad, because you'd start generally just a couple of minutes before launch, and launch itself only lasted eight minutes. So you could get through launch sequences pretty well, because they were only ten minutes. But then you had to turn around and do it again.

Entries took longer depending on where we started. We could do some that were just really short landings, or we could pick a point when they were still in orbit, do the deorbit burn, and come down. That took longer. We had a variety. When we did the orbit sims for eight hours all day, you would get several hours into it and it would change as they accomplished some procedures. Then, if the simulator crashed, we would have to bring it back up and then advance it to where it needed to be, check everything, put in all the malfunctions, and then start again.

It was an ugly process if you couldn't take it down and restart it out of sight of the crew and flight controllers. That was before they had the tracking and data relay satellite, so we had all these losses of signal from the tracking sites, and sometimes we'd be out of sight for an hour. This was later, but I actually could see things not looking right in the telemetry from the simulator. I knew that it was going to crash. I would look then to see when we were going to have a really long period of communications blackout, then I would tell them to take it down, restart it, and we'd bring it back up so it looked like it did when the crew and flight control team last saw it. Because if it crashed in front of everybody, they all just had to sit around while we went through that mess.

This is an interesting thing about that simulator—the operators of it had no insight into the health of it, so it actually was a major annoyance. I had to do a lot of arguing and blustering to get them to take the simulator down because they'd say, "It's working." And after I actually called a couple of heads above mine to get the message sent down to do it, they finally figured out I was right. They were quite upset that although I was sitting in another building, I could tell more about the health of their machine than they could.

We ran long-duration simulations, too. The first one we did was thirty hours. I don't know how this came about, but we had five different shifts over that thirty hours. I think my team took eighteen hours in three different periods, and Denny Holt's team had twelve in the other two periods. He was the lead of the two of us because he had started a year earlier and had done a lot of work on it.

There were a lot of malfunctions when you go for thirty hours because you have to put things in to keep every area in the Control Center occupied with something, so that they're looking for something. They're learning how to analyze problems and develop workarounds or tell the crew what to do. If something happened, the crew was supposed to go to a malfunction procedure. Well, they might not have a procedure for something that could happen.

So, over a period of time, to keep everybody interested, it turned into a whopping number of failures. In the end, we only did one thirty-hour simulation, and the rest were all fifty-six hours. We originally just thought we'd do a couple of the fifty-six hours, but STS-1 kept getting delayed, and they really could get better training, in some aspects, from the longer ones just because they would have to carry things over. If we just did one day, they didn't really have to plan for the next day, because they weren't going to have to live with the consequences of not doing something or of not doing it right. With a long one, they had to do something that made sense because they were going to be back the next day, living with it and carrying it on through, so we kept doing them.

They were getting better and better, and the crew was getting better—we were training for so long that the crew could handle anything. We were throwing ridiculous amounts of stuff at the crew to keep them from complaining they were bored. When we ran the long-duration simulations with the Control Center, there was a limit to how much we could put in because it all had to filter through one flight director. We still, I think, put about a one hundred fifty malfunctions in those simulations. It was a lot to keep track of. By the end of the sim we really didn't want that simulator to crash because it was harder and harder to restore where it had been.

Simulations were stressful for a lot of reasons. When we did the fifty-six-hour simulations, everybody who would be involved came to life, so we had the managers in the Mission Evaluation Room, who weren't necessarily there ordinarily, and the engineering guys who were on standby for support.[42] The program managers would come over and stick their two cents in, so there was a lot of outside help, or interference. I think they were even covered by the Houston newspapers. It was a big deal. Things didn't necessarily go well, and there are a lot of stories I could tell about all those simulations.

By the time STS-1 launched, they were over-trained.[43] Like I say, we had put in everything we could think of. We were having to put in more things. We never knowingly or purposely did anything that either the crew or the flight controllers could not get out of. We didn't do something just to show them that if you do this, you'll die. Never ever did we do that. There was sort of a camaraderie and a trust, so they understood that we only had their best interests at heart. We could, if we got mad, really do something to make somebody look bad, but nobody ever wanted to do that. They knew we had their best interests at heart because we were committed to the program's success and wanted everybody to succeed and look good in the end.

In fact, a guy I went to high school with was an airline pilot, and he really hated going to the simulator for training sessions because he felt like he was being tortured by the instructors. He told me this when he found out what my job at NASA was. I don't think the flight controllers or the astronauts ever felt like they were being tortured. I think they felt like they were being assisted in their learning and developing their proficiency.

The astronauts especially—because their lives are on the line—tend not to take anybody at their word unless they know without a doubt that they are the expert and know everything. If you give them an answer to a question, they'll go check with somebody who's more of an expert than you are to find out if that is right, and then they'll check with a third person. You can't ever guess or make up an answer, because you'll be found out and they'll never trust you again. There was a real issue of trust there, which was imbued in everybody that worked in that area, and it was passed on from one group to the next. When I hired somebody new into the section, I had a standard spiel I gave them and that was part of it.

I ran more simulations for STS-2.[44] By then I was section head, so that was the end of my actual operations or training experience working on a console in the Control Center or the simulator at JSC. Then I was a line manager. That came about when they picked some new flight directors. I think it was after the first flight or in that time frame. There were going to be people moving around and upward, and at that time there were supposedly going to be more flights coming more frequently. They picked several new flight directors, and my section head, John Cox, was one of them, which left his section open.[45] I think that was actually not long after STS-1, and so I applied for the job. Others did, too, obviously.

Nobody was named, and it went on and on. It turned out that the branch chief had selected me, and the division chief had okayed it, but it was stuck up above, because somebody didn't want me to be section head. It was either George Abbey or Gene Kranz, and I figured out it was Gene.[46] So this went on for two months, where neither side would budge.

My former section head, who had moved up, really wanted me to have it, and Kranz really didn't, I guess. Finally he gave in and I got it, but my branch chief, Bob Holkan, said that he really had to go to the wall for that, and he couldn't do it again.

As a sign of the times—this is another thing I want to add, because probably nobody else will provide this—back in the early 1980s when I was a section head, a law was passed about sexual harassment that defined it and made it unlawful. Training was supposed to be provided as part of the law. NASA hopped on it very quickly, because it was just a few months after it was passed that JSC had its first seminar on sexual harassment. And in MOD, the Mission Ops Directorate, the administrative assistant, Cecil Dorsey, who was there for decades, called me up to go to the very first one.[47] And I thought, "Why me? Has somebody complained about me, some of the guys that worked for me?"

Dorsey said, "Well, everybody is going to have to go." I later found out that a lot of the guys were claiming scheduling conflicts and didn't want to go. So I went ahead and went, and, honestly, that seminar could have been the subject of sexual harassment charges because of the behavior of one or two of the guys that went. They made rude and crude remarks about women and why couldn't they take it and what was the matter with saying this and that. The fellow who conducted it was excellent. He was a consultant from California. I haven't heard or been taught anything since then that changed anything he said. He had a really good handle on it. His interpretation of it has proven out over time.

The interesting thing was that I got called a second time to go, and I said, "I've already been. No thanks." Then they called me a third time, and I had to say, "If you call me to take this again before everybody else at the center has taken it, I'm going to consider filing my own charges," and I never got called again. I thought it was really interesting that for the very first session on sexual harassment in that time frame, they picked the *only* female supervisor in the directorate to go to it.

So I became a section head, and I really enjoyed that. It was an interesting time, because we were getting into more of a routine, where we weren't just doing unique, one-of-a-kind things for the first time for the first crew or the second. It was starting to look like more of the same. We had a lot of flight crews named in the pipeline. We were having to get them through the curriculum. We needed to standardize the training so that we made sure everybody saw the same basic set of things.

It was a time, really, to get better organized and also learn how to train our instructors, because we were having some turnover. As we increased the number of instructor teams at the simulator, we needed more new instructors to do the simpler training and have somebody in the pipeline. So we developed the instructor manual for all the things we trained in our section, and that included not just the data processing system and navigation, but also the star identification training. We had the Spacelab computer system training, the remote manipulator arm, and rendezvous.[48]

We had to put the Spacelab stuff together from scratch because that was new, to get it ready to train that crew for the first time. Same thing with the remote manipulator system. That was an unusual approach to training, because we actually did a lot of the training on higher fidelity, more engineering-type things, where they'd get a better response than we could from our simulator.

We had a lot of juggling to do because we had a lot of places to train. We had two bases at the simulator, the fixed base and the motion base.[49] Then we did some training at the engineering simulator. We even sent some of them up to Toronto for a while for the training on the arm, the remote manipulator system.[50] We had a lot of these little single-system trainer things that we did, and we also put together workbooks.[51] But the scheduling was a mess for all of that.

While I was section head, which started toward the end of 1981 and went on until the end of 1984, there were changes. They would take something out of my section and give me something else to do.

They gave me the scheduling for all of the training and crew activities. It had been done by a group—they had one guy assigned to each crew, and then they had one assigned to various facilities. And they were just frantic, running around.

They had one master schedule for the crews that they wanted to get out on Friday afternoon so everybody could see what their schedule for next week was, but if one scheduler was having trouble getting something nailed down for his crew, it held up the whole schedule. They gave the scheduling to me because they said it was broken. They said they didn't know if it could be fixed or not, but if it could be fixed, I could fix it. They told me to fix it, which I did, but it caused some ruffled feathers at first before it settled down.

I rearranged it so that each crew got their own schedule because the only people who really cared about the crew schedule were that crew and the simulator team that was assigned to do all their training. The single-system training we did by facility. The rest, for everybody that wasn't assigned to a crew, we did it by day and then marked it out by facility. So, if somebody only cared about one thing, they could just look at that part. We separated it out into enough different schedules that everybody got what they needed in the format they needed it.

It worked fine, except that the schedulers thought it was pretty revolutionary to do it this differently. The first time we did it, they ran up to John Young and said, "She's turned the schedule upside down, and you're not going to like it."[52] He called my division chief, who wandered down and said, "So what have you done with the schedules, Anne? I know it's going to be a good story, but I need to know what it is." I explained it all, and he said that sounded reasonable to him. We just had to make one modification, which was to give John Young a copy of all of the crew schedules, and it worked fine. The schedulers all of a sudden didn't have to be this close to a nervous breakdown all the time every week.

At some point along there they advertised for flight directors. The previous time when they added some, they'd just picked them. This time they did an open competition, and I applied. Everybody expected me to. A number of flight controllers did, too. I filled out the paperwork and had an interview with Gene Kranz. He was determined he was going to be totally fair about this and objective, so he had some checklist he was going by. He rated us according to all these things.

When they made the announcement, I didn't get it, which surprised a lot of the flight controllers. In fact, a couple of flight directors that got it thought I would have gotten it before they did. The reason he gave me was communication. He didn't think I communicated well enough. That was it. Everybody thought, well, that was just his reason for not picking me.

So I knew that avenue of advancement through Operations was closed off. If I didn't get it then, I never would because I was no longer working on the console. I was at my peak, in those terms. Sometime later my branch chief moved up to division chief and a branch opened up. I applied for that because I was expected to. I knew then I had no chance, because by then things had split up, and Kranz was now at the top of the directorate. I knew the promotion wasn't going to happen, but I applied because I was expected to, and I didn't get it. So I knew that advancement was foreclosed.

Looking back, I'm most proud that I took on so many things that I had never done before and didn't necessarily know how to do, but I figured out what to do, how to do it, and got it done. I think that's what I have the most sense of accomplishment about—I actually did accomplish things, things that hadn't been done before, or where people weren't sure what to do, or that were broken and needed to be fixed. They gave them to me, and I figured it out and got it done, and generally with less support than somebody else would have had.

Lisa M. Reed[53]

About the time STS-71 rolled around, when the Shuttle finally had the docking system, I moved into more of a specialty instructor role.[54] I trained the crews on the Orbiter docking system for a lot of the Shuttle-Mir flights.[55] After that, I got promoted to training lead. As a training lead, I am the maestro over a group of instructors. We work predominantly in the simulator to train a Space Shuttle crew for a mission. When we get assigned to a specific mission, we follow that crew from the time they're assigned until the time they fly. We do all of their training in the simulator.

For a Shuttle flight, a training team is comprised of five people: one training lead and four core discipline instructors. You train the crew on the different systems, whatever they may be, and how they work. You will work in the single-system trainers. You'll give them briefings. You'll do what we call tabletops, which is sitting down and discussing things with them.

At the same time, you will begin training them in the simulator. It's a stair-step approach. You will take them from the lowest level up to the highest level by the time you get to the simulator, where you're not really doing a whole lot of teaching, you're just having them practice what you've taught them over the years and, hopefully, they've learned it well.

The individual instructors are responsible for different discipline areas on the Shuttle: control, propulsion, data processing systems, environmental control and life support systems. They come in with the crewmembers prior to going to the simulator and train those individual systems in the single-system trainer. They teach them how they work, normally, and then we also have classes that throw in malfunctions so they can practice malfunctions in the single-system trainers. When we take them across the street to the simulator, we throw it all together, all the different disciplines, and that's where we practice flying for the mission.

Classes in the single system trainer are typically two hours long. Classes in the simulator are four hours, five hours, eight hours, ten hours, all the way up to thirty-six hours. Just prior to a flight, they'll do what we call a long simulation, which will take thirty-six hours, usually focusing on whatever the critical objective is on the flight. For example, on a docking flight, we take the window just prior to docking and after docking, and we will work all the way through that timeline and practice doing everything that they're going to do. It's the big dress rehearsal. All of those things tend to happen in the last twelve weeks of training.

As far as training for a docking flight, that encompasses more. It's all those same things. You still have the same instructors, but you have more specialty instructors that come in because trying to bring together orbiting bodies in space is a major task, so you'll need what we call a rendezvous instructor. He will teach the crew how to actually fly the vehicle—in this case, to Mir. Then we added a docking instructor. Once we had close enough proximity, that's where I would take over. I would teach them, once we had actual contact with the Mir, about the docking mechanism and capture, how to bring the two together to create an airtight seal so that we could eventually open the hatches.

As training instructors, we learned about the docking module any way we could. There were not a whole lot of drawings available to us. The ones that we first saw were in Russian, and none of us spoke Russian or read Russian, so it was interesting. We would attend tests of the mechanism. We would read anything we could get our hands on because it was a Russian system, and that was kind of unique in itself. They have a totally different design philosophy from the way we design mechanical systems here in America. Not that it's bad or anything, it's just different. We're used to working one way with electrical buses, and they may have it designed a different way. So we had to go through a lot of learning.

There were a lot of tabletops among the instructors where we would sit down and just try to figure out, "If you turn this switch on, what happens? A power source goes here. We think it's going to do this, but we're not really sure." We actually didn't have a good idea of how everything would work until after 71 flew and we actually got to see real flight data. It was an interesting time.

Hoot Gibson, commander of STS-71, and Greg Harbaugh, who actually worked the docking mechanism on 71, would laugh sometimes that we were all learning this together, because normally the instructors come in and they already know the systems and procedures.[56] But in this case, whenever we found out some new information, we'd pass it along to everybody else.

Real teamwork and communication were involved, and I think every-body realized that. With the teamwork, we also realized that there's no holding of information. This was a dangerous feat we were trying to pull off, if you really think about it, so the crew had to be well trained. We also had to understand how the mechanism worked so that we could figure out how to break it in simulations. If we didn't know how it worked, we couldn't give them realistic scenarios and teach them how to potentially get out of a hairy situation should it arise. That didn't actually happen in orbit.

That's part of my job, to come up with these situations. A lot of my friends chuckle at my job. I say, "I have a really great job. I don't have a whole lot of stress because I can take it out on the poor astronauts." Actually, the astronauts like to joke that we give them a really hard time, but I think they know we're their biggest cheerleaders. We try to throw malfunctions in, and as they go through the training flow it will get a little more difficult and a little more difficult. We're actually down there going, "Yes!" when they get it right, when they figure out what we've done to them and actually get themselves out of a sticky situation.

Lisa Reed at her instructor console in the Shuttle Mission Simulator during STS-47 training. Image courtesy of Lisa Reed.

We also train the cosmonauts. I think the biggest thing that was an obstacle in training the cosmonauts that first came here was language, obviously. I worked on STS-60, which was the precursor to the actual Mir Phase One flights.[57] They sent over Sergei Krikalev and Vladimir Titov, who have subsequently flown on the Shuttle.[58] This was November 1992, and I had just completed training my first crew, which was STS-47, and had been assigned my next task, STS-60.[59] STS-60, as it turns out, came about just after the Clinton Administration had made the agreement that we would work with the Russians.[60] The first step was going to be, "We'll fly a Russian cosmonaut on the Space Shuttle." They sent over two, because one would fly and one would be a backup. That was not decided right up front, so we had to train them both.

I remember we were all laughing because they came downstairs and said, "These two cosmonauts will be joining the training soon. The STS-60 crew will be training in February. We need to get these guys very smart on the Shuttle from November to February."

We were all going, "Oh, my goodness. Do they speak the language?"

They said, "Well, one of them speaks some, and one of them doesn't speak quite as well, but he speaks some English."

When they got over here, I remember they brought us all in to introduce us to these two cosmonauts. Everybody was feeling a little weird because we had never done anything like this. We've always worked with Americans who speak English. Everybody is at a certain level of knowledge when they come down for training. I remember everybody being very quiet. They didn't know what to say to us, and we didn't know what to say to them. We couldn't speak the language. There were these translators. All the normal icebreakers and jokes you try to use, they had to go through translation. That was our first experience of many in working with translators in training some of the cosmonauts. The neatest thing about that was they evolved, and we all become really great friends. Sergei eventually was chosen to fly, and flew on STS-60.

In order to get those guys trained by February so they could join the rest of the crew in the training flow for the simulator, we, the instructors, spent eight hours a day with them. For example, Monday would be electrical power day. I would start in the morning with a briefing for two hours. I'd have a translator sitting with me, and Sergei and Volodya—Volodya was the nickname for Vladimir—sitting across from me, and we would talk about the electrical power system. It was difficult, because, as you can see, I can just roll on talking here in English, but you have to pause with the translators. If it took an hour to teach it to an American, it took three hours with the Russians because you had to translate everything.

So, by the time we would get out of the briefing, we would go right into a single-system trainer class or a simulator session demonstrating what we had just learned. I actually felt sorry for those guys because they were getting a lot. It must have been like drinking from a fire hose for them. Every day it was a different system, but they rose to the task and did very well. It was fun.

We ran into a lot of problems with slang. We Americans like our slang. We don't realize we use it as much as we do. I remember one day I was teaching Sergei about the hydraulic system. I was explaining to him that when we come back from orbit, they'll start one Auxiliary Power Unit prior to the de-orbit burn.[61] It's basically just to make everybody feel good that you have one running, because you don't really need it. The term that all of the instructors here use is "warm fuzzy." I remember he stopped, and he looked at me, and he asked, "What is this 'warm fuzzy'?" And I couldn't explain it to him, because I tried to separate the words. "Warm," to him, was like "near hot." Then "fuzzy," we got off into talking about teddy bears and fur, and it just degraded from there. I don't think he ever quite understood what a "warm fuzzy" was. Maybe he does now. He's been over here for a while.

Our training team lead at the time was a guy name Henry Lampazzi, who is just a jovial person.[62] He's very nice, and he greets everybody walking in the hall. One day, I walked in to teach Sergei and Volodya a class. They'd had their English class prior to that; their English instructor had been up there. You would go in the room for class, and things would be written on the white board. You could tell what they were covering in the lessons that day. I saw the word "howdy" phonetically spelled out on the white board. I chuckled because I knew that Henry's greeting to everybody walking down the hall is "Howdy." He would see them, and he would go, "Howdy." They had never learned that. They probably learned the formal English greetings, "Hi. How are you?" They did not know what this "Howdy" was. They would ask their instructor to explain all these things. I'm sure we confused them thoroughly. We eventually got to a point where we tried to watch our slang.

They were good troopers for having to learn as much as they did in such a short amount of time. The STS-60 crew, it was just one of those really magical moments where the crew got along tremendously well. The training team got along tremendously well. We all got along really well together, so we had a whole lot of fun while we were training. We would laugh, and we would cut up.

I remember Charlie Bolden, commander for STS-60, would come in.[63] We only have four seats in the simulator. For a while, until they decided who was the prime cosmonaut for STS-60, he would have to flip a coin to see who would get that fourth seat. He walked in one day, and he goes, "Headskys or tailskys?" Sergei just laughed at him.

I wouldn't say the trust that built up between us came naturally. It was really interesting, because all of us had grown up in the era of the Cold War. Sergei was an engineer; Volodya was a military pilot. On the 60 crew we had, obviously, a military commander and military pilot. That was their background. Then we had mission specialists.[64] It was weird to think you're actually working with these people that you had been brought up all your life, and especially those with a military background, to think they were the enemy. I think the biggest lesson I learned out of that is that we're not all terribly different, no matter where we're born. I think they learned that, too. But there was a little apprehension up front because it's ingrained in you. Over time, I think it was more when we would go out and socialize, where you actually get to talk and meet everybody's family, that the trust built up. Then you'd come back and that would be reinforced in training. The more we got to know each other, it just built over time.

Training evolved between STS-60 and STS-91.[65] During STS-71 and all the subsequent missions, they had a prime and a backup Mir crew. At some point during the training they would bring those cosmonauts over and we would have to train them. It was a little bit different than 60. They weren't actual active participants on the Shuttle like Sergei and Volodya were. They would be visitors. Once the hatch was open and the two vehicles were connected, they could come over on our side.

We have to plan for contingencies in case something should happen, like a cabin leak. Realizing that they can't read all of our warning signs, we were concerned that they wouldn't, maybe, understand the master alarms and the caution and warning tones that they might hear. So we had to bring them in. I actually taught many of them and made up a special class on how you recognize the sound for fire, for cabin depressurization, and any other alerts that might say you need to get back to Mir in case we have to do an expedited undocking to separate the two vehicles.

That was interesting, because I didn't spend as much time with those cosmonauts—actually day-to-day—but they would come in with the translators, usually three or four at a time. I think my fondest memory was when they came in here with cameras. They wanted to take pictures. They were talking to the translator, "Can they get a picture with you?" I'd be standing there with three cosmonauts, and we'd all have our arms around each other, taking pictures. While I was instructing, one of them would hop up and want to be taking pictures of his buddy getting training in the Shuttle simulator. I thought that was kind of cute.

The other thing that amazed me, once I would explain to them, for example, rapid cabin depressurization and I'd show them on our CRT displays where that might occur, once they knew what it looked like here, they could go pick it out. They would point, "Oh, yes"—once I'd put the malfunction in—"that's where it's dropping." You could see that they were just as smart as our astronauts.

It's really funny. I remember someone saying that when trying to talk to foreigners, people tend to shout at them. But "they're not deaf; they just don't speak the language." You had to be very careful because you would find yourself, if you couldn't get a point across, saying it louder. And it doesn't get across any differently. It's the language; it's not the level of the volume.

I realized at that time that these folks were the best of their best, just like our astronauts here are. Now that some of our astronauts have gone over there and trained, I imagine it's the same feeling for them. It's got to be frustrating to be a highly intelligent individual and outstanding in your field, and then to be constrained by language and not be able to get your points across. So we tried very hard to help alleviate that for them. We had some fun moments. They were all really good folks.

I've been up close and personal with the Shuttle-Mir crews from day one. I worked STS-71, 76, 79 for a little while, and then I left, went to another job, and then came back.[66] I did STS-84 and STS-89, and 86 was in between there, too.[67] I lose track, but I did all of those missions for docking. In some cases, I was also their systems instructor at the same time.

We are assigned to a crew anywhere from a year and a half to flight or nine months to flight, depending on what the objectives are. I guess for most of the Mir crews, I worked with them about a year. As an instructor, you say you can always train them more, but most of the people that we have chosen in our astronaut program are just phenomenal folks. We set out our objectives, what we want to teach, almost at the beginning, and we go after that. They learn it very well. It's really amazing. You can always go deeper. This is such a complex mechanical system and flight machine, the Shuttle. Then you add all the other complexities of the docking system and using the arm to put the docking module on top of the docking mechanism, and then docking it with Mir. It's pretty complex stuff.

When STS-71 actually docked—we have what we call a Sim Control Area over in the Mission Control Center that is where we hook up the simulator with the flight control team in the Mission Control Center and they practice like we're on orbit. We sit in this room and we monitor the teams. We put in malfunctions for the flight controllers to see, so we can develop that teamwork. That's the last step in the training flow. Before a crew flies, we do these integrated simulations.

Well, during missions, that room is still there, and we can go see actual mission data. I remember sitting there the day that Hoot maneuvered the docking system in and called down, "Contact and capture." I had a lump in my throat for the rest of the day, I really did. It was just like, "This is so historical." The last time we docked Apollo-Soyuz, it was a different time, it was a different place, and we weren't necessarily friendly. This time it just had an overwhelming impact on me. I was amazed at my emotional response at the time. To this day I can still remember it. The rest of the day, the training team, we just walked around in a daze, because we had trained these people to do this and they had gone up and done it perfectly. It was truly amazing. All of the hard work paid off. We did grumble a little bit as it was going on because some of the things were hard to do. "What are we going to do?" There weren't any procedures.

All of the subsequent dockings that I watched, every one of them amazed me. They're bringing these two huge vehicles together at a slow rate, basically bumping them into each other, and having people work together from two different countries in space. That's really amazing.

It's really hard to describe my feelings when I see the hatch come open and the people that were once in this room traveling back and forth from one vehicle to the other, because it's almost surreal. You've trained for so long. We actually train for the hatch openings and the welcome ceremonies. They're not official, of course, in the simulator, but we go through the flight plans. We almost know step by step what they're going to be doing. After months and months of training, it's ingrained in our memory. To actually watch it, you almost feel like, "Wait a minute, we're not in the simulator. I'm not there. We're not talking to them. It's these other people talking to them. What's going on?"

You realize these folks are actually hundreds of miles up in outer space, and we trained them to do this. It's an emotional response. It's hard to describe. When the hatch opened on 71, I teared up. I really did. I've gotten control over myself for the other flights, but that one was just— there was never a question in my mind, if there was any crew that could dock, I knew that crew could do it. But we hadn't done it before, so it was a first. There was always that little bit of apprehension. Then when they do it, and they do it so well, you're just really proud of them. I guess it's a little bit, probably, like being a parent and seeing your kids do really well, because here we try to give them the tools and teach them things that will help them do their tasks. We also try to give them the tools so that if something goes wrong, they can save themselves and the vehicle and get home safely. When you actually wave bye to them after that last simulator session, it's like kids going away to college for parents. You hope you gave them what they needed to be able to do the job right.

The training teams are chosen for the missions way in advance. A lot of times we'll know what mission we're working before a crew is assigned. Usually it will be a couple of months before the crew actually hits the training flow where you're working with them a lot. You get a couple months' break in a lot of cases, a break from flight-specific training, that is. The instructors still train nonassigned astronauts during this down time to give them proficiency until they are assigned.

But for me, with the Mir docking flights, they were coming one after the other. For a while it got interesting trying to juggle the schedule, because I'd have the 84 crew training in the simulator that day, and I would have the 86 crew training in another trainer that actually does docking systems. It's the equivalent of our single-system trainer. Then I'd have another crewmember that was doing a joint integrated simulation. They were working with Moscow and the Mission Control Center. So you'd have three things you were trying to juggle. Working my schedule during that time was really interesting.

When we did do integrated simulations with Russia, we would pick a time where people across the ocean were on part-night and part-day, and people on this side of the ocean were on part-night and part-day. Sometimes we were coming in at one in the morning, and we would do a simulation until one p.m. the next day. It got humorous after a while, because at a certain point, especially if you've been working a normal day, and then all of a sudden, the next day you've got to come in at one o'clock, you'll get really sleepy about three or four in the morning. We'd all be sitting in there trying to wake up. We had to be, well, as we'd say, "Not stupid." You get dumb when you get sleepy.

There was a lot of coordination that had to be done with the instructors in Moscow, along with the instructors on this side, to come up with the cases that we would run. What we would try to pick were things that got both crews involved. Typically it was, "We'll throw in a cabin leak. They'll dock, and there will be a leak in the seal between the two vehicles." Or we would have one of the vehicles have a failed jet or something, or they had a leak and they had to expedite the process and then dock. You had to be awake for that to make it all happen in the simulator. It was not uncommon to see some of these instructors trotting up and down the hall outside the instructor station, just trying to wake up before we got into the really heavy part of the simulations.

I did not travel to Russia. It's really funny, I don't know how I've managed to avoid going. I think they needed me here most of the time for training. But, over the years, as we began to establish a presence with US astronauts on board Mir, people were beginning to have to go over there. A lot of our simulation supervisors would have to go over and work with their equivalents in Russia to come up with these cases.

There's a whole method to doing simulation. You don't just walk in and go, "Oh, well, I think I'll put in a cabin leak." You have to plan it all out. In other words, we don't want the simulation to go in a direction that we haven't thought about. "Okay, we can put in this malfunction and here are the four possible ways it can go." So we have to plan because there's a lot of money involved, a lot of time, and a lot of people involved, especially when you're connecting two control centers, Moscow and Houston.

To me, that was actually a pretty amazing technological feat that we were able to do that, and even more on their part. The Russians don't have, I guess, a Control Center like we do, where you can split it up. We can simulate while we're running a mission in another area. Their guys are running it out of the same—this is my understanding—Control Center, where they were actually being controllers for the Mir, so it was pretty amazing. That was where the simulation supervisors were, so a lot of them would go over and spend weeks coming up with cases. "Does this work for you?" They would call us sometimes.

Occasionally the Russians would call here, and we would do teleconferences. I would be in on those. They would come up with some great idea of what they wanted to do on the docking system, and we would see if that was all right. We would discuss it and come up with a really good case. That was kind of fun.

Another aspect of training involves protocol, what to do when you cross into the Mir. Astronauts had been told what they needed to do, so over in the simulations we would practice it. The reason being that it's going to happen one time and everybody's going to be watching, so it's like a dress rehearsal, if you will. I know they talked about it a lot, every crew that I ever trained for a Shuttle-Mir mission, because there were things that they would do. Obviously, the commanders would be the first to the hatch. You have to understand this little airlock that they're in is not very big. The airlock is probably—well, I want to say maybe six feet in diameter. It's very tight quarters. I guess the only blessing for them is that in zero-G they get to use all of the volume.[68]

They have to position somebody down there with a camera. Then they've got another person with a cable for a handheld microphone. It's one hundred feet long, so if you see any of the pictures of them going through the Mir and dragging this microphone so they can talk to Houston, that's it.

These are the things you don't see on TV. Prior to the commander actually opening the hatch, the docking targets are placed on the hatch on the Mir—somebody would have to go up and put those there. They need them for undocking, so they've got to make sure that they put them somewhere where they can find them later. There are two lights used during the rendezvous and docking to illuminate docking targets on the Mir. They would come across the hatch on our side, so they had to pull a pin and take those out. Otherwise, you couldn't get through. It was a lot of coordination just to get to the point where the commander could go, "Hi." A lot of things that had to be done. Obviously, you had somebody with a still camera, somebody with a video camera, somebody with a cable, somebody getting the targets and the lights.

I did have an occasion to go down to the Kennedy Space Center to actually train the crews. This was one of the things about the docking system. We didn't have any mock-ups here at the time that we could train with, so a lot of times the crew's first look at the hardware with an instructor would be at what we call the crew equipment interface test, which is about a month before flight.

I guess the funniest one, I went down with the 84 crew one of the times. They wanted some training on all of the duct work. The airlock contained air ducting that would provide air circulation from the Shuttle to the SPACEHAB, and it was docked to the Mir as well.[69] In addition to having to put these two vehicles together, you're bringing together two atmospheres, two different pressures, and different volumes of oxygen and nitrogen and carbon dioxide, based on how well their carbon dioxide scrubber may be working versus ours. There was a big duct that we had to connect and drag across—this was another coordination thing, for the greeting ceremony—to the Mir air revitalization system, so we had to train them on that. We had nowhere at JSC really to do that, so I used a lot of pictures, and we went through procedures and did lots of those tabletops.

I went down with the 84 crew. I'll never forget sitting in this little six-foot-diameter airlock on the floor. We were all sitting with our knees bent up to our chests, every one of us. There were about five of us in there. I was pointing things out, like this is what they're talking about when they say, "Disconnect this clamp and connect it to this duct," and actually pointing the stuff out to them. They all had their procedures out and they were making notes, because they do that.

At one point in time somebody didn't have a pencil. I remember Charlie Precourt, commander of STS-84, saying, "Has anybody got a pencil?"[70] Eileen Collins, the pilot, did this contorted motion, because we were all just kind of crammed in there.[71] She pulls out this pencil, "Here, I've got one." We were just amazed that she was even able to move in there. Trying to fit five people on the floor of that little airlock and trying to do some training was rather interesting. In zero-G it would have been much easier. I kept trying to get them to take me with them, but they wouldn't do it.

Training them on Earth for a microgravity environment is always a challenge. If you think about it, we practice launches. But I can't imagine going from our simulator, which I hear is fairly realistic, to actually knowing that you're sitting on top of all of this rocket fuel and it just lit. That's got to make you leave your brain back on the launch pad for just a moment. I think they have similar problems.

After every flight we have what we call a training debriefing. "What could we have done better? What did you notice that was different? Were there any,"—we call them "gotchas"—"any gotchas that you ran into? How can we change the procedures to make them better?"

All of the docking procedures and all of this duct work and what we call tunnel operations—that whole area back to the docking system we called the tunnel—we revised that for every flight. It got better every time. By the last three flights we pretty much had it down. We're actually going to be using this docking mechanism for our International Space Station flights. I think we've learned a tremendous amount of lessons from the Shuttle-Mir Program, and we will use them in the ISS flights that are going to be coming up in the future.

Each Shuttle-Mir flight was unique. I guess 71, it was a first. It was a "let's see if we can do this" kind of thing, and it was also us trying to learn this new system that we had very little information on. I think with 76 there was an extravehicular activity involved. Now we're getting a little more fancy. They actually went outside and were practicing some ISS assembly-type tasks while docked to the Mir.

Then on 84 it was just the crew. I think what I remember most about that crew was they were sort of representative of the changes happening in the Astronaut Office, the multicultural backgrounds that we were coming in, because we had Charlie Precourt as the commander; we had Eileen Collins as the pilot; we had Ed Lu, who is of Chinese descent; we had Carlos Noriega, who is of Hispanic descent; we had Jean-Francois Clervoy, who is from France. We also had a Russian, Elena Kondakova.[72] So it was not uncommon to be sitting around the table and hear Russian, French, Spanish, and English going on between all these astronauts. I remember that was a really, really neat group, as far as that aspect was involved.

Charlie is an amazing person. He's one of these very talented linguists. He just picks languages up. He would be in the cockpit, and he might explain something. He would speak to Elena in Russian. He might say something to Jean-Francois—who we called "Billy Bob," that was a nickname he got on a previous flight—in French. And he might be talking with Carlos in Spanish. It was really amazing. It was fun.

I hadn't really thought about doing anything else besides crew training, other than going into space. If they could just take me with them. As an instructor, I have been very blessed, or very lucky, I'm not really sure which. I've just had some really good assignments. I got to work on a lot of missions that were considered firsts. Even now I'm on a first, which is a first female commander.[73] So I consider myself very, very lucky. I don't have too many regrets about things I didn't get to do. I would still like to go to Russia, but at the time I was just too busy training them here, so I couldn't go there. They couldn't necessarily cut me loose from the schedule.

Senior Management

More senior managers at JSC are now female, a striking change from leadership in the 1960s, when NASA landed men on the Moon and began working on the Space Shuttle. Women in this chapter broke through the glass ceiling and achieved the highest positions in their chosen fields. They led some of the most important offices, without which the day-to-day work of the Johnson Space Center would grind to a halt. The teams, under their management, came up with innovative ways of managing the workforce, procuring items, and keeping morale high in spite of program cancellations.

NASA Johnson Space Center's senior managers learn more about Project Morpheus.

Debra L. Johnson started working at the Johnson Space Center as a cooperative education student from Texas Southern University (TSU). Hired full time in 1975, she completed her master's in business administration at TSU three years later and worked in the Office of Procurement in a variety of positions. In 2004, she began serving as director of that office, an organization that handles six hundred contracts averaging about six and a half billion dollars; regularly oversees numerous source evaluation boards; and manages the center's bartering, reimbursable agreements, and Space Act Agreements. No other NASA center comes close to matching the contracting activity JSC sees.

Hired by NASA as a human resources (HR) representative in 1987, Natalie V. Saiz has more than thirty years of experience at NASA. She was named director of HR in February 2004, shortly after President George W. Bush presented his Vision for Space Exploration, which included a plan to retire the Space Shuttle in 2010 and take the nation back to the Moon and on to Mars. Saiz faced the monumental task of transitioning the center's workforce from an operational program to a new research and development program known as Constellation. When that program was cancelled, the infrastructure and strategies she helped to put in place to transition Shuttle employees were applied again. She retained that leadership role until 2015, when she moved to the center director's staff as a special assistant to the center director for organizational change.

In 1978, the first year NASA selected women as astronauts, Peggy A. Whitson, a recent high school graduate, decided she wanted to be a spacefarer. A few years later, American space scientist Dr. James Van Allen at the University of Iowa tried to discourage her, saying that the Shuttle Program and the need for astronauts would not last long; robots would soon take over the task of human spaceflight. She disregarded his comments and applied to graduate school at Rice University, about twenty miles north of the Johnson Space Center, in 1981. Upon earning her doctorate in biochemistry and completing a postdoctoral position at Rice, Whitson accepted a National Research Council fellowship at JSC. Soon after, she became the supervisor of the biochemistry laboratories, where she conducted research on renal stone risk in astronauts. Whitson later became the project scientist for the Shuttle-Mir Program. In 1996, after many years of hard work at JSC, she achieved her dream of becoming an astronaut. She went on to fly three times on board the International

Space Station (ISS), serving as its first female commander in 2007. Two years later, she broke new ground when she became the first female and scientist to head the Astronaut Office, a position previously filled only by male test pilots. After returning to Earth from the ISS in 2017, she held the record for spending more time living off the planet than any other American.

When the Space Shuttle *Columbia* lifted off the pad for the first time at the Kennedy Space Center in the spring of 1981, Ellen Ochoa was a first-year graduate student at Stanford University. On the path to becoming a research engineer, she became interested in the space program when some of her friends applied to the astronaut corps. She decided she could combine her interests in engineering and research and do both in space. Before then Ochoa had not considered becoming a space flyer, but she recognized she needed to complete her PhD to be a competitive applicant. In 1985, she completed her doctorate in electrical engineering at Stanford and went on to work at the prestigious Sandia National Laboratories and at the NASA Ames Research Center. Five years later, she became an astronaut candidate and the first female Hispanic to be selected. She went on to spend nearly one thousand hours in space during four spaceflight missions. In 2013, Ochoa became the director of the Johnson Space Center, a position she held until her retirement in 2018.

Debra L. Johnson[1]

When I came to NASA, I had the option to rotate through the Budget Office, the Human Resources Office, or the Procurement Office. What always intrigued me about the Procurement Office is that you get to spend other people's money. So I knew that was the career for me. I went to Budget for one tour, then when they offered me a job, I selected to do contracts.

Official portrait of Debra Johnson.

As a student I came in, and I was in Building 419—that's when Procurement was located in the back of the center.[2] I stayed out there for a couple of years buying basics: pens, Xerox paper, those types of items. It was a great place to learn the process. I'm so thankful for that now, that I came in and started at the back of the center. I learned about the whole process, from the time that you buy items, to the time that you receive them, to the time that they send an invoice to make sure it's what you actually bought. I actually started at the bottom and really learned my way all the way through.

After we moved to the 225 Building, we thought that was really moving up in the world to be right at the back where the Printing Office is, because we were closer to the front and to Building 1.[3] I actually got to know a lot of people in the printing plant because we were right next door to them. It really paid off. That's when I started realizing the benefit of relationships at all levels. Most people target the top, but I learned the value of having relationships throughout all levels of the organization. I was able to utilize those relationships when I came over to Building 1 and started doing large procurement projects. I needed somebody to help me get the printing out, and it was great to walk up there and know everybody. They would take care of my products and move them up a little bit in the line.

I took over as director of Procurement eight years ago, on April Fools' Day 2004. I characterize us as not just a procurement organization but deal makers. We don't just write federal government contracts—we do write those—but we write commercial contracts, and we have for many years. In fact, we handle all of the international contracting that's done at the agency at the Johnson Space Center. We have a contract with the European Space Agency and with JAXA in Japan.[4] We negotiate all those contracts here at the Johnson Space Center.

We've done the first Space Act Agreements with SpaceX and Orbital Sciences for the Commercial Orbital Transportation Services.[5] Those were done here at the Johnson Space Center. Even with the Commercial Crew Program that's now located at NASA Kennedy Space Center, the deputy program manager is located here.[6] Kennedy sends people here for us to do the procurements and to look over our shoulders to learn because we've become experts at that. It is a matter of looking ahead to see where NASA is going and then planting yourself there.

We have the ability to do government contracting if that's what we want. We can do commercial contracting. We can do Space Act Agreements. We can do barters; we can do hybrids. When I first got here, we were strictly federal government contracting where you went by the regulations, and that's all we did. But we have learned that it takes all of that now to exist in this economy so that's what we do. We can do it all.

Things began to change when I came to Building 1. While I was working out at the other buildings it was just the Shuttle Program, followed by the International Space Station Program. There were different needs. A lot of the things that we wanted to do were no longer available from people that did government contracting. Sometimes they would say, "I don't want to sign a government contract." So we had to find other ways of doing things, like bartering for an airplane. Let me tell you that story.

The people that support the Flight Crew Operations Directorate (FCOD) out at Ellington Field with all the T-38s were actually doing the work.[7] I was the Procurement deputy at that time. There was a plane that NASA wanted that was owned by a foreign company. We couldn't go directly to them. We had to go to someone to be the intermediary to get the plane from that company, make it American-owned, and then we went to get it from them. We had to set up an agreement to make all of this work. I love doing those things. I love it when somebody walks in and they have a unique problem. If you just want to buy paper, anybody can do that. But if you come in and you have a unique problem that we have to figure out, I love it.

I love making deals. We had to learn to use skills other than just what was required by the regulations to be innovative and creative. If you were going to go buy something on your personal account, you would ask questions and try to cut these deals. We have the ability to do that now.

I love the fact that we can take government regulations and use those when we have to, but we can step away from those and start an agreement with a blank sheet of paper. We have some of those agreements also.

We have evolved in Procurement, and that, I think, is a result of the last couple of center directors. We're a player at the table at the beginning, and they make sure of that. Leadership starts at the top. Mike Coats values Procurement. Before him General Jefferson Howell valued Procurement.[8] They made sure that we were at the table at the beginning of the conversation. They don't plan it all and then call Procurement and say, "Go make it happen." We're there at the beginning, and that helps a lot.

The other day Steve Altemus called me up to talk about strategizing.[9] As the director of Procurement, a lot of times you don't get the opportunity to do that because that's what the staff does. So when they call up and say, "I want to talk about strategizing a little bit," that is the beautiful thing about contracting, going out and trying to make it work.

I think we have enough flexibility within our regulations now to do almost anything we want to. Congress says we can't contract with China. That is a "thou shalt not." We can't get around that. We send people out benchmarking all the time.[10] We benchmark DoD (Department of Defense), and we benchmark traditional customers like Shell. I'm on a group of chief procurement officers across the city of Houston, and that includes people from Shell, ExxonMobil, the hospitals, and the city of Houston. I listen to what they're doing, to how they're doing things, and I have found out there are very few things they do that we cannot do. We just haven't thought of them that way. So, I think our regulations are opening up and that allows us enough flexibility to do what we have to do to meet the new needs of the agency.

The agency is moving toward more commercial contracting. It's moving toward doing nontraditional things. As you know, we have contractors in our facilities now. When I first got here back in the 1970s that was unheard of, because these were government-only buildings. We didn't have a pool then, but when we built the NBL (Neutral Buoyancy Laboratory) that was for our astronauts only.[11] Now we have a lot of commercial companies utilizing our pool. It took our ability to write those agreements to make that happen. That means we have to think outside the box, because companies are not interested in doing a government contract. They're not interested in waiting a year to get one. They want it done right now, so it took us thinking differently.

But we also need to be able to do the traditional contracting when we need to do that. We can do it all; that's what makes us so flexible. There have been several times that I came up with an out-of-the-box suggestion. I love trying new things, especially if I've benchmarked them at other organizations. When I was in Information Technology, which was called ADPE (Automated Data Processing Equipment), one of our technical guys had found some equipment that was on the excess list, and we could still use it. Another government agency had it, but it was perfectly good equipment that we could use. The regulations required that you have to put it out for notice, let other people have an opportunity to purchase it. We wanted it. So we got with Legal and devised a new process to say, "We're ready to take it." There were some shipping costs that we were able to pay. Legal said, "No, you can't do that because regulations say you have to do this, this, and this."

Then the next question was, "Well, how do you get around those?"

Getting around regulations is probably not comfortable for Legal. We had to do an alternate approach for them to say, "Well, that doesn't work. I can tell you another way of doing it, but I can't get around the regulations." We had to change our terminology, but they assisted us. We got the equipment and got a good cost savings award. The technical organization, the Procurement Office, and Legal Office were all given awards for cost savings. We figured out a way to work that within the system, not get around it.

As deputy director of Procurement, I was able to move the organization toward some of these trends. I had a good director that allowed me to be a participant and manage an organization. That's when I had the leeway to sit back and look at what we were trending toward doing and start preparing training opportunities to think outside of government contracting. I think it really started then. I was deputy for seven years.

Before this concept became part of our efforts, we were not with our customers. Programs would go about planning and never think of us. We were an afterthought. We were brought in when they wrote the purchase request and sent it to Procurement. They would throw things over the fence to us, because we were over in another building. Now we sit with our customers. I think that makes a big difference. We have somebody sitting right in FCOD's suite with them. When they start talking about a plan, we're right there. I think that makes a big difference in that we don't have to go in and clean up at the end or tell them where they made mistakes.

Nobody wants to be invited to the party if you're the policeman. That ended up what we were more likely being, only because we were invited so late. We had to correct what they had already done. Now they realize that we're a team player and that we can help them through it rather than stop them. We can't do it their way, but we can do it another way, or let us try a different way rather than just saying no. It took a while for the NASA culture to change, just not at this center, but NASA procurement culture period. It had to change from, "No, the regulations won't allow you to do that," to "Yes, let me find a way to help you do that." I think that culture change has happened across Procurement.

Did JSC put some of those trends in place? You just want me to brag! I think JSC leads the agency in lots of things, but I'll talk specifically about procurement. All procurements over fifty million dollars that are competitive have to go through a source evaluation board process.[12] Most people think it takes forever, but normally it's about fifteen to eighteen months. We've streamlined a process where you can get a procurement that's over fifty million dollars in significantly less time. We're talking twelve months. We've done it at the center several times without NASA Headquarters' permission because it was under fifty million.[13] Once it's over fifty million we have to get their permission. They're letting us do a pilot now that all of the other centers are waiting for us to complete so they can use it. We've done that with a streamlined procurement approach; we've done it with the award fee approach.

This center has more award fee contracts than many of the other centers.[14] When Bob Cabana, who's now the center director at KSC, was the deputy center director here, he was the fee determination official in most of those contracts.[15] The approach, for which he helped us, let us develop a great process for award fees. The rest of the centers have used that. As soon as he moved to KSC he had the Procurement Office there call me. He said, "Send me all your award fee stuff." We've actually piloted a lot of different things in Procurement at JSC that we're not shy in using. We will let other centers use it, the same way we benchmark other agencies to get information. As soon as we develop something, we send it out and let everybody use it. They do the same with us.

Transparency has been important for procurement. One of the things I found in working with small businesses is that a lot of them say, "I would love to compete on this; I would love to get a contract, but I don't know how." This was all before we had websites. They would come in, and they'd have to take away hard copies of documents and go and read and figure out how to do it. It became wonderful when we started with the web. We can just put things online. I applaud this center for being one of the most transparent as far as the procurement process and being involved in procurements that come up. Mr. Coats and Mr. Howell before him opened their door for contractors to come in and see them, so they set the standards.

When the other technical organizations have procurements, contractors can walk in and talk to them. It's not the same way at different agencies and not the same way at different centers. It's time-consuming, but it's a learning process: the investment that you make up front in talking to them about your ideas, explaining to them what you expect for them to propose, and how you're going to make the selection. At the end, we give them a copy of the selection statement that says why we did or did not pick them and what the other person had that they didn't have.

We've seen better proposals, better contracts, and better relationships because we are transparent up front. It all starts with the ninth floor and the center director saying, "This is what I want the center to look like."[16] We get kudos from contractors all the time that we're very open. When we make changes to any of our processes, I have a meeting with all the contractors. I call them in and say, "You used to send it this way. We're going to change things." When we adopted the new award fee process, we had an open forum with all the contractors. They came in with pencils and pens, because that's part of their dollars. They need to understand how we're going to evaluate them. So we have a meeting with them, and we get feedback. Sometimes we don't always get it just right. They'll say, "You're not doing this quite right." We look at it, and we implement some of those changes.

I have about 136 civil servants and probably another 50 contractors in Procurement. We have experts that specialize in negotiating with the Europeans, with the Russians, with the Japanese. It is wonderful to get to know the different cultures because you can't negotiate until you understand their culture. We actually have classes that we go to out here. We not only learn the languages but learn the culture of negotiations. We walk through those different things.

We also have specialists that talk about negotiating with our commercial contractors, with SpaceX and Orbital, because they do things differently too.[17] So it is learning what works best in each situation and being adaptable and having in your grasp the ability to negotiate with any of those at any one time. Whether you're talking to DoD, which is strictly federal government to the extreme, and then having the commercial company that says, "I can take one page. Just sign it right there. That's my contract"—you have to be adaptable all along the way. The fun part about it is learning each one and being able to master each one of them.

Procurement is a partner with Legal. When we go to strategize at meetings, Bernie Roan, the chief counsel, comes with me. People always ask me, "Where's Bernie?" We come as a pair. I think that really has helped our young people in knowing that they can walk over to the Legal Office and it's not a bad thing. "I want to throw out this idea. We're thinking about doing this." It's great to know up front that they will set the framework. I love their perspective, too. They're not there to tell us "no," they're there to help us to find a yes. That's the approach they have. It's wonderful to go down the hallway and say, "I was thinking about—" and talk that over with an attorney.

Bernie has also had to groom his organization to support the type of procurement organization that we have. If we want to sit around and talk about what ifs, a lot of people say that's a lot of time invested, but it's better to catch me on the front end before I go design this new creature than you tell me at the end that it's not the right one. So they come in at the very beginning. As soon as we're at the table, Legal is also at the table with us.

I also have helped write new regulations. One of the first things I did in setting myself apart was coming up with a process that's called the screening process for small businesses. Previously, if I wanted to buy paper, I would call SBA (Small Business Administration) and say, "I want to buy one thousand reams of paper." They would come back and say, "Here's three companies. Get quotes from them."

Well, we don't know the companies. We don't know anything about them. SBA has just recommended them. I said, "Let's start another process. When you give me your three names, then let me bring them out here and interview them. Let me give them a statement of work. Let them make a presentation about the company. Let the technical people and me get in a room, and then we'll pick the one we want rather than you giving me one."

We started that screening process, and now it's part of the regulations. Any procurement that we have from SBA that's not at the competition level—where we have to go through formal competition—instead of the SBA giving me one name, they have to give me several names. We bring them out, and they actually do a presentation to the technical organization as well as Procurement. We pick one based on that.

I did not like the fact that we were given somebody we knew nothing about. It sets everybody up for failure. The technical people don't have any confidence in them. They don't know us, we don't know them, and you just set them up for failure. But once you bring them in, and the technical guy can pick, then they've invested in that company and they will help them succeed.

After that our small business program really became positive, because the technical person felt, "Well, I picked them, so I'm going to help them succeed." That's what I implemented. I wrote it up and put it in the newspaper at Headquarters. They're using the small business program all across the agency. That was many, many years ago.

A lot of other things I've done have been more on the "soft skills" side. When I came in there was only one person in the Office of Procurement that was a minority. In fact, it was a minority male. Females were not in leadership positions, because back in the 1970s most of the women were in administrative positions. There were none in leadership positions. When I came in I had a degree. There were several other females that came in with degrees, and we started working together informally. We worked together, finding out what all the unwritten rules were—and there *are* unwritten rules. There are ways that women have to approach positions. Back in the 1970s when we came in with IBM, we had our little bowties, we had the suits, and we dressed like men. The only thing that we were taught in school was the way that men handled things.

When we got out here and started looking at getting promoted, we realized that as women we could not use the same path men did. So we got together and started looking at ways that we could advance ourselves.

All of us ended up being managers. One of the ladies that was in our group ended up being the head of the Office of Procurement at NASA Headquarters and went on to be the head of OFPP, the Office of Federal Procurement Policy. It was great; she's done wonderfully.

All four of us ended up being managers. We credit that to us getting together and going through the list of what does it take. There are unwritten rules. There are things you have to do to make yourself attractive to people, to be mentored and sponsored. Without a sponsor there's no way you're going to get a promotion. When the door closed back in the 1970s and early 1980s, the people who decided who would get a promotion were all men. There was nobody in the room to speak for a woman unless she had already been sponsored by a man. We had to figure out how to get the men to sponsor us.

If we're ever going to get in that room and be managers, because that was the main thing—back then we called it the "key to the men's room." How were we going to get that key to the men's room? Back then when we came out here, some of the buildings didn't even have—literally— female restrooms. I was in Building 225 and one of the other ladies was in 36, one of the science buildings. There were no restrooms for females in that building. As women became more of a population out here, they converted some of the men's rooms to female, but a long time ago there were no restrooms for women.

When we came in there were no rules to get promoted and you didn't know what it was going to take. Back then I started as a GS-5. To get from a 5 to a 7 to a 9, that was unwritten.[18] Right now we have a roadmap, and I applaud us for doing that. That is still in place, iterations of it, different levels of change, but we started that. To go from a GS-5 to a 7 you should have negotiated a contract. You should have bought fifty thousand things, whatever it takes to move up. We wrote that down because nobody told us. We learned through observation. We found out how those guys got to the positions they were in, and we started doing those things. That's how we got promoted.

Then we started telling everybody that came in. We passed that roadmap on. Now it's a formal process. What it takes to get ahead is actually written down so that it's not a secret. When they went behind closed doors, the name of the individual promoted came out. That's the way it was in the past. Now people know what goes on behind those closed doors.

As I mentor and sponsor people now, I tell them one of the things you need to do is to set yourself apart. Everybody can do a great job, but you need to do something that nobody else wants to do. You need to take on that job, the ugly job, and then be successful at it. People will always remember you for that. But you need to be careful to avoid being the only one that does that forever, because if you are, they'll always keep you there. So as soon as you become the expert, you need to train someone else so that you can plan your exit. I've seen a lot of people that became specialists in areas. They were so unique that they couldn't lose them, so they were not available for promotion or doing different things. It's great to do something different. I took on a couple of ugly jobs throughout my career.

One of them was small business. In order to do small business, you have to learn the regular procurement regulations. Nobody wanted to do it. I took that on, learned it, and became an expert at it, so I was always called on when they had a question. No matter where they were, they would always say, "Well let's go ask Debra about small business." That set me aside and gave me a different aspect than everybody else.

Then I started training other people, and I moved on to the next thing, which was ADP (Automated Data Processing) or IT (Information Technology), which nobody wanted to do because that required a lot of other work, too.

Do I think being a minority female could have possibly held me back? I think you can look at it in different ways. I looked at it as an opportunity. Because there was not one before me, I was going to be the first. I've been the first at a lot. I was the first CO (contracting officer), the first branch chief, first division chief, first deputy, first procurement officer. When all the procurement officers meet across the agency, I am still the only minority female. There are other females now, but I am still the only minority female.

That gave me a lot of opportunity to make changes. I've actually had a wonderful career. It has been wonderful in that everything I wanted to do I got to do. It doesn't mean I didn't have to work for it, but I got to do it.

If I was coming through again, I would go back and rotate out and do things like going to the Education Office, doing outreach, taking a tour in the budget area and human resources, or one of the program offices. I didn't then. I stayed in, because that's what I was required to do. After I got to be a manager, I didn't think it was time for me to go out. I would do that now, because it really broadens you. Back then, we would just have to make sure we knew what was going on and study or get with our customers, and if we knew somebody else in that area, go meet with them and learn about it. I encourage a lot of people, when I talk to them, don't just stay in your field. Learn what's going on across the whole agency, not just JSC, but the whole agency.

I also have training for my staff so they are prepared to do everything. The first thing we do is organizational training, team building. We do a lot of leading at the speed of trust. I just set up a seminar where I'm going to have women talk about their voice. Women speak with a different voice than men, and women cannot take the tone or approach that men do. That's going to be a telecon. At least those things are available today. Back in the 1970s you couldn't find those things. That came across my desk, and I set that up.

I require every one of my division managers and deputies to take at least one training class every year, to participate in fellowships and take training. They have to have at least one retreat with their organization a year. It helps build that team. I set up training measures for my direct reports, and then we get together as a team and set it up for their organization. It's taking advantage of a lot of things. It's negotiating, it's how do you become a professional at work. It's how you're taken seriously. All of those things help people become well-rounded and not just a procurement person, but a good businessperson that can make a deal.

If you look at it that way, we're not just procurement people. I train people to have that broader perspective and that when you leave Procurement—and one day you will leave—you can go do other things and not just procurement.

Lessons learned? The ones I think that I'm going to take away with me are the ones about helping other people. When it comes down to it in the end, it's not about which contract I wrote or what the deal was or how much money we made or saved, it's about the people we touched. I want to be able to know that I helped somebody else, so I spend a lot of my time grooming. I groom all the managers that work for me. I spend time with them talking about the things they need to do, the mistakes I made, and the things I would do differently. I talk to them about what happens at senior staff meetings, what they're looking for. The little tidbits that would make a difference in somebody's career is what I really apply myself to right now, helping the next generation.

The new ones that come in, I meet with every single one of them and talk to them about it. We've hired some attorneys that are here. We've got some great young folks that are here. I talk to them about getting connected. You have to know you cannot make it on your own. I don't care how smart you are, you're going to get to a point where you're going to need somebody else to help you. You need to start grooming yourself now so that a senior leader can say, "You know what, that person has got something. I want to sponsor them. I want to see that they make it to the next level."

I talk to people a lot about doing things so that they can be seen, like going to the National Management Association lunch where Mr. Coats was speaking. I send that out to all of my employees and say, "This is a good opportunity to hear what's happening at the center but also for senior staff to see you." I encourage people to take advantage of the different team-building activities we have, the Christmas lunch, and then the picnic. A lot of people in this generation like to go home. They love their time, but if you invest a little bit of it so that people will see you on the outside, it goes a long way. I do think when I came in the 1970s one of the things that helped me break the glass ceiling, or the concrete ceiling, or whatever you want to call it, was the fact that people got to know me as a person. You have to put yourself out there. You have to make yourself available outside of your particular office so that people can see that you're a real person.

I developed friendships when I was buying pencils and paper that I have been able to take advantage of in this particular office. It's amazing that those people will always remember you. They'll remember how you treated them. It's nice to be able to go back and capitalize on some of those relationships.

One of the big things I learned is that at one time I didn't value the people in the office, for instance, the administrative officers or secretaries, when I first got here. Then I woke up and learned that they're the most important people at the center. They can block your way forward, or they can help you. When that lightbulb came on, that's when I think my way became a lot easier. I tell people that right up front in the beginning: don't underestimate your support in your organization. Learn to treat them well, because they will be the ones that will either smooth your path or close the door. Your telephone messages will never get delivered; you'll never be able to get on somebody's calendar. Treat everybody the way that you want to be treated, and it pays off. Sometimes people hear that and say, "Oh, that's just a cliché." But I have learned that it's powerful.

I think the biggest challenge I've struggled with is being a part of the team. Just in the last eight to ten years, we saw Procurement become a part of the team. I think that was the biggest struggle. When I first started in Procurement, we were the roadblock, and everybody said, "Oh my God, I've got to go to Procurement. I know it's going to take you all forever"— that was the attitude. A lot of people really took it personally. They said, "It's not us." It's really the whole mission support team, whether it's Budget or Procurement or HR—we stand in the technical guys' way. So when they have to come to us, some of them were screaming and hollering. "Get this, I need it now!"

I think that has been the biggest challenge, overcoming that and becoming a part of the team, and being valued as a part of the team. That really has made a difference in our success.

My greatest contribution is going to be the people: the people I've hired, the people I've mentored, and the fact that I've been here in my position as a minority female hopefully will open the doors for others. Things are changing. I think one of the best things you can do in your position is to leave a good legacy. That's been my objective, to make sure that when I leave they don't have to remember my name. They'll remember I was here, of course, but the fact that I helped somebody else. You don't have to remember what I have done. Given the fact that I've helped other people, my legacy will never end because it'll always be alive through them.

Natalie V. Saiz [19]

I was going to school in Albuquerque at the University of New Mexico in 1987. I was interested in human resources, but I didn't really know how to get into the field.

Official portrait of Natalie Saiz.

I had been reading about it and I recognized you really needed to have experience, so I went to our career services office at the university. I was looking for opportunities, and I found two. One was here in Houston for the Johnson Space Center, and one was at IBM in New York. IBM at that time, in the mid-1980s, was just a huge company, a very powerful company, unstoppable, that kind of thing. I was thinking, "Wow, that would be awesome to work for IBM." But it was really, really far away. I came from a small Hispanic family. I didn't think there was any way my parents would let me go to New York. At the same time Johnson Space Center was an amazing place, and I didn't really think I could work here because I wasn't a scientist or engineer. I really thought that NASA was mainly for technical people.

I had been on a trip to Johnson Space Center when I was a sophomore in high school. I was involved with Distributive Education Clubs of America (DECA). I made it to the national competition that year, and nationals was in Houston. We came on a tour of Johnson Space Center as part of the DECA trip. The tour opened my eyes a little bit. I had been here, and it was familiar. I still wasn't sure NASA would hire me because I didn't know they had anybody other than scientists and engineers.

I never really dreamed of working for NASA. I never really thought about it at all, but the fact that I came here on a tour, I think, was something that made it a little bit more possible. So I applied. I was very, very happy I did, twenty-five years later.

I started in the Human Resources Office and was an HR rep. I tried to work with a lot of different organizations. I worked a lot with Space and

Life Sciences, so I got to know the employees and the organization very well. At the time, we only had one female director of an organization, and that was Dr. Carolyn Huntoon.[20] I supported her as her HR rep. I was very excited to do that because I felt like she was a real role model for everyone, particularly women, at the center. I think there was a little bit of a kindred spirit because we were both women. Yes, she was technical, and I was mission support, but at the same time I really feel like she took me under her wing to help me out.[21] When she became center director, that was really exciting for me as well.

I worked with other organizations such as Space Station Freedom and Engineering. I worked a lot of the staff offices as well, so it gave me a really good perspective of the entire center and the different organizational cultures and the different people. To this day when I see people at the café or out in the mall, I think back to when I was an HR rep working with them on different issues.

I was a group lead in Engineering, and then I became supervisor of the Human Resources Management Branch for about five years. Then we created a different branch within HR, the Human Resources Operations Branch, and I became the branch chief of that organization. I was selected about a year later to be the assistant director for Human Resources. I was essentially the third person in the organization.

As soon as that happened, someone ended up retiring from the organization. Then my boss moved to another position because there was another opportunity for him. At the time, our center director was General Howell, and he asked me to serve in an acting capacity as the HR director. I had only been in Building 1 for five weeks. Being at this level for such a short period of time, I felt it was very much a sink-or-swim situation. For me I felt it was very high risk. Sometimes I would go into a meeting, and I wasn't really sure how everything was supposed to happen at the senior executive level. I, fortunately, had a couple people that guided me as far as helping me understand the culture and the norms and how to work at this level, because I really hadn't had that training yet.

At the time I was part of a leadership development program, and I had just been selected a few months before I was asked to serve as acting HR director. I needed to go to NASA Headquarters for a minimum of six months to a year. But General Howell was relatively new as our center director; he'd only been here a short period of time. I wasn't sure if this was the right thing for me to do, plus I was acting director. The job wasn't mine yet. I went into his office and I said, "I'm part of this leadership development program. I'm supposed to do a rotation at NASA Headquarters. Would you like for me to withdraw from the program?"

He said to me, "Well, tell me why you want to go to NASA Headquarters." I decided in that moment it was really important to complete the program. I wanted to go to NASA Headquarters.

I said, "Well, I want to build relationships at Headquarters. I have relationships with certain levels, but I don't really have them at the levels I need to in order to be an effective HR director. I think once I get up there and have this time to really understand more about the agency perspective and establish those relationships, I'll be a more effective HR director."

He listened and then he said, "I think you should go ahead and go to NASA Headquarters and do your rotation."

I said, "Okay. Thank you. Thanks for your support." I was hoping he was going to say, "And your job will be waiting for you when you get back." But he did not do that. It was really risky. I tell people, when I talk to different leadership classes, when you take rotation assignments or when you need to go off and do something you need to do, there are no guarantees. You have to just step off on that limb and see what happens.

I did what I told him I was going to do, and that was establish relationships. When I came back, I had a really good agency perspective. I saw all the centers and had insight into their cultures. I learned that there are a lot of really dedicated and passionate people all over NASA. Every center has that. Sometimes you tend to think you're the only center that has that passion and dedication, but it's everywhere within NASA. That's an important lesson to share.

A few months before I came back, General Howell was going up to Headquarters, as he did frequently, and he contacted me and asked if we could meet. I said, "Well, absolutely." So we sat down and talked. At that point, before I came back, he asked me if I'd be the permanent HR director. I was very happy he offered me the job, and I said yes. It's been eight and a half years since I became the HR director.

When I was an HR rep and an office chief, branch chief, I never really aspired to be the director of HR. I knew that only men had held the job before. I was intimidated by the fact that I wasn't sure how much time it was going to take, and I wanted to make sure that I was also committed to my family and not having my family sacrifice because of my job.

I really had aspired to be the deputy director. I really wanted to be the deputy. But then because of everything that happened and how it transpired, here was my opportunity. I realized at some point that if I really wanted to have some big center-wide impact, then I needed to actually be the director, because as director you decide what projects to work on, the priorities. I really wanted to emphasize flexible work schedules, more of a work-life fit, having that family-friendly environment and culture. So we advocated having more flexible schedules here at the center, which I think a lot of people are using, and more telework. We think we're using a lot of that as well.

I also wanted to spend a lot of time working on diversity and inclusion. That's been very much a follow-on, if you will, to the whole Shuttle and Constellation transition effort, because it's all about transparency. It's about transparency of people knowing what opportunities are out there. Can they put their name in the hat for certain jobs? It's all about having that engaged team, so we have done a lot.

The diversity initiatives have been focused on the engaged workforce, the engagement teams. There's also a team looking at rewards and recognition. There's another team working on barriers to innovation. It's not just out of our office. A lot of people in the center are working on that. We have an Inclusion and Innovation Office in Human Resources.[22] It's only comprised of three people, but it's the organization within HR that keeps all the stuff going.

Now we have employee resource groups (ERG). Employee resource groups are something that's relatively new. It's been in the last year or so that we've instituted these employee resource groups. We've had something similar in the past, but there's a little different focus now. We did a lot of benchmarking with private industry to find out how to put these in place, Texas Instruments, Georgia Power, different places that are really focused on your business and providing business solutions. We have an African American group, we have a Hispanic group, an Asian group. We have an LGBT (lesbian, gay, bisexual, and transgender) group called Out and Allied. We have a human/systems integration group as well; it's more technically focused. We have five ERGs.[23] Those are, I think, going really well.

I should add that the Johnson Space Center Human Resources Office is unique. When I started working here, I realized that it was a very different human resources office. One of the big differences I saw immediately was that there was a real value in trying to learn the customer's business and talk their language and really be involved with them, to the point where HR attends staff meetings and knows the people. You're not just this organization that's sitting on the side waiting for things to come to you, but reaching out there. We have always strived to be partners with our customers. The way we've put it is we want a seat at that strategic table. That's something we have tried to emphasize. We have heard many people say that our organization does do that, and we are very involved with, I think, almost all the big initiatives.

One of the things our current center director, Mike Coats, has told me is that when he wants to get something done, he'll talk to Human Resources because he knows we'll make it happen. For example, our inclusion and innovation work at the center has been a huge effort that we've been intimately involved with.

Also, the Program and Project Management Development Program (PPMD) was a leadership development program we started at JSC but ended up becoming an agency-wide program that we ran for about three years; it was a huge initiative that has really paid off.[24] We had a technical advisory panel that was helping us with PPMD, and we worked with the real legends of NASA. We had Arnie Aldrich, Tommy Holloway, Glynn Lunney, and Jay Greene, people who were really strong advocates and real partners with us trying to help us with the curriculum.[25] I think we have always had that mindset of engaging the customer and working as partners, and I think that's made us more effective as an organization, particularly a human resources organization.

Technology has been extremely helpful because we can get information out a lot more quickly. We have consolidated and have information in one location. I have a small group of people here at JSC in my Human Resources Office that work information systems for HR. They do different applications and things that are very unique to the field.

As a matter of fact, most of their development efforts have been picked up by other federal agencies. For example, we automated the benefits statement that all federal employees can see online now. That was something we started here at JSC. They've done a fabulous job of combining the technical requirements with the HR application and bringing those together.

We have a unique opportunity here and service that we're providing the agency. That's something that I think has been very different in our office as well.

HR is in charge of the civil servants here as far as HR services. The contractors, our team is twelve thousand people or so. We have a group of people called the HR principals, and those are HR directors from the other contractor companies. We've been meeting with the HR principals for several years. It started off with my predecessor. We kept that going, and it's really evolved now where we have a rather large community that gets together about once a month or once every six weeks, for about an hour meeting. We usually meet at the Gilruth Center, and we have different topics that we cover.[26]

One example of when our HR principals came together was right after Hurricane Ike.[27] We called a meeting with the HR principals. Frankly I wasn't expecting many people to show up. I thought we'd have five to seven people. It was, I think, a couple of days after the hurricane. We walked in, and it was standing room only. We had another meeting that afternoon, and it was the same number of attendees but with different people. We realized that we had a lot of support and we had a big network that was already established, so we felt that was really beneficial.

Hurricane Ike, I think, really made us realize that we needed to rely on that community through the transition efforts, too. That's where we really started focusing on our Shuttle transition and Constellation transitions.[28] We did a lot around Constellation and the Space Shuttle transition, trying to get a lot of communication out, working collaboratively on job fairs and what could we do for the contractor community. They were the ones hit the hardest with layoffs after Shuttle retirement. So we did a lot with workforce transition.

Before the retirement of the Shuttle, in 2007, there was a tragic incident in Building 44.[29] That was definitely a very sad time for the center. I remember I was in a meeting in our center director's office at the time, and my phone was buzzing. I found out later Pat Pilola, who was the division chief of Avionics, was calling me.[30] I was meeting with the center director, so I didn't think I should answer because I was with my boss.

But at that point the director of Engineering walked in and said there'd been a report of someone with a gun in Building 44. Mike Coats and Steve Altemus said, "Let's go over there." I think the SWAT team had already been called, so I came back to my office. I immediately got in my car and started driving over there. They were all stationed down by the fire station area. We went over there. The EAP (Employee Assistance Program) counselor was there.[31] We were waiting to see what was going to happen. They were trying to account for the employees and see who was there and who wasn't. Fortunately, there weren't as many people in the building as there could have been because they were having an all-hands meeting in Building 2, but there were definitely still some people in the building.[32] The division happened to be going through a reorganization, and they were announcing the new leadership of the organization.

One of the things I really appreciated about Mike Coats is that he surrounds himself with a leadership team that he relies on for advice and consultation. I remember very distinctly that a spouse was at the gate wanting to come in. Someone had called her to tell her there was a shooting in her husband's building. Of course, she needed authorization to come in. I remember there was a small group of us standing around Mike Coats, probably about six people. The question was, should we let the person in? Most of the people were saying that they didn't think we should let her in because we really didn't have much information.

We didn't know anything definitive at that point, so should we let spouses in? Was she just one of many that were going to come? We didn't know. Most everybody was saying, "Well, I don't know if this is a good idea; we probably shouldn't do this." I remember I was the only person that said, "You need to let her in, and you need to bring her over here."

I was very pleased that Mike Coats then looked up at security and he said, "Yes, let her in." Now, he probably was going to decide that anyway. I'm not saying that I changed his mind, but I appreciate that he was getting advice from his staff, not just the SWAT team people. They did a fabulous job—I'm not trying to criticize them in any way—but there are different perspectives that people have. I think HR does bring a unique and valuable perspective to the table.

So she came in. She was in a room with the EAP counselor. As difficult and as sad as the situation was, it turned out to be a really, really, really good decision to let her in, because she did not have to hear about this on the news. The center director went in there and told her that her husband had been killed. I just remember watching Mike Coats. I'm getting emotional. I had so much admiration for him, because he was just so caring and compassionate. That was a true mark of his leadership.

Fortunately, a lot of us — including myself — had already been through the Critical Incident Stress Management team training that Jackie Reese does through the EAP Office.[33] It's a way to bring people together and talk about what happened and then help them. It's not extensive counseling or anything, but it's a way to talk about an incident and then identify those that may need more help. So we used that. It was a very good skill we had because we could talk with them about those things.

One of the things we talked about is that people really need to know what their avenues are. What are their avenues for raising issues, raising concerns? What are the complaints? Interestingly enough, we had already been working with different organizations, with the EAP, with Legal, with Procurement, the union, and the ombuds on resolving issues and displaying the avenues on our website.

We developed something within a couple of months right after the incident, and we put that out as much as we could. We still utilize it, and we still talk about it quite often in leadership classes and all-hands meetings. Mike Coats is a really firm believer in everyone having an avenue to seek help when needed.

Going back to that day, I remember a lot of different emotions, anger, sadness, frustration, all the things you see in a death and in a catastrophic event. One of the things that was helpful is I already had an HR principal community. I called the HR principal community together very quickly, and we got information out, and we talked. We tried to get them information, because the whole team wanted to know what was going on. I felt like my role was about communication with the HR leadership of the contractor companies.[34]

Because of my involvement on the Joint Leadership Team, I also knew the head manager of that particular company.[35] The company wanted to have some sort of gathering for the employees of that building. They needed a place to come together and grieve. I remember that I was very supportive of that. There were a lot of people that worked in that area and needed to have some closure, needed to have some time together.

There were very intense emotions for some people who were very angry that the contractor company was having a separate gathering, but it was for the coworkers who needed a place to grieve. It was held at the Hilton Hotel. I went as well, and I thought it was very well done. It was very short.

A few people spoke. I noticed that the employees that did attend really appreciated having a place to go to try to cope with this tragedy. I think HR has to be very objective and very open minded and has to bring that to the table and give that other perspective. It's very easy for some people to get on a track where they're just like, "Okay, I just want to see this one side of the story." I thought that was a very good thing the contractor did for their employees as well.

Hurricane Ike was another example of how we used the Critical Incident Stress Management team training. Many employees had lost their homes, so this training was very useful. That was a really difficult time, but it was very rewarding. One of the things that our center director asked as soon as the storm passed is, "How are we going to account for everybody?" We started working with organizations to make sure we had identified everyone to ensure they made it through the hurricane safely. It was a lot of manual work.

We got the HR principals together. They were all reporting on their companies. We were in the process of not only trying to account for them but also trying to identify how much impact they had. We decided to categorize employees as red, green, or yellow. Red basically meant that they had lost their home and they couldn't live in their current condition. Yellow was a temporary situation, like they didn't have power. Green, they got through the storm okay.

We also came up with a concept of employee recovery advocates. We call them ERAs. A trusted person within the organization would volunteer to be an employee recovery advocate who would go to the individual that was considered a red category and try to see how they could get their basic needs met.

For example, Duane Ross was a recovery advocate for one of our employees within Human Resources who had a lot of flooding in their home.[36] He went over there and took them a meal and provided water and said, "What insurance companies can I call, or what kind of help do you need?"—even charging their cell phones and doing anything that needed to be done. Obviously, the person going through all that has a lot on their mind. They have too much to do at once. Later on, Boeing wrote it up as a best practice in their newsletter sent across the entire nation. That was really neat, too.

Then we set up a rest and relaxation place out at the Gilruth Center.

I went down the hall to Mike Coats's office, and I said, "So many people have lost power, but Gilruth has power. So we want to set up a rest and recovery place for people to come and use their computer. IRD (Information Resources Directorate) took a bunch of computers out there.[37] We want to put computers there so people can look up their insurance companies and make phone calls and find places to come in and work on their houses."

We had the big Gilruth ballroom open with the big screen; we had some TVs in there. We had some basic things like water and a few snacks. We opened the showers, and people could take showers there. They could come and rest and get a break from the heat, because it was hot. It was in the middle of September.

I was really, really proud of our organization because we came together. We didn't have a plan necessarily for all those things I just mentioned. Now we do. We started trying to figure out what do we need to do; let's do this or that. It was really nice. After we went through all that, Mike Coats recognized the organization with a Center Director's Gold Coin award.

It was a bad situation but it turned out to be very, very rewarding. We've followed up with that concept after Ike, and we developed a system called NASA Cares, which is an automated employee accountability system. It's now turned into the employee notification system, ENS. We've been testing it. It's now been picked up across the entire agency. There's an iPad application, and you can push information out, like Center Ops: "This is what's happening."[38] Then employees have to log in and say whether they are safe or not safe or whatever their situation is. It shows who's in certain zones based on their ZIP codes and where they live, and you can identify the red, the yellow, and the green people. We're hoping we don't have to use it, but we have worked really hard since Hurricane Ike to develop this automated tool to make it a lot easier. That was developed out of my office with IRD.

A few years earlier, in 2004, President George W. Bush had announced that the Shuttle would retire in 2010. Several years before that deadline hit, we knew it was going to take several years to really transition the workforce. There was a balance with really needing to keep the right people and the right number of people on the Shuttle Program as we were downsizing to eventually transition everyone.

We had a really strong partnership with the Shuttle Program management team with John Shannon, and before him Wayne Hale, and one of my staff members, Sue Leibert.[39] About five years before retirement, I met with a leader in the Space Shuttle Program and explained what I wanted to do. I said, "We know this is a partnership for the transition. I'd like to have one of my employees work in your organization half the time and then half the time in our organization to bridge that gap between our organizations, because we need to be working hand-in-hand on this. Obviously, you have mission-related needs, and we have the people skills as far as processes to transition everyone."

We already had a good working relationship, but we established more of a formal partnership early on. We wanted to do that because we know it was a shared responsibility, so we worked very closely with the Shuttle Program. We had a Human Resources rep and a Human Resources development rep that would meet with the management very frequently and came to the table with different ideas, like interviewing everyone, talking about their plans and what was their timing and when would they be interested in moving. We wanted to communicate as much as we could. We developed a Shuttle transition website and we had a lot of information there. Almost everyone on my staff was working Shuttle transition. We had a few people starting off, and then we added more people and more services. The last year or two it was almost the whole office in different capacities. Every year we were figuring out how we needed to step it up as far as our Shuttle transition.

The person that I mentioned, Sue Leibert, she was the one that was working half-time on Shuttle, half-time in our office. She actually got a more expanded role as well where she was the HR lead for all four human spaceflight centers. That also helped too because the Shuttle Program is comprised very heavily from four centers in particular and had workforce in other areas too, but mainly at Kennedy Space Center, Marshall Space Flight Center, Stennis Space Center, and JSC. So that was very, very helpful to have that partnership.

We did a lot from that level but then we also did a lot at the center level. On the website I mentioned, we had something that became very, very popular called "Fact or Fiction," where people could write in rumors they heard. Are we going to have a buyout?[40] Are we going to have a reduction in force?[41] I heard this is happening, that's happening.

We made a commitment to answer every one, and we did it promptly. We didn't run our answers by Legal. We didn't stop and say, "Legal needs to review these."

Another center called and said, "We understand you have this. Our center director wants something similar." They were asking us the same thing. "Are you running your answers by Legal?"

We said, "No, we want to respond quickly and with what we know. We'll address 97 percent of those. The 3 percent or even 1 percent, the ones we get wrong, we'll have to deal with those." We just felt like that would slow us down. We have great legal staff. I'm not saying that. It's just that if you have to put it in legal language, then it's going to slow you up.

We wanted to err on the side of being responsive, so we had a team that worked on the website. There was a lot of work put into that, really putting different features there. Then we developed something called the Job Opportunity Bank (JOB), where we wanted everyone at the center to post their vacancies, even if they were at the same grade level, because one of the themes that we wanted in the transition was transparency. We wanted everything to be transparent, because what we didn't want is people who were working Shuttle to the very last mission to feel like someone had forgotten them or that someone else was taking the new jobs. "I'm going to be the last one here and I'm not going to have a good job when I'm done." We wanted to have a lot of transparency, and that's why we did the JOB tool. Of course, a lot of this was a result of the strong partnership between the Shuttle program management and our organization. They gave us the support we needed.

We could develop all these great tools, but if no one used them then they wouldn't be very effective. It wasn't easy. There were some organizations that were hesitant, that didn't necessarily want to put their jobs in. We got a lot of buy-in from the center and made the case about why this was important, and why transparency was so important to this transition effort. We're still using our JOB tool, but it's been used for different positions that we have now.

The other thing that happened is Constellation came in, the new Vision for Space Exploration, then very shortly thereafter it was canceled, too. We had been working for five years for Shuttle transition, and we were really shocked about the Constellation transition. Fortunately, we had a lot of the tools already in place that we could quickly apply to the Constellation Program.

The thing we didn't have was identifiable new work, because a lot of our Shuttle transition efforts were focused on the Constellation work. That was going to be the follow-on program. At the time, we weren't sure if MPCV (multipurpose crew vehicle) was going to survive or not.[42] Finally, I think a whole year later, maybe longer, it was decided—but at the time we didn't know. There was a lot of uncertainty.

I think one of the things that made it really work well is we had strong support, as I said, but, also, we communicated. My philosophy is you need to just get whatever you know out there. People are educated and they're going to read about it in the paper, so you might as well be the one telling them what you know or don't know. If you don't know, just say, "We don't know yet." I think that really went a long way. I tend to forget what a big deal that was, but it really was a big effort.

Another initiative I was involved in was the Center Director's Program Manager Forum. That's not a forum we're using as much these days, though we did for several years. When we had three large programs here (Shuttle, International Space Station, and Constellation), those three program managers reported to Headquarters, not to our center director—yet they're located here at the center. There were so many issues that we found ourselves constantly saying well, what does the Shuttle program manager think or what does the Constellation program manager think, what does the center director want to do. We had all these competing priorities and different views. When Constellation was staffing up, Shuttle and Station were very worried about pulling their key leaders out of those roles because they had missions to achieve, too. I had suggested to Mike Coats that he, as center director, get together with the leadership of the organizations and talk. They started out by talking about key leadership positions that needed to be filled in Constellation, but then it evolved into different topics.

So we set up those meetings about every other month. They would sometimes be held at seven in the morning because you couldn't get everybody together at the same time otherwise. The meetings included the three program managers, Mike Coats and his deputy, and me, because I was in charge of the agenda. I had great insight into the programs and personnel because I was part of working the agenda and organizing the meetings. Oftentimes it would just provide huge amounts of information that I could work with as far as doing follow-on work.

It got to the point where the program managers just would not miss the meetings. They would come in from vacation or they would ask to adjust the meeting date because they really wanted to be there; it was an important topic. That went on for quite a while, until Constellation went away. Honestly, we don't utilize it as much anymore, but that was how it first went into play. I was really proud of that, too, because that was a good way to bring the leadership together.

One of the areas that I try to emphasize as director of HR is relationship building, whether it's the Equal Opportunity and Diversity Office or the union or other centers or our customers, we start with relationship building. Working with others is the only way you're going to get anything done. This forum goes back to the importance of relationship building, because you've got to get together and talk about the issues rather than being frustrated with each other.

Since I have been at JSC there are definitely visible changes in opportunities for women at the center. Women have a lot more opportunities now, as you can see from the demographics. A lot of the men that became leaders, they had daughters. Their daughters were trying to break into professional fields, and they were seeing some of the struggles of their own children. I think then they started paying attention to their own work environments. Mike Coats talks a lot about his daughter, who is a high school counselor. General Howell would talk about his kids. I think they started realizing we need to look at our own environment. I think they were a lot more willing and a lot more receptive to those kinds of flexibilities, and just that overall philosophy.

When I first became the director, General Howell would have a meeting every morning at seven thirty. Five people would come to it: the deputy center director, associate director, HR, and I can't remember who the other person was now. It was always really a challenge to get here at seven thirty. It was a struggle because I had three kids; I had small kids. I thought, "How am I going to do this?" It was very hard and challenging. I'd make it there, and I could tell none of them had those stressors that I did getting in to work.

I was the only woman in the group. I could tell none of them were thinking about, "Ooh, did I pack that baby bottle, and did I remember the pacifier?" But I didn't want to say anything because I didn't want to be disinvited to the meeting or told, "Well, if you can't make it don't worry, you don't have to come," because then I would have been left out of key decisions and I wouldn't have that seat at the table.

After about six months, I did feel like I had earned enough credibility that I met with General Howell and I said, "Is there any way you could make your meeting at eight? Because I have to get my kids to daycare, and it's just a little easier for me—a lot easier for me."

I remember he said, "Natalie, why didn't you tell me before? Sure, we can move this meeting to eight o'clock, no problem."

The thing is, though, it never ever entered his mind that that would have been a challenge for me. Finally, I guess it helped having the courage to bring it up and not just accept that I'll just not go, because it would have been fine if I didn't go. It wasn't like I was a key person, necessarily, but he still cared enough to make that change. I really appreciated that.

Another person that was really helpful to me was Leonard Nicholson, who was the director of Engineering.[43] For several years I was an HR rep supporting him, and then prior to that he was the Shuttle program manager for many years. When he was the director and I was supporting him in Engineering, I had just had my first child. We had adopted our oldest, and then I had my first child really close together. I came back from my leave; I was going to be working thirty-two hours a week. We had a weekly meeting, and I asked if we could start the meeting at nine o'clock because of the different daycare issues. His deputy was sitting there, and his response was really interesting. I love him too, but he said, "Well, I don't know. Is it possible?" He wanted to push back, because obviously that was going to impact him.

Leonard just stopped him and said, "No, we will start the meeting whenever you want to, Natalie." I remember just being so affirmed by that, because he was very accommodating, and it only lasted for six months, and then I came back to work full-time. Then I could go back to resuming our meeting at eight. I'll never forget that he was very supportive.

I've had a lot of people that have been very flexible, so I wanted to make sure when I got into the position like I have now that I would institute some of those things and sell those ideas to the directors and say this is why it's valuable. If you give people a lot of flexibility in their jobs and give them the ability to manage what they have to at home before they get here, they're going to be more focused when they get here. They're not going to be distracted and thinking about all the things they didn't get done. You build commitment in your organization. They may not want to leave, because if they go to another organization that doesn't have those flexibilities they're going to be more inclined to stay where they're working.

I think that's how you keep good people, building that commitment. I try to tell people that and managers that when they get a little nervous about workforce flexibilities, that you have to trust your employees and that's why. I think it's worked well.

It's tough to say what my greatest accomplishment at JSC is. I think it's probably workforce flexibilities—and it's not that I've done this by myself, because there's been a lot of help along the way. Technology has helped tremendously, and you see the younger generation wanting some of these flexibilities. Also, I think with society allowing fathers to be much more actively involved, they want flexibility, too. It's not just the woman's responsibility to do things at home or to take kids to the doctor's office. I think that all these changes in society have really helped. I think that really trying to have an impact with a work-life balance and allow people to have those opportunities have been something I really focused on.

The other thing is inclusion and innovation, primarily the inclusion piece, trying to build trust with your supervisor and having good dialogue and giving feedback for people that are not like you. It's sometimes intimidating to give feedback to people that are not like you. Sometimes people are afraid to do that. Those would be the two areas that I hope would continue.

As far as a tangible initiative, the Shuttle transition would be the biggest thing I'm most proud of as far as our organization as a whole working on, in close collaboration with the leadership in the Shuttle Program.

My greatest challenge? It's sometimes difficult to be a mission support person, a human resources professional in predominantly a technical organization. Everything here is measured by technical expertise, and technical expertise is very valuable and very important. So, I think that having to always prove your credibility or to have that credibility and to prove your value has always been a challenge.

It's been something that I work on every day to add value and not be a hindrance, or not be that bureaucrat. I really feel proud that our organization has done that very well. It's about staying connected with your customers, because the ones who are not connected with them, then I think they lose their effectiveness.

I think I'm probably the only center HR director that is not located with their employees; my employees are in Building 20, and now they're moving to Building 12.[44] Some people that come here, visitors, and other center employees, they're surprised that I'm on the same floor as the center director, that I'm on the ninth floor, that this is where our office is. I think it's an example of how much we're relied on, because I go down the hall at least four times a day for a meeting or for something that I need to talk to the center director or the deputy center director about.

Peggy A. Whitson [45]

When Brent Jett got up and named me as the new chief of the Astronaut Office at an all-hands meeting, I think there was some shock that day.[46] Even the folks that had been proponents of mine were, I think, surprised that I had been selected since there had never been a nonmilitary chief. That announcement was more stressful than going to chair the Monday morning meeting for the first time.[47] I had been at and chaired many Monday morning meetings as deputy chief when the chief was out of the office. This was a bigger step for our office, to transition from always having military guys to now having a nonmilitary chief of the office. That was in some folks' mind a little harder. I think they thought my leadership style would be different because I was a civilian. But they were unsure. In the end, I don't think my leadership style was that different than anyone else's, so I think folks adjusted. But I think it was a little bit of a shock to everyone that, "Hey, it's a nonmilitary person. Is it going to be different now?"

I think being a female chief of the Astronaut Office was probably less of a step than being a nonmilitary and nonpilot astronaut. Being the first female, I happened to be in the right place at the right time. I've been lucky that way a number of times in my life. I'm thrilled to have been there and been able to take those steps. But again, I don't consider the female part as big as the nonmilitary, nonpilot part.

Because it is a pretty visible role, the position does need some approval from above. In my case, Brent Jett—who was at that time the head of Flight Crew Operations—was a big proponent of mine and selected me. Immediately after my second spaceflight he started talking to me about my preference in terms of a position, and we discussed that option. At the time, there was some discussion about whether or not I would fly another Shuttle flight or take this position.[48] He told me he wanted me to take the position of chief. I ended up selecting this position.

Because I had a history of spaceflight and was pretty well known, hopefully it wasn't too hard of a sell for him to convince folks to have me be chief of the office; I don't think it was. I think he thought it was important to transition to somebody that had Station experience as opposed to Shuttle experience. At the time, I had the most Station experience in the office and had served as a commander on board the Space Station.[49] I think he felt like that was a good transition for where NASA was heading in the future.

The Astronaut Office and its chief have many duties. We support and provide the crews that are trained and ready to fly in space, and all the things that go behind that. This involves supporting crewmembers in training, doing procedure development, conducting hardware fit checks, and providing capsule communicators for the crews that are on orbit.[50] In addition to actual on-orbit operations support, we also have crewmembers who provide their expertise in areas for future vehicle development, so we have folks involved with Orion and Constellation or exploration programs.[51] For the new Commercial Crew providers, we've had people integrated into their organizations as they develop their new vehicles so they have our experienced crewmembers' insight.

So we have a pretty large organization from the perspective of supporting all those activities. The size of the office, from an astronaut perspective, is determined by the manifest and the number of crewmembers that we need to provide each year and also five years out, taking into consideration how long the training flows are, for instance, for International Space Station training. It's a two-and-a-half-year training course, and we have about twelve people in training at any one given time.

It's challenging because you don't always know how the manifest is going to change or the direction of the programs. You have to have that capability to surge when needed, to provide all the resources that NASA needs in terms of astronauts and be able to select the right mix of crewmembers with certain skill sets. What skill sets are needed is obviously driven by the mission. Selecting crews for a Shuttle mission, we need a commander, a pilot, and then whatever the tasks are, whether it is robotics or extravehicular activities (EVAs), the spacewalks, or other specific transfers, or anything else that we need a specialist in. For a Shuttle crew, you pick and choose the experts based on that. For most crews, you want to interleave some experienced people with some inexperienced people so that you continue to have that flow of experienced people available as the older folks attrit and leave the agency or leave the Astronaut Office. We want to maintain that capability.

For ISS crews, it is a little bit more complicated. You have to have people support different activities, so one person has to potentially be the commander, the EVA specialist, the robotics specialist, perform the scientific activities, and carry out the maintenance procedures. We have to find in one person the capability to do all these things, which I think is a little more challenging. Then, if they're going to be the left seat crewmember in the Soyuz, the copilot version of what we have in the Shuttle, they also have to have a very high-level Russian capability, because everything is written and spoken in Russian. That adds another level of complexity.

When putting the crews together, I had to think about an experienced and inexperienced crew mixture. I wanted to ensure having a strong enough EVA crewmember potentially matched up with a less strong EVA crewmember. Because the Soyuz launches only three crewmembers at a time, I had to plan crews based on the two-to-four-month overlap between each of those launches. For on-board capabilities, EVA, robotics, and the like, I had to write a plan such that it was independent of the Soyuz launches. It actually gets very complicated at times, because one decision is going to bias how you make the next one and then the subsequent one after that. That intricate linkage is a challenge, but that makes the job interesting.

When it comes to selecting crews for flights, the office has staff that help us try and make sure everyone's trained. If there's somebody that has a deficiency in an area, then we try to get them up to a certain level. I took inputs from my branch chiefs and my deputy chiefs on people's performance. Being somebody that's not the most highly skilled person in the world, hard work counts in my book, too, so that was a factor in who I might give preference to a flight for. If everything else came out even, that was a factor.

In the Station training program things would change over the course of very short periods of time, so even though I might have a plan that went out a year and a half, invariably I'd have to change it and come up with another plan based on some change in either what was going to be happening during the mission or crewmember changes. It was always a very dynamic process. Again, that makes it interesting. It can get a little frustrating at times, but I wasn't bored.

All the Shuttle crews were selected when I was named chief, but then we added an additional Shuttle flight. So I did get to select one Shuttle crew, the STS-135 crew.[52] Because it was only a four-person crew, it needed to be four experienced people. They had to do the job of what's normally done by five or six crewmembers, so we were selecting a very talented crew. Also, because it was the final crew, I wanted it to be a group of folks who could handle the PR associated with being the final crew.[53]

They needed to have that capability in addition to all the technical ones we needed to get the job done on the mission. I ended up picking Chris Ferguson, who was at the time my deputy, to be the commander. He had previous flight experience, and his experience working as my deputy chief of the Astronaut Office proved that he was extremely capable.[54] He just did a phenomenal job with the public relations and the outreach, making sure that all the Shuttle employees knew how much they were appreciated. He did a fantastic job.

Doug Hurley was his pilot, and he also had a lot of experience. I had no questions about him at all either, and he was a good compliment to Chris Ferguson — Fergie. Sandy Magnus and Rex Walheim were also selected.[55] I needed one strong EV crewmember for contingency scenarios, and that was Rex with his previous EVA experience. I also wanted an Expedition crewmember because there was going to be a lot of transfer involved, so I wanted a previous Expedition Station crewmember in order to make the process most efficient. Sandy has some amazing organizational skills, and she was the best choice for that. All of them together, an extremely talented group, were able to accomplish all the tasks that normally a five- or six-person crew would have to do. They were very efficient workers and just great people. I think they were a good choice.

When I came in I did move around a few folks in leadership positions, which is not all that uncommon in the Astronaut Office. The rationale for the changes depended on what I needed. In some cases, I wanted to get to know people a little better because I'd been in two Station training flows and had been out of the office a lot, so I didn't know some of the people very well. I moved some people to positions where I would have more potential to interact with them. Others I knew had what I thought would be a very good capability, and that was needed in some of the more complex branches. I moved them into some of the heavier-hitting branch chief roles, and got to know folks in a different way than I had known them previously.

We move people around a lot. We have to be followers and leaders when we fly in space. Learning about people's leadership and followership capabilities is important. So that's somewhat unique within our office — we do move folks around to have them demonstrate these capabilities and try and find them those opportunities so that we can understand the strengths and weaknesses of everyone. It's important to know when you're trying to assign someone what might make the best mix crew-wise. It's much easier to do, obviously, if you know the people.

My leadership style—I had been the deputy chief when Rommel was the chief of the office.[56] I think one of the best characteristics that I observed from him, that I tried to carry over, was his ability to get people to talk to him, explain things, and ensure that they were sharing their opinions and ideas. At first, I was a little surprised he asked questions. I knew that he knew the answers to many of the questions he asked, but he was trying to get folks to be involved, engaged, and make sure they were providing their positions and voicing the issues that they had. I think that was a good talent that I learned from him.

He also is a very optimistic guy; I think leaders need to try and show a positive side. We were in a very challenging time with the Shuttle transition coming to an end and then with Constellation being canceled. I thought it was important to maintain some optimism, although that had to be tempered with the fact that people don't want someone who's walking around with rose-colored glasses all the time either. They want leaders who understand and know the realities but can try and find a positive way to address the problems, issues, or deficits if we have them. So I tried to do that. It was a time of dramatic changes and as a result there was a lot of uncertainty out there that worried people. When you have that many changes going on, it really makes people uncomfortable.

When I started out as chief, I think we had close to eighty astronauts, and when I finished three years later, we had fifty-three active astronauts. We transitioned quite a few folks and attrited a lot of folks in the post-Shuttle era. Losing folks, as hard as it was, was the right thing to do. I knew that. I knew where we needed to be in terms of numbers in the office. I was grateful when we lost those folks and they stayed in the space industry, because that meant we were still going to have some influence. It was important, I think, to recognize where we were and where we needed to be two or three years down the road.

Communicating with the folks in the office about the necessity of downsizing was something we talked about at staff meetings and town halls. Also, individually, I would talk to folks about it as well, people who were trying to decide what was the right thing for them to do—which direction should they choose. It involved a lot of talking with individuals to find out their desires and how to best fit them within the future of NASA and our office. I hated losing that many people, but at the same time I knew we needed to get our numbers down. It was a considerable process to manage that.

I think the biggest thing I did to boost morale was communicate. That sounds easy, but when there aren't definitive directions and goals, sometimes that gets challenging. When the agency's path was not clear, it made it more challenging to convey that optimistically to anyone.

I think probably one of the most challenging and satisfying parts about being chief of the office is a lot of these folks that worked for me were a lot smarter than me. While this is not a bad thing, it can be a tad bit intimidating. I think it was really important to draw on as much of that talent as possible to help try and find solutions or define directions.

We had a big review by the National Research Council (NRC) of the Astronaut Office's size and its significance in the post-Shuttle era. That was instigated by the OMB (Office of Management and Budget) putting a line-item requirement into the NASA budget that required this review.

It was a big challenge to try and educate and address the concerns and misconceptions that were out there about the role of astronauts and our training mechanisms. I brought together a lot of the folks in my office to help come up with the rationales we needed for the presentation we had to give to the NRC. That process took over six months to develop all the information and present to the commission on three or four different occasions, plus the additional information that was required after they presented a number of questions. It was challenging to try and convey that in a way that the NRC members could understand, and then portray back to NASA the importance of what we were doing and how we were doing it.

It was a very challenging time. We had a lot of folks dig in and help with different aspects of the presentations and documentation. It was a great team effort. In the end, we were obviously very successful. The NRC came out and not only supported the process we were using but said maybe we were pushing our margins a little too close in terms of the number of people that we had for out years.[57] I was very pleased with the results from that. I couldn't have asked for a better outcome, but it was a huge team effort to get all these pieces arranged in a coherent, logical story. We had people making videotapes of certain aspects of training and developing lots of different charts. We developed white papers to go along with it just to provide all the detailed background of how we determine the size of the office and what we need in a crewmember and why.[58]

The office, unfortunately, has to do versions of this every two or three years. We have to justify our existence, because we are a very public organization in the sense that we're the face of NASA and at times the target. We come under the gun pretty frequently, particularly relative to the size of the office and how we train, whether we need to use T-38s or how we could do this cheaper.[59] It happens frequently, unfortunately. I was hoping this review might last for two or three years, at least, before we had to do another one.

When it came to the new commercial ventures, we developed a white paper because we knew there was lots of talk about whether there was a need for an astronaut corps, even though we are still flying on ISS.[60] It takes two and a half years to train people to fly a specific mission on board the Space Station. No commercial venture is going to take over that two and a half years' worth of training and integrating with international countries through intergovernmental agreements that we have established. We tried to lay out in our white paper what we thought the ideal process would be. We do think that the commercial companies are going to want to have their own pilots, but during test flights we want to be involved. It is similar to the way the military does a lot of their development—they'll have what they call a joint test force. We envisioned that we would do something similar.

Eventually our crewmembers would be trained to launch and land those vehicles, to take them to and from the Space Station. The commercial providers may have other pilots that would do other missions with tourists or other people, but missions to the ISS would be flown by NASA astronauts. There are a lot of unknowns, but we just tried to lay out a plan: Here's what the astronaut requirement is going to be, and here's how we envision you could potentially make it work. There was also talk about rental car versus taxi. Since the taxi driver version would require additional launches to carry the driver, we didn't consider that too feasible. In any case, the ISS crewmembers still need to be able to perform the emergency descent. There are trades to every aspect. I think in the end it will be most beneficial and efficient for our crewmembers who train the two and a half years for a specific mission to also include ascent and entry training to their plan.

As chief of the Astronaut Office I also worked with MOD (Mission Operations Directorate) and FCOD (Flight Crew Operations Directorate).[61] The Astronaut Office has a very important role to play in integrating with these organizations. Working with MOD is probably the most obvious one because they're responsible for doing mission operations. Luckily, because I'd done two long-duration spaceflights, I knew a lot of the flight directors. Actually, the head of the flight directors was a guy that had worked with me during my first expedition, John McCullough, so it was easy to develop a very close relationship with MOD during the course of that timeframe.[62] Even if we didn't always agree 100 percent, we always went together to the program or to anyone else. When there were issues or concerns, we always wanted it to be a united front from Mission Operations and the Astronaut Office. We strove to make it that way. It required a lot of communication and expressing concerns both ways so that we had our stories ready and intact. It helped us, I think, a lot.

We also work with the Mission Operations Directorate's training team a lot. We have folks that are directly tied in with the training organization. Obviously, with a two-and-a-half-year training flow, there are always issues that come up, so we work closely with them as well. It involves five different space agencies and has to be coordinated. The other organization we spend a lot of time with is the robotics and spacewalking organization, the EVA group. We met and coordinated with them frequently and had teams interact at a lower level to ensure that our operations experience is integrated into their training and into their procedures.

We also work with the ISS Program a lot. I had worked with Mike Suffredini on a couple different levels before I became chief.[63] That gave us an experience base from which we were able to work well together.

I guess those are the big groups in terms of organizations that I interacted with. It was a real honor, I think, to be able to work with so many different organizations and groups in my role as chief of the Astronaut Office.

When it came to missions, the Astronaut Office also provided information and data for the flight readiness reviews for Shuttle and Station.[64] Typically, the Flight Crew Operations Directorate, our bosses, represented us on the boards for most of those activities. I went to Kennedy Space Center for all the launches and almost all the landings of the Shuttles that were flying there. There were some conflicts where I would be in Russia for a Soyuz launch, and so my deputy and I would divide up responsibilities. In those cases where there were conflicts, one of my deputies would go to one and I would go to the other. We were also involved with the return of the crew from Kazakhstan after landing.

By the way, the direct return of crews upon landing was developed when I was chief of the office. This faster return process enabled crewmembers' access to the facilities and the data collection here sooner. Since science is one of our primary objectives for ISS, getting more data was an important reason for this change. I would fly out on the helicopters to pick up the crews at the landing site with the Russian search and rescue team and bring them back to our NASA aircraft to fly them back. It was very interesting getting to be a part of all that.

When I'm asked about memorable launches or landings during my tenure as chief, I think each one is memorable, each individual one. From the Station side, the individual crewmembers provide so much flavor of what their mission was about. When they first land, they want to tell you everything in the first ten minutes; they want to share that excitement of their particular mission. It's special to be a part of that. Of course, STS-135 is probably the most memorable Shuttle landing.

When I was head of the office, I handled some crises, one of which

Expedition 29 Commander Mike Fossum comments to Peggy Whitson, chief of the Astronaut Office, about how strange the effects of gravity feel just a few hours after he and his crew landed in their Soyuz.

involved astronaut Mark Kelly.[65] His wife, Congresswoman Gabby Giffords, was shot in Tucson in 2011.[66] Initially we didn't know what condition Gabby was going to be in, and whether or not Mark would be available to command his flight. We pretty quickly named a backup commander in case we needed it. C. J. Sturckow, my deputy at the time, was named as the backup, just in case.[67] Because the mission was, I believe, six or seven weeks away from launch, we needed a very experienced person to be able to just step in and take the reins, if necessary. Over the course of the next few weeks, we were waiting to see what would happen with Gabby and to make the decision on whether or not we'd go with C. J. or Mark.

I think for the mission success, Mark was the right choice, from his experience over the course of his training flow during the year or so that he'd already been assigned to the mission and in training. He knew the mission best. However, it was obviously a time when there were a few other things on his mind. That made it a very challenging decision. Whether to go with someone with less overall understanding of the specifics of the mission or someone who could be focused 100 percent on the mission. It was a big decision, a very challenging one. We ended up continuing with Mark on that mission. I do think it was a challenge for lots of folks for lots of reasons, especially because of the extraneous activities that were required as a result.

I had one other instance—I think it was a few weeks before—where I had to replace a crewmember because of an injury. It was also too close to his flight for him to recover from the injury. That one was very challenging. That was Tim Kopra.[68] He had been training with his crew for a year, and he was obviously very disappointed.[69] He took it as a professional, because he understood the decision, but it was obviously very, very disappointing. I got input from various folks and looked at how to split up the crew training. We ended up selecting Steve Bowen, and then offloaded some of the other responsibilities that Tim had.[70] Tim had been MS-2 (mission specialist-2), which is the flight engineer on board the Shuttle, and plays an important role.[71] He was EV-1 as well. So, it wasn't an easy replacement in a short period of time, but Steve Bowen had the EVA expertise. The EVAs themselves had already been worked out by Tim and Al Drew.[72] The details had been pounded out, so it was just a matter of practicing the EVA. We made Steve the lead EV crewmember, but we didn't make him the lead in terms of the flight engineer responsibilities. We split them up between the people who had been the MS-1 so that we could more evenly distribute the training load in that last six or seven weeks before the launch of that Shuttle.

Those were the two instances I had in a very short period of time trading folks out. It happened pretty commonly when I was deputy chief in the ISS arena, where we had to change out a lot of crewmembers. Typically, it was some last-minute medical issue. As chief, I changed the timing of the medical certification process to better ensure that crewmembers who were not medically qualified were not assigned to an expedition. That worked out better, such that we didn't have nearly as many ISS crew change-outs.

When I had those challenges, the office tried to be as supportive as possible. We have a family support group in cases where there are issues, maybe somebody's in the hospital, whether it is a crewmember or their spouse. The family support group consists of the astronaut spouses that work to ensure we have the right support for the crewmembers or their families as needed, depending on the situation. The spouses' network will get together and provide food for the kids and make sure the kids get to school—whatever they need.

I met with them a couple times. Typically, just to share information. We have a liaison in our office who works and interfaces with the family support team, sends out emails to let folks know what's going on with the various missions and different aspects during the missions to keep families informed. I've had to worry about that and use that more often than I would like, but it's great that folks come together and do that.

Since I've been chief of the office, we've also had some new astronaut candidates in. Typically the chief of the office is the deputy chief of the selection board. I was actually the chair of the selection board for the 2009 class, filling Brent's role basically, because I was not chief of the office at the time. So I played a key role in their selection and knew them from the selection and those interactions during selection.

I felt like they were my kids. I was very much more, I think, invested in how their training was going, where they needed help, and making sure they got help where they needed it. I had several little social events where I'd bring them over to my house and talk to them about what my expectations were and make sure they felt like they could talk to me if the need arose. I wanted them to not feel too intimidated, to come and talk to me if they had issues or concerns. I worked pretty hard to maintain a relationship with that class, in part because I felt responsible, having chaired the board that recommended them for selection. I had much more ownership of them in my mind, a feeling of responsibility. I paid a lot more attention to that group.

Ellen Ochoa[73]

I was asked if I would become the deputy director of the Flight Crew Operations Directorate in 2002. It was a lateral move, so it wasn't a position that was advertised and for which they needed to do a competitive placement. It was a few months after I had gotten back from my fourth Space Shuttle mission.[74] I was one of the senior astronauts in the office and had gotten to

Official Portrait of Ellen Ochoa.

fly several times. Because I was involved in the senior leadership of the Astronaut Office, I knew that there was a big push to make sure that we were flying all the astronauts. The assembly of the International Space Station got delayed a little bit because we were waiting for Russian modules. And then, as we got into assembly, it was very focused on doing spacewalks; there was a lot of emphasis on making sure we had experienced astronauts doing some of that work. So, while we were flying new people all that time, we weren't necessarily flying a lot of new people on each mission.

We had a fair number of astronauts who had never flown, and one of the big goals of the office was to make sure we got these people flight experience, both as a matter of fairness, wanting to get them into space after training, but, also, they become much more valuable to the office, to all of JSC, once they have the flight experience. Then they can come back and provide their expertise to the programs, to training, to everything that goes into planning future missions. They're much more knowledge-able once they've had that opportunity. Because of that big push I knew it wasn't likely that I would get reassigned to a fifth mission any time soon. I was well aware of that even before I flew my fourth mission. I felt lucky to get assigned.

I had been doing some other jobs in the office since coming back. Troubleshooting roles where they had a particular issue they were work-ing on with some of the international partners and they wanted me to get involved. I had done a lot of the senior roles in the office already, so when they approached me about becoming deputy director of FCOD that was obviously quite a different step for me and a chance to do something different, a new challenge. I thought about it for a little bit, and I felt like it was the right time for me.

FCOD manages the Astronaut Office and the Aircraft Ops Division (AOD). Within every directorate you're in charge of all the policies, the personnel, budget, and making sure that your directorate provides what it is supposed to provide to all the human spaceflight programs and to the agency as a whole. You need to provide trained crewmembers. You need to provide the right aircraft resources and the people to fly those aircraft. Obviously, the astronauts have a big role in public outreach, so you need to make sure that you are providing that role, in addition to the operational role that astronauts provide, and everything that goes along with doing all that.

About six weeks after I became deputy director, we had the *Columbia* accident.[75] It was really soon after I came to the office. In November and December of 2002 we got a new director of Flight Crew Ops, Bob Cabana, a new deputy—Bob selected me as his deputy—and a new head of the Astronaut Office, Kent Rominger.[76] We refer to him as "Rommel." So we were all new in our roles for that mission, essentially.

Normally the director of Flight Crew Ops goes to Florida for the launch and also to Florida for the landing, and the deputy would be in Mission Control for both of those as the Flight Crew management rep in Mission Control. We both went to Florida for the launch. That gave me a chance to shadow Bob in terms of what the director of Flight Crew Ops does during the whole launch run-up. There are certain meetings that you attend, including the prelaunch Mission Management Team (MMT) down there.[77] Then you're actually with the Mission Management Team in the Launch Control Center during launch.[78] Your role there is, of course, following along with everything that was going on. In any part where there's a decision that needs to be made, or a go for launch, you're representing the flight crew and saying whether or not the flight crew themselves are ready to go, and whether, in your judgment as the manager of the Flight Crew, you feel everything else is ready and you're willing for the flight crewmembers to get on the vehicle and actually go launch.

There were a number of missions planned for the upcoming year. Bob had said, "Hey, I want to go do that for the first couple launches, but then we'll alternate that role."

I said, "Well, why don't I come down with you to Florida for this first launch? I'll get to see how everything unfolds prelaunch. Then I'll feel comfortable later in the year going down and doing that role myself." So we were both in Florida for the launch. That was neat for me because I knew Rick Husband really well.[79] We had flown together before. I actually got to see him suit up on launch day and give him a hug before he headed out of crew quarters.[80] Of course, I knew all the astronauts on the flight, but I was closest to Rick because we had flown together.

Then Bob and I came back to Houston. Naturally both of us are following along in the mission, but he said, "I'll attend the mission management meetings," because there's always lots of other things going on as well. So I did not attend any of the Mission Management Team meetings during that mission.

January is when you start the budget process for the upcoming year, and I was going to have to learn that from scratch. He wanted me to be involved in preparing the budget for the upcoming year. It was completely new to me. I can remember spending time getting briefings from our resources person that supported our directorate, and then starting to talk about what's really involved in putting a budget together. I was focused on learning how the whole directorate operated and keeping tabs on the mission, but Bob was the one going to the Mission Management Team meetings.

I, of course, heard about the fact that there was an impact on the Orbiter. I remember seeing the email with the video of it and watching it. I had some discussions with our FCOD personnel and the chief of the Astronaut Office, trying to determine if we were asking the right questions in the MMT meetings.

I was not aware of the great amount of discussion that was going on at a lot of different levels that I probably only found out about as the investigation proceeded, particularly when I ended up reading through the whole report that came out from the CAIB (*Columbia* Accident Investigation Board).[81] They of course had compiled all the different conversations that had gone on in hallways and in emails that had been traded between flight controllers and engineering people who were concerned. There was a whole level of things that I didn't know was going on and only found out after the fact. But I had some knowledge of the event in general.

Obviously, I think many of us didn't have a full appreciation of what this meant and how worried some people were. That didn't make it to Bob and to me. Some of these conversations that were going on were showing people were clearly extremely worried about that event.

I was in Mission Control for landing. As I mentioned, the director goes down and sees the landing and is there with the families. The deputy is normally in Mission Control at the MOD management console. I was sitting next to John Shannon and Phil Engelauf, who were representing MOD management.[82] That's just a longstanding role for the deputy. Obviously, the main reason to be there is in the event of a contingency, but I wasn't really worried. Again, I didn't fully appreciate the seriousness of the impact on the Orbiter. I wasn't aware of so much of the work that had gone on, so I wasn't going in there thinking about that.

A lot of things have to happen for a safe entry, but we had always had a safe entry. However, I reviewed the contingency book, because I knew that was my job. I had looked through it to see what my role was if something happened. I had the big book with me, but certainly was not at all expecting to ever open the book that morning.

There's so much that happened after we knew the Orbiter and the crew were lost that I could spend days just talking about what happened in the following three hours. I, of course, pulled open my contingency book. I did bring out the notebook I kept that day to remind me of things. We do have what we call an FCOD Contingency Action Center that's activated on launch days and on landing days. It's over in Building 4 South, where the Astronaut Office is. I talked to them within five or ten minutes of the contingency being officially declared.

The deputy CB chief was there.[83] Of course the chief was down in Florida also. So that's where the deputy Astronaut Office chief is, plus a couple of assistants and support people. We talked. They have a list of things they need to do, and we started to talk about who we needed to contact. Each astronaut on the crew has what we call a CACO, a casualty assistance calls officer. Before flight each astronaut designates another astronaut in the office who will work directly with that astronaut family should there be a contingency. One of the first things we needed to do was call those astronauts and put them on notice that they're about to take over this role.

That started a whole chain of events. One of the things we wanted to do very quickly was inform the crew on board ISS what had happened.[84] I did leave the Shuttle Control Room and went across the hall into the Station Control Room. We brought in Beak Howell, who was the JSC director at the time.[85] We brought up a private air-to-ground loop to talk to the crew. Beak talked to them first to let them know what had happened. Then I actually got on and talked with the commander, who was Ken Bowersox at the time.[86] We didn't have very much information at the time, but we told them as much as we could and that we'd keep in contact with them, because we wanted to make sure they were informed.

After that I called up Bob Cabana. He had, I would say, by far the toughest job that day, because he was at crew quarters with the families, and had to inform the families. The contingency, I think, was declared around eight fifteen a.m., because that was supposed to be the time of the landing on Saturday, February 1. At eight forty-one, we informed the ISS crew. At eight fifty-eight, I talked with Bob. There was a mission evaluation team meeting at nine o'clock. Then our first Mission Management Team telecon, which essentially turned into a mishap response team telecon, at nine thirty that day. I attended that as the Flight Crew representative.

That was the first time they had everybody together to talk about how we would respond to the mishap. Ron Dittemore talked as the Shuttle program manager.[87] Linda Ham was the head of the Mission Management Team, so she called it all together.[88] We talked about the very first steps, what do you have to do to protect information, protect the debris, and what the next steps are. There are a number of teams that get kicked off. One of my jobs being the senior Flight Crew rep in Houston—obviously again Bob and Rommel were really dealing with the families at that point—was to make sure we had crew reps on all the teams that were forming. There are a couple teams for which you predesignate crew reps. There's something called a rapid response team that would go to the actual site. Our crew rep was Jerry Ross on that, so he came in that morning.[89] There was at least one other team that I think had predefined crew reps.

The engineering team pulled together to come up with a very detailed timeline of everything that we knew; we wanted to put a rep on that. It just ballooned out from there. Every time something would come up where we were going to form a special team for it, then we tried to make sure we named a crew rep to that.

Pretty early that day, before noon, I got a call. We had a bunch of astronauts out at Aircraft Ops that wanted to head up to East Texas.[90] JSC was starting to receive calls from people up there with reports of debris on the ground. Very quickly the whole team had to try to figure out, "Okay, what do we do, how are we dealing with the debris, how are we logging the calls, how are we pulling all the information together into one database that we can then respond to as a team?" We of course wanted crew reps involved in all that. We talked about how we were going to get crews out there. We wanted to involve our medical folks because they knew how to deal with this issue better than we did. We got all of them involved; that was one of the first things that happened that day as well.

Then, of course, there was an independent commission that was formed, and Admiral Gehman chaired the *Columbia* Accident Investigation Team. We needed to have a crew rep on that team. Those were all things that were forming pretty early on in those next couple days or so.

Those were long days. I left home—I don't know exactly, maybe around six or six thirty in the morning Saturday. I can remember telling my husband, "Well, I'll be home for late breakfast." It was about fourteen hours later when I got home, and that was pretty typical over the next couple of weeks, maybe longer. There was just so much going on, so much happening. We were trying to make sure we were involved in all aspects of the mishap response, but, also, we had special activities associated with the astronaut families and dealing with all of that.

We had no shortage of volunteers, clearly. Everybody in the Astronaut Office wanted to help. Certain people, because of their particular position in the Astronaut Office, were logical people to select. We have branches in the Astronaut Office; there's a branch that supports the Shuttle Program. I think Dom Gorie was heading that, so he was very involved in the investigation.[91] There were other senior astronauts that were logical people to put on these teams. Some of them were just based on what job they currently had in the office. Others were based on their seniority and experience. Maybe they had supported the Shuttle Program Office itself at some point or had previously been head of the Shuttle Branch in the office, so it made sense.

Then you just start to tick down the names as new teams came up. Of course, it wasn't just me. It was me working with Rommel, and there were four of us at the time—Bob and I as the director and deputy of FCOD, and Rommel and Andy Thomas as the chief and deputy of the Astronaut Office—who would get together and talk about who does it make sense to put on these teams.[92] There was a level then where I wasn't even involved; Rommel was involved as you got another layer down.

Everybody wanted to be involved. The hardest thing is not to have a specific role after that, I think. Not only did we have a need for a lot of people, but we actively tried to get people involved as much as possible in some way, supporting some team, because the worst thing is to come in and not have a specific role to do, and not really know when you would.

Of course, in the immediate aftermath there was just no telling when we might get back to flight. The whole question starts to come up in the media, if nowhere else. Should we even continue the Shuttle Program? There's a whole uncertainty factor now thrown into the whole future of human spaceflight that you have to deal with in addition to the more specific response to this particular mishap and what it might take to understand it, respond to it, and get back to flying. Assuming we were allowed to do so.

Within the next day or so we were trying to work on the memorial that was going to happen here at JSC on the following Tuesday. I don't remember at what point exactly, but within twenty-four hours we knew the president was coming.[93] That kicked off a lot of activity around the center. We certainly needed to have someone representing the crew, and someone in charge of notifying all former astronauts about the event and helping to have astronaut escorts for crew families that would be coming in.

Certainly, I was not one of the prime people working on the specific plans for the memorial, but the office was supporting it in every way possible. Sylvia Stottlemyer, who was the administrative officer for FCOD, was extremely helpful.[94] The astronaut that we assigned to it, Mike Bloomfield, was another key person in thinking about all the ramifications.[95] We needed to understand again what roles astronauts would play in that memorial. In a lot of cases they were escorting the families of astronauts coming on site. We, of course, made a special point to talk to the families of the *Challenger* accident and of the Apollo 1 accident and invite them.[96]

If their spouses or families were coming, we had a current astronaut who would meet them here and bring them on site and be with them during that time. You can imagine that could be an extremely difficult time for them, in addition to the current families who were involved.

We had a flyover as part of that, and because Aircraft Ops is part of our directorate, that's something that we were involved in planning.[97] Of course, we can primarily hand it over to Aircraft Ops; they know exactly what to do. We just made sure that we had plans for that in place, and that they were included as part of the memorial.

Then, because a lot of former astronauts were coming—with not a whole lot of notice, but for a lot of them it was really important for them to be here—we wanted some kind of reception so that they didn't just show up, attend a memorial, and then walk out. People wanted a chance to get together and talk about it, so we wanted to provide that as well. Those were the immediate concerns we were thinking about.

We wanted to make sure we were reaching out to all the current astronauts and their families as well, trying to make sure that we had periodic all-hands meetings. I don't think we actually planned an all-hands with astronauts and their families prior to the memorial because there wasn't time. But I think by the end of that week we did. Then we tried to do that periodically in the first few weeks after the accident to make sure they were getting the information that we knew about the mishap, what caused it, and how we were responding to it.

We were also thinking about counseling and support for JSC's team members, although it wasn't necessarily FCOD's particular responsibility. I think that was something they were really thinking about at the center management level, because, clearly, while FCOD is greatly affected, so are so many other people at the center—both because so many supported the Shuttle Program, but also because so many people knew the astronauts personally. They helped train them, or they were neighbors, or their kids were on the same soccer team, or they went to the same church. We knew it would have a huge, huge impact on the whole center.

The clinic was open all day Sunday. The Employee Assistance Program counselors were in the cafeteria. I know they told the folks in MCC (Mission Control Center) that they were going to make counselors available to them. Then we did the same thing for sure in FCOD for the astronauts and families. So that was something we thought about not only in FCOD, but they thought about center-wide, of trying to provide that. That continued for quite a while, extra support in that area.

I did have the opportunity to go up to East Texas. I didn't stay up there for days or weeks like some of our crewmembers did, but I did go up a couple of different times. They actually had a memorial service up there one week later, and that was my first chance to get up there. In addition to attending the memorial service there, we had a chance to talk to everybody in the incident response center they'd set up, not only our folks but folks that had come over. Dave King from Marshall Space Flight Center was heading that up, so we talked to him.[98] Jim Wetherbee was one of our astronauts who was a primary person up there, and Jerry Ross.[99] We had astronauts that would go up for several days, and then we'd switch out. We essentially did that. We had that specific astronaut support until everything associated with our crew had been located.

I got to just be there that day and talk to a lot of different folks and see how that was all headed. It was hard, but it was really neat, too, in some way to see everybody pulling together and to see the support from the community. We had really incredible support from the community where you had people cooking and providing food and bringing it in. That went on for weeks. People doing laundry for our folks—just really, really incredible.

We got to not only be at the main center in Lufkin, but we went out to Hemphill and I think one other place, and we talked to a couple of the satellite centers. Just with the support from everyone there, as well as we ended up having, I don't know, two dozen different state and federal agencies that were involved. Huge support from the US Forest Service. I went back at one point; this was after the Forest Service folks—the people that are really experienced and go around and provide wildfire support—came in to support the actual debris collection. They had set up this huge camp. They already knew how to come in and set up a whole tent city, and that's exactly what they did here. I came in one night and just did an astronaut outreach, talking about Shuttle missions and what we do and gave them a chance to ask questions. We did that with various astronauts. I went by, so I got to see that whole operation.

Then, as you may unfortunately remember, there was a helicopter that had a mishap during the recovery, and we lost a couple folks there.[100]

I went up for that memorial, which was just really tragic. It was almost right at the point where people were starting to get over the initial shock of *Columbia*. To have this happen was just a real downer. But again, being up there gave me a chance to see our folks that were up there, and everybody else who was supporting, and go around thanking people. That was an amazing, amazing operation up there.

The accident and recovery weren't the only things going on. We had our day-to-day tasks. Bob and I managed FCOD by talking through it every day all day for fourteen hours. As every new thing came in Bob would either say you go do this or I'll go do that. It was just literally minute by minute. We worked very closely, as I mentioned, with Rommel and Andy. We were just literally tagging up every few hours because so much was going on.

We still had to do this whole budget preparation too, which, again, I was learning from scratch. NASA as an agency has a deadline that you have to meet every year with preparing all your plans. You have to prepare five-year plans. Of course, that was even harder when you had no idea what your plan was, much less coming in new to the budget anyway. That was all going on that spring as well. All I can say is I was learning something new every minute of every day.

Bob and I were just joined at the hip, meeting constantly and talking constantly about now here's something that we need to do, how are we going to do this. Of course, we called on everybody that worked in the FCOD Office, and everybody that worked in the Astronaut Office and AOD as much as possible, all of our support people. Everybody really stepped up; you know, it's what JSC does. You do what you need to do.

I cannot recall when we knew that NASA would be flying the Orbiter again, so we could move forward with our five-year plan. I believe it was January 2004. It was about eleven months later when President George W. Bush came out with the Vision for Space Exploration. I flew in for that announcement, when the president came over to NASA Headquarters and rolled out the Vision for Space Exploration. He said we're going to continue to fly the Shuttle, but we also have an end date. We had to really change the approach to the assembly of the International Space Station, cut down on the number of flights that we were planning, because we weren't going to be able to do our original plan.

There was a huge replan effort in the ISS Program then. In the immediate aftermath of the accident they had a huge issue. We depended on the Shuttle for our consumables. The Progress does come up, but the bulk of them come up with the Shuttle.[101] Anything we need to get down comes on the Shuttle. So now we had no way of getting anything down, and we had a very limited way of getting things up. Not even knowing how long that might go on, they immediately had to go into what are we going to do, can we even keep the Station crewed? Of course, as you know, we went down to two crewmembers, I guess, by April. Until the Shuttle started flying again, we stayed at two crewmembers but were able to keep them supplied. Part of our job was working with the ISS Program on the plans for keeping it crewed and how we were going to work on that.

We were working very closely with both programs—the response to the Shuttle and then how we actually responded on the ISS side. With this new vision, knowing that we would eventually get back to flight, of course you still don't know exactly when that's going to happen. The program obviously responded and figured out how to assemble the ISS in fewer missions. We supported that from the crew perspective as well. Of course, every week the programs have control board meetings and go through their whole process.[102]

I have no idea how we came up with the budget that year. The bulk of the budget in Flight Crew Ops is really for our Aircraft Ops. There's actually very little budget associated with the Astronaut Office itself. We have to know how many people are in there so we know how many civil servants to account for, but there's very little money that is provided to the Astronaut Office itself. What you mainly have to figure out is how many flight hours you're planning to fly in each of those years. Then you have to guess about gas prices and spares. We had to guess how many flight hours we thought we would need to fly both for the T-38s, which are the bulk of our flight hours, but then also the Shuttle training aircraft and things like the Shuttle carrier aircraft, which are much fewer hours but then more expensive per hour to fly.[103]

When you don't know exactly how many flights you're going to support, it was difficult. We had to come up with some scenarios and some assumptions about those flight hours. That was one of the major things we were trying to do. At the time, we were also advocating for an upgrade to our T-38s in terms of their avionics, adding some safety features, and upgrading the ejection seats. So that was another added wrinkle to the budget. That's certainly not something we're doing every year. This was something we were specifically requesting funding for from the Shuttle Program.

We ended up having to get an independent review of what we were planning to upgrade, and how we were planning to do it, to satisfy the Shuttle Program management that we had a sound plan, a cost-effective plan. I know that was one of the major things I was working that was independent of the whole response to *Columbia*. That was eventually important to our continued operations.

Did things return to normal after the accident? Well, there was certainly a point where the number of hours I was putting in at the office got back to a little bit more of a normal schedule. I don't know that I could tell you how long that took, several weeks at least, probably a few months. Clearly we were very focused for the next two and a half years, really up until Return to Flight, once we had a good idea of what actually caused the accident, of trying to lay out what it would take to get back to flight.

9

Center Management

The center director is the top government official at NASA's Houston field center, the Lyndon B. Johnson Space Center. The director wears many hats and oversees all center management, strategy, operations, development, testing, and facilities. The center director implements programs, carrying out plans and policies drafted by NASA Headquarters in Washington, DC. In addition, JSC's leader strives to maintain the center's technical expertise in human spaceflight. Without the vision and courage of these men and women, the twentieth and twenty-first century spaceflight successes and advances might not have been achieved.

Dr. Carolyn L. Huntoon came to JSC in 1968 as a National Research Council fellow and quickly became the authority on human space endocrinology and metabolism. Huntoon pursued flight research from Apollo through the Space Shuttle Program, when Spacelab missions made conducting extensive research on humans possible for the first time since Skylab. Over the years, she successfully combined her scientific interests with her administrative and managerial skills as head of the Biomedical Laboratories Branch, JSC associate director, and director of Space and Life Sciences. In 1976, Center Director Christopher C. Kraft asked Huntoon to serve on the selection board for the first class of Shuttle astronauts, a position she regularly filled until she headed the center. In 1994 she became the first female center director at JSC and within the agency.

Center Director Dr. Carolyn Huntoon introduces JSC employees to Houston Mayor Bob Lanier, US Senators Barbara Mikulski, Christopher Bond, and Kay Bailey Hutchison, and NASA Administrator Daniel S. Goldin. Lanier, Mikulski, Bond, Hutchison, and Goldin were visiting JSC.

Carolyn L. Huntoon [1]

I spoke with NASA Administrator Dan Goldin on several occasions about the possibility of becoming the JSC center director.[2] When he first brought it up—I never asked for the job—I told him I was reticent. JSC was mostly engineering, and there had never been a scientist managing the center. He said, "You know, there are a lot of engineers that work at NASA, and you have many great ones at the Johnson Space Center. If you can be a manager, you don't have to be an engineer."

Official portrait of JSC Director Carolyn Huntoon.

He thought my background, my education, my talent for getting along with people was what the center needed at that time. I understood the priorities we had with the Shuttle and the Space Station. So he came back to me again, and I talked with several people, including Dr. Chris Kraft, Aaron Cohen, Gerry Griffin, Pete Clements, and others whose opinion I valued, and they all encouraged me because I was qualified, and it was a remarkable job.[3]

My management style, in the broadest sense, did not change when I became center director, because I like to deal with people. I like to get information. I want people to tell me things and show me with numbers or graphs. I'm not into micromanaging, but I cannot understand magnitudes of things if someone doesn't show me the before and after or the effect. I guess that's the science background I have and that hasn't changed very much through the years in the various jobs I've had. The idea that I want to see how things work when something is an issue, that hasn't changed. Having served in most every level of management at the center gave me the understanding of the magnitude of the job.

Being center director was a bigger job, certainly, and the center budget—when compared to the budget in Space and Life Sciences—was larger, and it was spread over more programs. There were more issues than I had had before, which was one of the changes. The job of center director required more involvement with the various political offices in Houston and Washington.[4]

I certainly had given many technical speeches in Life Sciences and, because of my involvement with astronaut training and selection, I had given numerous general speeches about the Shuttle Program and crews. But I really got into public speaking more as center director, because I was called upon to represent the center much more in Houston as well as nationally.

I think the delegation of work was also something I accomplished. I had to depend on people I thought were telling me the right thing, and I had to learn to rely on their judgment. But my intuition about work and people is pretty good, and I still stick with that a great deal.

I learned early on in managing Life Sciences, you always wait until you hear at least two versions of the story before you act, because everyone has a different view of an event. People want you to make decisions when they come tell you something. The lesson I learned was to listen and say, "Well, I'll look into it," and find out the other side of the issue or the rest of the story. Only then would I make a decision. As I matured in management, I learned more about making timely decisions but not being too quick to jump to conclusions.

Soon after becoming center director I initiated a major reorganization at JSC. We had not had one for some time. We could have managed the center with the organization that was present, because it's the people that are important. The right people can be in almost any job title reporting most anyplace, if they're the right people and doing the right job.

What we tried to do was streamline the organization and make it so there were clear lines of responsibility, because over time these meshed together. We were really trying to clean up the lines of responsibility and take advantage of the timing, because the Shuttle-Mir Program was drawing to a close.[5] Many people had been devoted to Shuttle, and we needed their expertise to do Station work. So it was just the right time to tighten the organization.

My primary goal as center director was, of course, to run the center in a way to accomplish the work for which we were responsible. The work was operational human spaceflight and every aspect of it. The Johnson Space Center has expertise in almost every field of engineering and science. The center was formed with three major contributors (engineering, operations, and science), and I wanted to make sure that that was the way we continued our work with Shuttle and built the Station. We successfully accomplished eleven Space Shuttle missions and initiated the center's support of the International Space Station.

One of my goals was to recruit well-qualified engineers and scientists. At this time, employees who had worked hard on Gemini and Apollo were retiring and leaving some huge voids. We were able to bring onboard many excellent engineers and scientists.

We have an excellent Personnel Office at the center, as well as an outstanding Business Office and Facility Operations. I wanted to make sure we supported those people, because they are the ones that kept our operation running smoothly. Often when there's a tightening of the budget, it was our support offices (the Center Operations, the Personnel, and Business Office) that were affected the most. We had to assess and increase support of those organizations. It was a period of assessment for the center, of rebuilding some of the organizations that needed the attention and focusing on what the change in the agency would be with the Shuttle and the Station becoming the primary focus.

I also opened the Technology and Transfer Commercialization Office.[6] From the time I came to the Johnson Space Center, technology always was a big issue. The fact that we were, not just in science but also in engineering, doing innovative work that could be used other places had always been a focus. There had not been a specific technology office for some years.

NASA Headquarters was paying more attention to technology at this time, and we needed someone to not only interface with our scientists and engineers, but also to work with the Washington office to make sure that the money and the time we were spending on development was being put into use wherever it was needed in the private sector or in other government agencies.[7]

Throughout my career I formed networks and partnerships with the community and with academic institutions, industry, and commercial operations. It was quite evident that we had a tremendous job, and we were in a unique place in history in our country. We had funding to accomplish things, but we certainly did not have all of the knowledge at the center.

So, in order to reach out and get the best people in the country thinking about some of our problems and helping, we formed quite a few relationships with the private sector. During my postdoc, I formed an advisory group of scientists from Massachusetts General Hospital, the University of Miami's medical school, and several from California, all outstanding men and women in my field. They came to the space center several times a year, reviewed the work I was doing and made suggestions, or they would go home and do things to help. It was a model I used even after that, to reach out to get people to address our problem areas.

We also did that in a technology sense. When we needed a piece of equipment, we would call on everyone who was working in that area and tell them what we needed and see who would respond. When I became center director, I wasn't the only one at the center working with the community on technology efforts, but it was not a coordinated effort. We expanded it into an office, where the engineering as well as the science aspects of technology could be managed.

One of the facilities that opened while I was center director was the Sonny Carter Neutral Buoyancy Lab.[8] JSC originally had the WETF (Weightless Environment Training Facility), which had been built toward the end of the Apollo Program.[9] The round centrifuge was removed from the building and a big tank put in its place. All of us thought that that was probably as much as we would ever need for EVA training.[10]

As we were preparing for the Space Station, and actually a little bit before, when we were doing missions like the Hubble repair, we realized that the WETF was not large enough.[11] It was scheduled two and three shifts a day because there were so many EVAs on the Shuttle, and it was not large enough for all the training on payloads, Shuttle repair, and Space Station construction. So several studies were done, and the first people that came forward were the engineers from the Man-Systems Division, which was in the Space and Life Sciences Directorate.[12] They presented their case. The WETF was not going to be enough to support the Shuttle flights, the Hubble repair, and build the Station. We needed more capacity.

The Hubble repair EVA training was conducted at the Marshall Space Flight Center because they had a Neutral Buoyancy Simulator.[13] Still, we did not have the time or space in the WETF tank to accomplish our work. They presented all this to me, and I laughed them out of the room the first time. I said, "You know this agency is not going to build another tank and certainly not one that size."

They left, went back and calculated their data some more, and came back in again and convinced me. So I took it up to the center director, who at that time was Aaron Cohen. The first time we told Aaron this, he reacted the same as I had and told us there was no way the agency was going to buy or build another tank.

We went around the country and looked at everything you could think of that might be big enough for that work. People say, because they don't know, "Well, they make movies where they have ships that turn over in the water. They must have big tanks out there." Well, they don't. We found out that in Hollywood they flood a parking lot when they need to do an ocean scene, and that's how they have enough water. We also found there were large concrete tanks that had been built in, I believe, West Virginia, but were never used, so we could flood those and use them. We went there and looked, and they were not the right size.

We went to look at what the Russians were using, and we went to various countries that had some facilities, because the idea is, if it's just a matter of needing a big area with water, we don't need to go build that; we can find it. Well, we could not. The idea of using open water in the Gulf of Mexico was considered, but the amount of equipment and the downtime required because of weather and salt water caused us to put that aside.

The idea that we needed the facility, and that we needed to build it, happened about the same time that one of the facilities built for the Space Station Freedom out at Ellington Field was vacant.[14] It was a very large facility, so we examined why we could not do in that facility what we had done years earlier in the old centrifuge facility, which is put a swimming pool inside it, and that way we wouldn't have to build the outside structure.

I don't know of anything at the center, other than the Shuttle and Space Station Programs, that was examined as much by that many people. I recall there were so many against it until they saw all the numbers, and we ran the numbers so many times of how many crews you could train for how long on what missions. Realizing we were going to use the crew to assemble the Station was the big issue.

A continuing problem at the center was having the visibility and the resources to fund the next level of medical research that needed to be done for longer duration flights. The administrator decided to consider forming institutes for all the science areas within NASA, and I was asked to serve on that agency-wide committee. We looked at every kind of institute within the government. Then we considered different sciences—space science, Earth science, microgravity, and life sciences—to determine if they would benefit from an institute or not. The committee concluded that the life sciences would probably benefit the most from having an institute formed because it was having the hardest time attracting investigators into the program. Because of that, the administrator asked me to head that effort, the Space Biomedical Research Institute.[15]

It was a sizable effort to consider why and how we would do it. We finally decided that the only way it would be successful would be to make sure it was in the proximity of the Johnson Space Center. The other was to make it the broadest science involvement we could, and so we encouraged the institute to be not just one school or medical center but to be a combined effort. There were two groups proposed, and Baylor College of Medicine was selected for funding.

I think it was important to do it at that time. I had trouble recruiting scientists. In the life sciences community, the biggest funding source is the National Institutes of Health, and their yearly budget at that time was nine to ten billion dollars.[16] I think it's doubled in the past five years or so since then. They fund most of what goes on in medical research in this country. Therefore, we did not have the opportunity to have a community that was solely a NASA medical community looking at NASA problems. We could not get a consistent group of high-level people to pay attention to our medical issues. It's logical when you think of it. Our funding source was small, so the decision was made to form this research institute. The research institute would be a long-term effort that would follow experiments and the investigations over years and not just be short-sighted as far the funding goes.

As I was addressing that issue, I began to get more involved in work at NASA Headquarters. The administrator wanted me to be an assistant to him to do this institute, as well as some other issues in Washington. I moved to Washington and received an offer about that same time to go to the White House to work in the Office of Science and Technology Policy (OSTP) and conduct policy work that affected science in the whole nation.[17] I took that opportunity and worked there for several years. It was there that I worked with managers and scientists across the federal government to solve issues of mutual and national interest.

I hope there are many contributions I made over the years that had a lasting impact on NASA, OSTP, and the Department of Energy (where I served as assistant secretary for several years). Talking about the "I" part is hard for me, but it's also hard for anyone that's worked in an organization like NASA, because it isn't an "I"—it's always a "we." Many people are involved—it's a team effort. You form relationships with people that last, that move on into the next phase.

The relationship between the Station and the Shuttle that we set up when I was center director grew into a good way of doing business. Some of the people that we were able to recruit and retain certainly have contributed greatly. The methods that we set up for external involvement in our science and engineering programs have lived on, so there have been quite a few contributions.

One of my goals was to try to get women and minorities into the system as contributing members. I think we were able to accomplish this. Maybe it would have happened anyway, but I happened to be there at that time.

Sometimes the gender issue was overemphasized or it overshadowed the fact that I was an accomplished scientist. But I accepted that, because that was the community at the time, and I really didn't have a choice. You either accepted the ground rules, or you didn't play in the game. NASA in general did not discriminate, and NASA was very good to me as an individual, as it was to many people. There were individuals in the program that did discriminate, that did make life hard for me and other women. Luckily, they moved on. I outlasted most of them. I think the idea of women not being just like men was something that some people could not get over. They would hire women to do jobs, and the fact that they didn't operate just like men meant they couldn't do the job. Some managers weren't willing to give them a chance to develop their own style of doing things. I ran into that several times. Other women have too. It's not unique to me.

I decided that what I was accomplishing and what I was able to get done, in spite of all that, was worth it to stay there and do it, and that's why I did. Women have left not just NASA, but other organizations at mid- or high-level careers because they just didn't want to put up with what you had to put up with in that difficult environment. To me, it was worth it.

Afterword

Ellen Ochoa

When my son was five, we drove by the Johnson Space Center on our way to socialize with Astronaut Eileen Collins's family. At that time, I had been an astronaut for fifteen years and flown into space four times. As we went past the *Saturn V* rocket on the front lawn, he asked, "Can boys be astronauts? Or just girls?" I realized then just how far the United States space program had come.

Ellen Ochoa holds her son, Wilson Miles-Ochoa, at the Cape Canaveral Air Force Station Skid Strip following STS-96.

I grew up in the 1960s, when NASA sent men to the Moon and women weren't part of the narrative. Space interested me, but there were no female astronauts. In the late 1970s while I was at San Diego State University, that all changed—NASA selected the first six female astronauts.

As *Making Space for Women* demonstrates, NASA has many exciting positions, and these are not limited by gender. Being an astronaut is one of the most recognized positions in the agency, but it is not the only career option. The space agency needs scientists, engineers, mathematicians, and technicians, and just as important are the people behind the scenes. Some of these you read about in this book, but the list is exhaustive, and also includes writers, educators, software designers, graphic artists, attorneys, and many more. All are needed to help the agency achieve its vision to reach for new heights and reveal the unknown for the benefit of humanity.

For those interested specifically in STEM positions, education is important. Math and science are the building blocks of that training, but so are English, writing, and speaking skills. Being able to communicate effectively in this world is vital. Dedication and hard work, as you have read, are necessary. The women in this volume persevered and worked doggedly even when the situation was bleak. When they looked back, they described their experiences as worthwhile. My own career has been challenging and extremely rewarding.

I hope *Making Space for Women* has inspired you to consider exploring careers in STEM and to look to NASA for guidance on how best to achieve your dreams. With the passage of the INSPIRE Women Act, the future looks bright for women pursuing aerospace careers, and NASA will be enriched by their contributions.

Notes

Throughout the book, readers are directed to the JSC History Portal and NASA's Astronaut Homepage. Rather than cluttering the notes with URLs, links to these websites are listed here:
JSC History Portal: https://historycollection.jsc.nasa.gov/JSCHistoryPortal/history
/NASA Astronaut Homepage: https://www.nasa.gov/astronauts

Introduction
1. Linda S. Keklis, Regecca Wespic Ancheta, and Jeri Countryman, "Role Models Make a Difference: A Recipe for Success," *AWIS Magazine* 34, no. 3 (Summer 2005): 18.
2. See, for instance, Harriet S. Mosatche, et al., "Effective STEM Programs for Adolescent Girls: Three Approaches and Many Lessons Learned," *Afterschool Matters* (Spring 2013): 24, accessed September 4, 2015, http://niost.org/images /pdf/afterschoolmatters/asm_2013_17_spring/ASM_2013_spring_6.pdf; Elizabeth Dayton, "Factors That Influence Females' Pursuit of STEM Fields: A Literature Review Emphasizing Psychological Influences," 7–10, accessed September 4, 2015, http://sierraschoolworks.com/section/wp-content/uploads/2013/11/FACTORS -THAT-INFLUENCE-FEMALES%E2%80%99-PURSUIT-OF-STEM-FIELDS -E.-Dayton-for-Sierra-College-STEM.pdf; Catherine Riegle-Crumb and Chelsea Moore, "The Gender Gap in High School Physics: Considering the Context of Local Communities," *Social Science Quarterly* 95, no. 1 (March 2014): 266.
3. Well before electronic computers were in use, NACA hired women with mathematics degrees to compute and plot data for engineers conducting flight research.
4. Capsule communicators are the people who communicate with flight crews from Mission Control in Houston. The term is a holdover from the early piloted spaceflight programs.
5. Throughout the book I use the term "JSC" and "MSC" interchangeably, depending upon the time period.
6. Howard E. McCurdy, *Inside NASA: High Technology and Organizational Change in the US Space Program* (Baltimore: Johns Hopkins University Press, 1993), 17–20.
7. Margaret W. Rossiter, *Women Scientists in America before Affirmative Action 1940–1972* (Baltimore: Johns Hopkins University Press, 1995), 278–79.
8. James R. Hansen, *Engineer in Charge: A History of the Langley Aeronautical Laboratory, 1917–1958* (Washington, DC: NASA, 1987), 50, 63, 187, 206–207, 209.

9. Sarah McLennan and Mary Gainer, "When the Computer Wore a Skirt: Langley's Computers, 1935–1970," NASA *News and Notes* 29, no. 1 (First Quarter 2012): 28–29.

10. United States Department of Labor, *1958 Handbook on Women Laborers* (Washington, DC: GPO, 1958), 11–12, 22.

11. Nancy Grace Roman, "The Role of Women Scientists in the Space Program," Nancy Grace Roman bio file, Record Number 1821, NASA Headquarters Historical Reference Collection, Washington, DC (hereafter cited as Roman bio file); "Nancy Roman" bio, NASA Quest, Roman bio file; Nancy Grace Roman, interview by Rebecca Wright, September 15, 2000, accessed November 14, 2014, JSC Oral History Project, JSC History Portal.

12. "Professional Women Employed by Prime Contractors on Apollo and Gemini Programs," Record Number 008998, NASA Headquarters Historical Reference Collection, Washington, DC (hereafter cited as HRC).

13. James E. Webb to Heads of Offices, November 5, 1964, "Status of Women in the Federal Service," Record Number 008998, HRC.

14. Ibid.

15. Charles F. Bingman, interview by Carol Butler, April 9, 2000, transcript, accessed August 5, 2013, JSC Oral History Project, JSC History Portal.

16. Zoe Von Ende, "Women Not Barred from Space Program," *Denver Post*, March 18, 1966, Roman bio file.

17. Roman, interview.

18. That is not to say that women did not experience discrimination. They did, but the issue was not NASA but societal views of women's place in the workforce. In a 2018 interview with the JSC Oral History Project, Poppy explained her work as a contractor with TRW, and how the protectionary laws in the state of Texas in the 1960s limited the hours she could work, therefore impacting her ability to make a living wage as a computress. "The Personal Views of Eight Women Who Succeeded in It," *Life Magazine*, September 4, 1970, 20. Francis M. "Poppy" Northcutt, interview by author, November 14, 2018, transcript, JSC Oral History Project, JSC History Portal.

19. Judith Viorst, "Women Find Place in Space, Too," *New York World-Telegram and Sun*, September 1964, record number 008998, HRC.

20. Toni House, "Woman Space Launcher Comes Home," *Evening Star*, December 24, 1970, record number 008998, HRC; Peggy Mihelich, "Small Satellite a Big Success for Engineer Marjorie Townsend," American Association for the Advancement of Science, posted November 2, 2010, accessed November 25, 2014, https://www.aaas.org/blog/member-spotlight/small-satellite-big-success-engineer-marjorie-townsend.

21. Ivy Hooks, interview by Lauren Kata, Society of Women Engineers, April 9, 2003, accessed November 14, 2014, https://ethw.org/Oral-History:Ivy_Hooks.

22. Jeanne L. Crews, interview by Rebeca Wright, August 6, 2007, transcript, accessed November 14, 2014, JSC Oral History Project, JSC History Portal.

23. Will McNutt, "Girl Helped Bring Astronauts Home," *Free Lance Star*, April 28, 1970, accessed August 7, 2014, https://news.google.com/newspapers?nid=1298&dat=19700428&id=v9JNAAAAIBAJ&sjid=YYoDAAAAIBAJ&pg=7290,6562615&hl=en.

24. Ivy F. Hooks to Dr. Harriett G. Jenkins, December 14, 1977, JSC History

Collection, Center Series, Flight Crew Operations Directorate Papers, Box 3, University of Houston-Clear Lake, Houston, TX.

25. Crews, interview.

26. Carolyn S. Griner, interview by Mike Wright, March 2010, Audio Recording, Series 107, Digital Collection, NASA Marshall Space Flight Center Historical Reference Collection, Huntsville, AL.

27. McCurdy, *Inside NASA*, 49.

28. Sy Liebergot with David M. Harland, *Apollo EECOM: Journey of a Lifetime* (Ontario: Apogee Books, 2003), 88.

29. Eugene F. Kranz, *Failure Is Not an Option: Mission Control from Mercury to Apollo 13 and Beyond* (New York: Simon and Schuster, 2000), 303–304.

30. Crews, interview.

31. Email to author, November 18, 2014.

32. Shirley H. Hinson, interview by Rebecca Wright, May 2, 2000, accessed December 4, 2014, JSC Oral History Project, JSC History Portal.

33. Anne L. Accola, interview by Rebecca Wright, March 17, 2005, transcript, accessed August 16, 2013, JSC Oral History Project, JSC History Portal. Eugene F. Kranz started working for the Space Task Group in 1960. He went on to become a flight director and led the Mission Operations Directorate from 1983 until his retirement in 1994. His book, *Failure Is Not an Option: Mission Control from Mercury to Apollo 13 and Beyond*, details his time in flight control. Read more about his lengthy career by downloading his oral history transcripts on the JSC History Portal.

34. Rossiter found that the federal government was no haven for women scientists. Women received few promotions and were not well paid. *Women Scientists in America*, 302–303.

35. Carolyn Huntoon, phone call with author, November 19, 2014.

36. For more details on this phenomena, see Rossiter, *Women Scientists in America*, xv–xviii; Sue V. Rosser, *Breaking into the Lab: Engineering Progress for Women in Science* (New York: New York University Press, 2012), 33–34.

37. Kim McQuaid, "'Racism, Sexism, and Space Ventures': Civil Rights at NASA in the Nixon Era and Beyond," in *Societal Impact of Spaceflight*, ed. Steven J. Dick and Roger D. Launius (Washington, DC: NASA, 2007), 429.

38. Jay F. Honeycutt, interview by author, September 18, 2014, JSC Oral History Project (unreleased interview).

39. McCurdy, *Inside NASA*, 172–73.

40. Ibid., 50–57.

41. Ibid., 102–103; see also Roger D. Launius, "Public Opinion Polls and Perceptions of US Human Spaceflight," *Space Policy* 19 (2003): 163–75, https://doi.org/10.1016/S0265-9646(03)00039-0; McQuaid, "'Racism, Sexism, and Space Venture,'" 421–22; Amy E. Foster, *Integrating Women into the Astronaut Corps: Politics and Logistics at NASA, 1972–2004* (Baltimore: Johns Hopkins University Press, 2011), 81–82.

42. Foster, *Integrating Women into the Astronaut Corps*, 71.

43. McQuaid, "'Racism, Sexism, and Space Ventures,'" 431.

44. Ibid., 423, 434.

45. Ivy F. Hooks, interview by author, March 5, 2009, transcript, accessed June 26, 2013, JSC Oral History Project, JSC History Portal.

46. See Jessie Hendrick, "Women Find New Challenges at JSC," *Space News Roundup*, March 14, 1997, 4.

47. Women's numbers had barely increased since the fall of 1965. "Historical Progress of Women at JSC," March 21, 1990, JSC History Office, Houston, TX.

48. Ihor Gawdiak with Helen Fedor, comp., *NASA Historical Data Book, Volume IV: NASA Resources, 1969–1978* (Washington, DC: NASA, 1994), 66.

49. Sally K. Ride, interview with Rebecca Wright, transcript, October 22, 2002, accessed June 13, 2013, JSC Oral History Project, JSC History Portal.

50. Carolyn L. Huntoon, interview by author, April 21, 2008, transcript, accessed July 17, 2013, JSC Oral History Project, JSC History Portal.

51. In 1982, four years after NASA selected female astronauts, the Navy opened its test pilot schools to women. The Air Force followed six years later.

52. Matthew H. Hersch argued that the media and box office promoted the idea of the space pilot and portrayed scientists as inferior or incompetent. "'Capsules Are Swallowed': The Mythology of the Pilot in American Spaceflight," in *Spacefarers: Images of Astronauts and Cosmonauts in the Heroic Era of Spaceflight*, ed. Michael J. Neufeld (Washington, DC: Smithsonian Institution Scholarly Press, 2013), 35–55.

53. Alan L. Bean, interview by author, February 23, 2010, transcript, accessed July 31, 2013, JSC Oral History Project, JSC History Portal.

54. Foster, *Integrating Women into the Astronaut Corps*, 99.

55. Kathryn D. Sullivan, interview by author, May 10, 2007, transcript, accessed November 26, 2014, JSC Oral History Project, JSC History Portal.

56. Interestingly, three of the six interviewed for NASA oral history projects gave examples of how they had held back in certain situations. Huntoon, interview; Foster, *Integrating Women into the Astronaut Corps*, 100–101.

57. Rhea Seddon, interview by author, May 20, 2010, transcript, accessed July 22, 2013, JSC Oral History Project, JSC History Portal.

58. Stanley H. Goldstein to Distribution, "Pilot Course, Working with Women," November 4, 1980, JSC History Office, Houston, TX.

59. Jay H. Greene, interview by Sandra Johnson, December 8, 2004, transcript, accessed July 18, 2013, JSC Oral History Project, JSC History Portal.

60. Eugene F. Kranz, interview by author, December 7, 2011, transcript, accessed July 10, 2013, JSC Oral History Project, JSC History Portal.

61. Ibid.; "Limits to Inhibit," *Space News Roundup*, August 9, 1985, 1, 3.

62. Jack Knight, interview by Sandra Johnson, November 28, 2007, transcript, accessed July 18, 2013, JSC Oral History Project, JSC History Portal.

63. Ginger Kerrick, interview with author, September 4, 2012, transcript.

64. Judy A. Rumerman, comp., *NASA Historical Data Book, Volume VI: NASA Space Applications, Aeronautics and Space Research and Technology, Tracking and Data Acquisition/Support Operations, Commercial Programs, and Resources, 1979–1988* (Washington, DC: NASA, 2000), 465.

65. Since the sixties, NASA has housed its aircraft at Ellington Field. Before JSC opened, Ellington Air Force Base, now a city airport, also housed center employees. (The Department of Defense deactivated the base in 1976.) Throughout this book women refer interchangeably to both, depending on the time frame, or simply call the facility Ellington.

66. Robert C. Goetz to Assistant Administrator for Equal Opportunity Programs, "Deputy Center Director's Meeting with a Sample of Women Professionals at JSC," September 28, 1984, JSC History Office, Houston, TX.

67. Alan Boyle, "The Changing Face of NASA," NBC News.com, July 21, 1999,

accessed December 3, 2014, http://www.nbcnews.com/id/3077898/ns/technology
_and_science-space/t/changing-face-nasa/.

68. Julie Kramer White, interview by author, April 26, 2011, transcript, accessed July 26, 2013, JSC Oral History Project, JSC History Portal.

69. "Celebrating National Women's History Month," *Space Center Roundup,* March 10, 2000, 2.

70. Margaret W. Rossiter, *Women Scientists in America: Forging a New World Since 1972* (Baltimore: Johns Hopkins University Press, 2012), 41.

71. Emily Canal, "NASA Astronaut Recruits Come at a Time of Change for NASA," *Forbes,* June 25, 2013, accessed June 26, 2013, https://www.forbes.com /sites/emilycanal/2013/06/25/new-astronaut-recruits-come-at-a-time-of-change-for-nasa/#5c1b423547e4; Lori B. Garver, "Women's Empowerment-Women's Education," (remarks, Johns Hopkins University Women's Network meeting, March 14, 2012), accessed August 1, 2013, http://www.nasa.gov/pdf/655357main _garver_apl_hopkins.pdf.

72. "Celebrating National Women's History Month."

73. Hansen retired from NASA in the fall of 2017, and Ochoa retired as center director the following spring. Peggy Whitson retired in 2018.

74. Tu-Quynh T. Bui, email to author, February 5, 2015.

75. Tim Schrag, "Ex-Astronaut Shares Love of Knowledge with Educators," *Hutchinson News,* February 12, 2015, accessed February 12, 2015, https://www .hutchnews.com/article/20150212/News/302129883

76. Debra L. Johnson, interview by Rebecca Wright, August 29, 2012, transcript, JSC Oral History Project, JSC History Portal.

77. Pamela A. Melroy, interview by author, November 16, 2011, transcript, accessed August 1, 2013, JSC Oral History Project, JSC History Portal.

78. This is related to women's conversational style as discussed by Deborah Tannen in *Talking from 9 to 5: How Women's and Men's Conversational Styles Affect Who Gets Heard, Who Gets Credit, and What Gets Done at Work* (New York: William Morrow and Company, 1994), 153–55.

79. For example, see Laurie A. Rudman, "Self-Promotion as a Risk Factor for Women: The Costs and Benefits of Counterstereotypical Impression Management," *Journal of Personality and Social Psychology* 74, no. 3 (1998): 629–45.

80. Johnson, interview.

81. Melroy, interview.

82. Natalie V. Saiz, interview with author, August 15, 2012, transcript, JSC Oral History Project, JSC History Portal.

Chapter 1

1. As of February 1963, JSC employed 580 women, most of whom were secretaries. Robert B. Voas, "Ladies in Space," speech, February 1963, 8, Papers of Robert B. Voas, Record No. 5028, NASA Headquarters Historical Reference Collection, Washington, DC.

2. Nancy L. Fee, interviewed by Rebecca Wright, November 29, 2011, JSC Oral History Project and available on the JSC History Portal website.

3. Estella Hernández Gillette sat for this interview in September 2012.

4. Before the center opened, Ellington Air Force Base also housed center employees.

5. The GS, or General Schedule, is the pay scale for government employees. The scale, based on education and experience, starts at GS-1 and goes through GS-15. A GS-4 position is nearly entry-level.

6. At the time, Maxime A. "Max" Faget was the director of Engineering at the Manned Spacecraft Center (MSC), and he served in that position for twenty years. He is recognized as the designer of American spacecraft from Mercury through the Space Shuttle. Oral histories with the father of America's spacecraft are readily available on the JSC History Portal.

7. Betty M. Ensley worked as Max Faget's secretary at the space center from 1962 to 1977. She was one of the Space Task Group members who moved from Virginia to Houston.

8. Marion Loveless was secretary for the Structures and Mechanics Division from 1964 to 1965.

9. The Vibration and Acoustic Test Facility tests space hardware to evaluate its ability to withstand the vibrations and shock of launch and landing.

10. Gillette mentions Clear Lake, the region of Houston where NASA is located, which includes the surrounding areas of what was formerly called Clear Lake City, just off the shores of Galveston Bay.

11. Martin L. Raines started working at the White Sands Test Facility located in New Mexico in 1964. Four years later, he moved to Houston and eventually became the director of the center's Safety, Reliability, and Quality Assurance Office (SR&QA). He retired in 1987. Learn more about his career by downloading his interview on the JSC History Portal.

12. The Apollo Program was coming to an end, resulting in layoffs.

13. Kelly Services is a job placement agency.

14. John G. Zarcaro came to MSC from the US Navy in 1962. He started out in the Landing and Recovery Division and retired in 1977 as the manager of the Space Shuttle Program Assessment Office.

15. Clifford E. Charlesworth joined MSC in April 1962, where he served in a number of positions including JSC deputy director. For three years, from 1983 to 1986, he served as director of Space Operations, which included the Flight Crew Operations and Mission Operations Directorates. He retired in 1988. Apollo 11 was the first mission to land on the Moon. Read more about the mission at http://www.nasa.gov/mission_pages/apollo/missions/apoll011.html.

16. In 1962 Lois M. Bradshaw was an MSC stenographer; by 1980 she had worked her way up to manage the Program Administrative Office.

17. The Katharine Gibbs Schools were the gold standard for secretaries at the time.

18. RIFs are a permanent reduction in the workforce.

19. When there is a RIF, a civil servant is entitled to a job if s/he has a higher ranking than another civil servant holding a job within a federal agency. That job can be in the same or lower pay grade. To avoid a layoff, an employee can accept a position with a lower pay grade, backing into that position and bumping another person.

20. William E. "Gene" Rice began working at NASA in 1962 on Apollo propulsion and power systems. He retired in 1989, when he took a position with Northrop Grumman. Read his oral history and about his career on the JSC History Portal.

21. Glynn S. Lunney had a lengthy NASA career, beginning at the Lewis Research Center in Ohio in 1958. He later became part of the Space Task Group. With more experience under his belt, he became a flight director and managed the Apollo-Soyuz Test Project and the National Space Transportation System. After

retiring from NASA, he went on to work for several aerospace companies. Read more about his career and oral history at the JSC History Portal.

22. Christopher C. Kraft was JSC's second center director, a position he held until he retired in 1982. Just out of college he began a career with the National Advisory Committee for Aeronautics (NACA) in 1945, and then in 1958, after NASA was established, became a member of the Space Task Group. He is best known for his role in the creation of spaceflight operations and serving as NASA's first flight director. Learn more about Kraft and his time at NASA by reading his oral histories at the JSC History Portal.

23. In 1975, Marilyn Bockting became the first female division chief at the Johnson Space Center as head of the Program Administration Office. Bockting came to the center in 1963 as secretary to the Deputy Center Director James C. Elms. She retired in 1980.

24. George M. Low served in a variety of positions within NASA, including MSC's deputy center director, manager of the Apollo Spacecraft Program Office, and NASA deputy administrator.

25. In 1959, Richard Smith "Dick" Johnston began his NASA career with the Space Task Group at Langley Research Center in Hampton, Virginia. He remained with the agency for twenty years until he retired in 1979. Read more about his career and oral history at the JSC History Portal.

26. Gregory W. Hayes came to JSC in 1973 as a management intern and eventually directed JSC's Human Resources Office.

27. Gillette is referring to the number of applicants that became the 1980 class of astronauts. Applicant pools for the Space Shuttle Program that year were much larger than previous programs as NASA opened opportunities to a diverse group of people, not just test pilots, as well as women.

28. At the time, George W. S. Abbey headed the Flight Crew Operations Directorate, which oversaw the Astronaut Office.

29. Joseph D. Atkinson, Jr., was chief of JSC's Equal Employment Opportunity Office. Before coming to NASA in 1964, he worked at Kelly Air Force Base. Along with Jay Shafritz, he is the author of *The Real Stuff: A History of NASA's Astronaut Recruitment Program.*

30. Cyril E. Baker came to the Manned Spacecraft Center in 1963 from the US Air Force. He provided administrative support for the astronaut corps until his death in 1982.

31. "CB" is the Astronaut Office organization code.

32. John W. Young flew Gemini, Apollo, and Shuttle missions. He commanded STS-1, called the boldest test flight in history. Learn more about his career by reading his biography on NASA's Astronaut Homepage.

33. Building 4 housed the astronauts and flight controllers.

34. The 1978 astronaut class called themselves the "Thirty-Five New Guys."

35. Payload specialists were not professional astronauts. They flew onboard the Shuttle and managed a specific payload or experiment. Members of Congress, the military, and other nations flew as payload specialists.

36. Federal contractors work for companies that receive contracts with the federal government to perform a job or provide services or materials to the government. They are not government employees but work on government projects. Civil servants, by contrast, are employees of the federal government.

37. Travel agents relied on Semi-Automatic Business Research Environment computers, or SABRE computers, to look up, price, and book airline tickets, hotel accommodations, and car rentals.

38. For each flight, astronauts flew medallions as commemorative items for friends, family, and colleagues.

39. The STS-51C flight was a classified Department of Defense mission. For more information, see http://www.nasa.gov/mission_pages/shuttle/shuttlemissions /archives/sts-51C.html.

40. Loren J. Shriver and Ellison S. Onizuka were part of the 1978 class of astronauts. Both flew for the first time on STS-51C. Shriver then went on to fly STS-31 and 46. He went on to become vice president of engineering and integration for United Space Alliance and their chief technology officer. Onizuka, assigned to STS-51L, died during the liftoff of the Space Shuttle *Challenger*. Read Shriver's oral histories on the JSC History Portal.

41. NASA lost the Space Shuttle *Challenger* and her crew on January 28, 1986. For more detailed information about the space flyers onboard, see https://history. nasa.gov/Biographies/challenger.html.

42. NASA management and astronauts flew in and out of Ellington Field in Houston, Texas.

43. Carolyn L. Huntoon was then the JSC's associate center director, and is featured in this book.

44. In 1967, three astronauts perished at Cape Kennedy (now Cape Canaveral) during a test of the Apollo spacecraft.

45. Jack R. Garman was then deputy director of Mission Support at JSC, handling various data systems and software. He started at MSC in 1966 as a recent graduate of the University of Michigan's engineering physics program. By the time he retired in 2000, he was the center's chief information officer. To learn more about his lengthy career, download his oral history transcripts from the JSC History Portal.

46. Ronald L. Berry was then the director of Mission Support at JSC, which handled Mission Control software and support when the missions were on orbit. In addition, Mission Support helped to develop software for the crew trainers and provided computer support for the center. He started at JSC in 1962 and retired in 1995. To learn more, read his online transcript on the JSC History Portal.

47. Chris C. Critzos was a longtime NASA employee, having started out with NACA.

48. Brewster H. Shaw was an astronaut, who had—at that point—piloted STS-9 and commanded STS-61B. Before coming to NASA, he had been a combat fighter pilot as well as a test pilot. To learn more about his experience with the agency, read his biography on NASA's Astronaut Homepage and read his transcript on the JSC History Portal.

49. From 1982 to 1989, Mary Lee Meider worked in the Astronaut Office.

50. Francis R. "Dick" Scobee flew on STS-41C and commanded STS-51L. His wife, June, and the other 51L families came together to establish the *Challenger* Center. June Scobee Rodgers is the founding director. Learn more about astronaut Dick Scobee by reading his biography on NASA's Astronaut Homepage.

51. Paul J. Weitz was deputy chief of the Astronaut Office. He flew on Skylab 2 and STS-6. Read more about his NASA missions by downloading his biography

on NASA's Astronaut Homepage and reading his oral histories on the JSC History Portal.

52. CAOs helped the families as they dealt with the deaths of their spouses, siblings, and parents.

53. Lorna Onizuka was Ellison S. Onizuka's wife. Loren Shriver, who had flown with Onizuka on STS-51C, became the CAO for the family.

54. Donald R. Puddy started out working in flight control at the space center in 1964. He went on to become a flight director, acting deputy director of the Ames Research Center in California, and assistant associate administrator for Space Flight at NASA Headquarters. He served as director of Flight Crew Operations from 1987 to 1992. In this position he oversaw the Astronaut Office and Aircraft Operations at Ellington.

55. STS-26 was the first flight following the *Challenger* accident. Learn more about that 1988 mission by visiting http://www.nasa.gov/mission_pages/shuttle/shuttlemissions/archives/sts-26.html.

56. Harvey L. Hartman was then a deputy personnel officer. He started at MSC as a management intern with Personnel in 1966 and was director of JSC's Human Resources Office when he retired in 1999. Learn more about his career by reading his oral history on the JSC History Portal.

57. Jamye Flowers Coplin talked with the JSC Oral History Project in November 2008.

58. John Young and Virgil I. "Gus" Grissom flew Gemini III. Information on the flight can be found online at https://nssdc.gsfc.nasa.gov/nmc/spacecraft/display.action?id=1965-024A and in their biographies on NASA's Astronaut Homepage. Grissom died in 1967 during a spacecraft test at the Cape.

59. From 1963 to 1970, Sarah W. Lopez handled a variety of assignments for the Astronaut Office.

60. John J. Peterson worked in the Astronaut Office from 1964 to 1967.

61. Maxine Henderson worked in a variety of positions for the astronaut corps from 1963 to 1966.

62. Gay R. Alford was secretary to Alan B. Shepard, chief of the Astronaut Office. He became the first American in space in 1961 and went on to fly to the Moon on Apollo 14. Read more about his career on NASA's Astronaut Homepage and also read his transcript on the JSC History Portal.

63. Lynn M. Cross spent her entire career working for the Astronaut Office.

64. Charlotte A. Maltese worked for the Astronaut Office for a decade, from 1966 to 1976.

65. Antoinette J. Zahn worked for the office from 1965 to 1968.

66. Charlene Stroman started in Procurement in 1963 and then moved to the Astronaut Office, where she worked from 1965 to 1968.

67. Like Stroman, Mavis E. Study began working for Procurement. For four years, from 1965 to 1969, she provided administrative support in the Astronaut Office.

68. M. Scott Carpenter, L. Gordon Cooper, Gus Grissom, and Walter M. "Wally" Schirra were four of the first seven astronauts selected by NASA, known as the "Mercury Seven." You can learn more about their careers by reading their biographies on NASA's Astronaut Homepage or their oral history interviews on the JSC History Portal website. Carpenter was the first to be called an aquanaut for his

work underwater. Cooper flew twice in space, once in Mercury and then as commander of Gemini V. Schirra was the only Mercury astronaut to fly the first three spaceflight programs: Mercury, Gemini, and Apollo. James A. "Jim" McDivitt was part of the second class of astronauts, named in 1962. He flew Gemini IV—the first flight to include an extravehicular activity (EVA)—and Apollo 9. After the first lunar landing, he became the Apollo Spacecraft Program Office manager.

69. Martha I. Caballero started working at MSC in 1964 and retired in 1998. She provided administrative support to the Astronaut Office from 1966 to 1971. She earned a bachelor's in accounting in 1978 and went on to become a section and branch chief.

70. Goldie B. Newell worked in the Office of the Director of Flight Crew Operations from 1966 to 1971.

71. Coplin went to work in the Office of the Director of Flight Crew Operations, headed by Donald K. "Deke" Slayton. Thomas U. McElmurry and Donald T. Gregory both worked in that directorate. Sue M. Symms worked in the office from 1963 to 1973. She went on to work in the Astronaut Office until she retired.

72. Building 2, now known as Building 1, housed the center's management and director.

73. Wite-Out is a white fluid used to make corrections on typewritten paper.

74. Coplin is referring to the families of astronauts Stuart A. Roosa, Ronald E. Evans, Joe H. Engle, and William A. Anders. Learn more about each spacefarer by downloading their biography from NASA's Astronaut Homepage, and read the oral history transcripts for Engle and Anders on the JSC History Portal.

75. Astronaut Ron Evans flew on the final Apollo mission, Apollo 17; Janet "Jan" was his wife. Read his online biography on NASA's Astronaut Homepage. Learn more about the final Apollo mission by reading https://www.nasa.gov/mission_pages/apollo/missions/apollo17.html. An interview with Jan can be found on the JSC History Portal.

76. Apollo 10 was a dress rehearsal for the first lunar landing. More information about that mission can be found online at http://www.nasa.gov/mission_pages/apollo/missions/apollo10.html.

77. Donn F. Eisele, selected as an astronaut in 1963, served as the command module pilot for Apollo 7. Read more about his Air Force and NASA career by viewing his biography on NASA's Astronaut Homepage. Edgar D. Mitchell came to NASA in 1966 from the US Air Force Aerospace Research Pilot School. The lunar module pilot on Apollo 14, Mitchell retired from NASA and the Navy in 1972. Learn more about his career by downloading his biography from NASA's Astronaut Homepage and reading his oral history interview on the JSC History Portal.

78. Thomas P. Stafford, Eugene A. "Gene" Cernan, and John Young made up the Apollo 10 crew. You can read more about their careers on NASA's Astronaut Homepage; interviews with Stafford and Cernan are available on the JSC History Portal. Selected in 1962, Stafford piloted Gemini VI, commanded Gemini IX, Apollo 10, and the Apollo-Soyuz Test Project crew, the first international spaceflight. Cernan, named an astronaut in 1963, piloted Gemini IX with Stafford and was the last man to set foot on the Moon.

79. Apollo 9 tested the procedures for lunar rendezvous and docking in Earth orbit. Additional information can be found online at http://www.nasa.gov/mission_pages/apollo/missions/apollo9.html.

80. The December 1968 Apollo 8 mission was the first to circle the Moon. For more details, see https://www.nasa.gov/mission_pages/apollo/missions/apollo8.html.

81. The command module was the spacecraft that took astronauts to the Moon, while the lunar module landed astronauts on the surface and provided them shelter on that body. Apollo 10 did not land on the Moon.

82. In 1968, astronauts awarded—for the first time—Silver Snoopy pins to members of the NASA team who made important contributions to the Apollo Program and its missions.

83. Michael Collins graduated from West Point in 1952 and joined the Air Force. Stationed at Edwards Air Force Base in California, he served as an experimental flight test officer before being selected as an astronaut in 1963. He flew two space missions and left the agency in 1970. Learn more about his career by downloading his biography from NASA's Astronaut Homepage and reading his oral history transcript on the JSC History Portal.

84. Charles "Pete" Conrad, Jr., a naval aviator and test pilot, graduated from Princeton University in 1953 with a bachelor's in aeronautical engineering. Part of the "New Nine" class, Conrad flew four space missions: Gemini V, Gemini XI, Apollo 12, and Skylab 2. Read more about his astronaut experience by downloading his biography on NASA's Astronaut Homepage.

85. Neil A. Armstrong is best known as the first man to walk on the Moon, but before the Apollo 11 mission he had been a naval aviator, an engineer, and a research pilot. On Gemini VIII, he successfully docked two spacecraft in orbit for the first time. Learn more by downloading his biography from NASA's Astronaut Homepage and reading his oral history transcript on the JSC History Portal.

Buzz Aldrin, the second man to step on the Moon, was selected as an astronaut in 1963. He flew the final Gemini mission with Commander James A. Lovell and set an EVA record with his spacewalk. In 1971, he resigned from NASA. To learn more about his career, read his online biography on NASA's Astronaut Homepage.

86. NASA selected Frank Borman in 1962. He commanded Gemini VII and Apollo 8. After retiring from NASA, he worked for Eastern Airlines. Learn more about his experiences through his online biography on NASA's Astronaut Homepage and by reading his oral history transcript at the JSC History Portal.

87. Apollo 12 was the second lunar landing. Read more about that flight at http://www.nasa.gov/mission_pages/apollo/missions/apollo12.html.

88. Thomas K. "T. K." Mattingly flew on Apollo 16, along with John Young and Charlie Duke. More information on this flight is available online at http://www.nasa.gov/mission_pages/apollo/missions/apollo16.html.

89. Joseph P. Kerwin was a medical doctor and naval flight surgeon before becoming a scientist-astronaut in 1965. He flew only once, on Skylab 2 in 1973, but remained with NASA until 1987. Learn more about his career by downloading his biography on NASA's Astronaut Homepage and reading his oral history transcript on the JSC History Portal.

90. Skylab was America's first space station. NASA provides an excellent overview of the program online at http://www.nasa.gov/mission_pages/skylab.

91. Assigned to Apollo 13, T. K. Mattingly was removed from the crew when he was exposed to the German measles. John L. "Jack" Swigert flew in his place. Mattingly remained in the corps and flew three missions: Apollo 16, STS-4, and STS-51C. He resigned from the agency in 1985. Read more about those missions

and both of their careers by downloading their biographies from NASA's Astronaut Homepage, and read Mattingly's transcripts on the JSC History Portal. Swigert, selected as an astronaut in 1966, flew one space mission. He left NASA in 1977 and later was elected to the US House of Representatives, but died before he could be sworn in.

92. There were problems with the Skylab 1 mission from the beginning. The micrometeoroid shield and sunshade tore off during launch, and, once in orbit, the loss of the shade exposed the workshop to the sun and temperatures inside the workshop rose quickly. Because of these problems, NASA delayed the launch of the Skylab 2 crew.

93. In the 1960s, Americans became increasingly involved in Vietnam, fearing that the loss of the country would result in other nearby Southeast Asian countries falling to communism. Thousands of American troops went to Vietnam.

94. Houston's Sam Houston Coliseum was a popular indoor arena. In 1962, citizens of the city welcomed NASA to their city by holding a barbecue for MSC employees in the coliseum. As part of their welcome, the Mercury Seven astronauts, along with Robert R. Gilruth, Walter C. Williams, and John A. Powers became members of the Reserve Deputy Sheriff's Posse and received Texas-style cowboy hats.

Chapter 2

1. *Aerodynamics* is the study of how the forces of air interact with moving objects. In other words, how does motion, along with gravity or the thrust of an engine, impact how a spacecraft will fly? An expert in aerothermodynamics studies the heating of a spacecraft's surfaces during flight, especially those at the high Mach numbers experienced by a vehicle re-entering Earth's atmosphere.

2. Hooks managed and oversaw the Orbiter/carrier separation systems for the Approach and Landing Tests in California, a series of flight tests separating the Space Shuttle *Enterprise* from the top of the Shuttle carrier aircraft, a modified Boeing 747. These tests proved the spacecraft was able to safely land after a mission. She also helped to design the solid rocket booster (SRB) separation system. The SRBs helped propel the Orbiter into space, where it separated from the stack two minutes after launch.

3. Flight dynamicists study and model the motion of a vehicle in the air or microgravity.

4. Dottie Lee sat for this interview in November 1999.

5. George Gamow authored several popular books, including *One Two Three . . . Infinity: Facts and Speculations of Science* and the Mr. Tompkins series.

6. In the early 1950s H. Julian Allen of the Ames Aeronautical Lab demonstrated that blunt nose cones worked better for the reentry of missiles than did sharp points. Max Faget proposed using the blunt-body shape for the Mercury spacecraft.

7. *Trajectory* is the path a spacecraft or satellite follows as it moves in space or through the air.

8. Personal computers like those we use today did not exist at the time. When Lee started with NACA, computers were not ubiquitous.

9. A *Mach number* tells engineers how fast an object is going compared to the speed of sound (which is 768 miles per hour at sea level). At Mach 2, the object is going twice the speed of sound. Spacecraft travel anywhere from Mach 2 to 30 while reentering the atmosphere, while supersonic fighter jets at their fastest travel

between Mach 1 and 2. Commercial airliners like a Boeing 737 travel around Mach 0.8. A *Reynolds number* can help to predict and compare different fluid flow situations. For instance, this number can be used to help scale up what happens in a wind tunnel—where technicians use small models of a flight vehicle—to what will happen when the object is flying through the atmosphere at full size. It also helps predict how the flow will behave. For low Reynolds numbers the flow is smooth, and for high Reynolds numbers the flow is chaotic.

10. An *integral* is a mathematical operation in calculus that solves the area under a curve. *Hypersonic flow* (flow going faster than five times the speed of sound) requires solving many integrals, here having to integrate three times.

11. Maxime A. "Max" Faget is recognized as the designer of American spacecraft from Mercury through the Space Shuttle. He served as director of Engineering at JSC for twenty years. Oral histories with the father of America's spacecraft are readily available on the JSC History Portal.

12. At the time, Joseph G. "Guy" Thibodaux was a NACA engineer. He became chief of the Propulsion and Power Division at JSC in 1964, a position he held until he retired in 1980. Download his oral history interview from the JSC History Portal to learn more about his career at NACA and NASA.

13. Robert R. "Bob" Gilruth was head of PARD.

14. From 1960 to 1968, Floyd L. "Tommy" Thompson served as Langley Research Center's center director.

15. "MacDac" stands for the McDonnell Aircraft Corporation based in St. Louis. Robert O. "Bob" Piland started as a mathematician with PARD. Once in Houston, he worked a variety of Apollo positions and later in Engineering and Development, Science and Applications, and Space and Life Sciences. To learn more, read his oral history transcript on the JSC History Portal. Gs stands for the gravitational forces a pilot would encounter; excessive force can cause a pilot to pass out.

16. John B. Lee, Dottie's husband, started working with PARD under Bob Gilruth in 1948. Lee, a World War II veteran, held a bachelor's degree in mechanical engineering from the Virginia Polytechnic Institute. He went on to work on the Mercury, Gemini, Apollo, and early Space Station designs. Learn more by reading his oral history transcripts, available on the JSC History Portal.

17. Bob Gilruth formed the Space Task Group in 1958 to initiate a piloted satellite project.

18. A calorimeter measures the heat absorbed or emitted from an object.

19. Ablative heat shields burn away as they reenter the atmosphere and therefore are used only once, unlike the Space Shuttle, which was designed to be flown over and over.

20. To protect the Apollo spacecraft as it reentered the Earth's atmosphere, NASA used an ablative heat shield that consisted of a brazed stainless-steel honeycomb structure filled with phenolic epoxy resin. Workers injected the cells by hand using specially developed guns.

21. The Apollo 1 fire happened at Cape Kennedy in 1967. A cabin fire during a launch pad test resulted in the death of the three astronauts in the vehicle.

22. Charles B. Rumsey joined the NACA Langley Memorial Lab in 1942. After working for the Aircraft Loads Division for two years, he chose to enlist in the military. When discharged, he returned to Langley, where he studied aerodynamic heating and boundary layer transition. He retired in 1975.

23. *Angle of attack* is the angle between the oncoming air or relative wind and a reference line on an airplane or wing. Boeing maintains a page that illustrates the concept of angle of attack at http://www.boeing.com/commercial/aeromagazine /aero_12/whatisaoa.pdf.

24. Federal contractors work for companies that receive contracts with the federal government to perform a job or provide services or materials to the government. They are not government employees but work on government projects. Civil servants, called "engineers" by Lee, are employees of the federal government.

25. Marshall Space Flight Center also designed, developed, and tested the Shuttle's external tank along with the solid rocket boosters and motors.

26. Thermal protection systems protect vehicles from the heat of reentry. On the Space Shuttle Orbiter, they were reusable protective tiles; older vehicles had single-use heat shields.

27. The final design included delta wings, but Faget envisioned a vehicle with straight wings.

28. James A. Chamberlin was a Canadian who came to work for the Space Task Group. An engineer, he made many valuable contributions to NASA's human spaceflight programs, from Mercury to the Space Shuttle. Avro, a Canadian aircraft manufacturer, scrapped the CF-105 Arrow fighter airplane in February 1959, giving NASA the opportunity to offer jobs to engineers like Chamberlin.

29. *Convective heating* is the transfer of heat from one place to another by the movement of fluids. In order to reenter the atmosphere, a spacecraft must travel several times faster than the speed of sound. In doing so, it creates a shock wave — the shock wave is an intense way to get all the air out of the spacecraft's way by creating a spike in the temperature and pressure. That really high temperature air transfers heat to the spacecraft.

In the high-temperature gas surrounding a vehicle during atmospheric reentry, atoms/molecules and electrons interact in a way that radiates energy to the surface of the vehicle, which is known as *radiative heating*.

30. *Laminar flow* is a type of fluid flow where the fluid travels smoothly or in regular paths. In aerothermodynamics, this type of flow convects less heat to the surface of a vehicle than turbulent flow at a given condition. *Turbulent flow* is a type of fluid flow where the fluid mixes and has irregular fluctuations; it is chaotic. In aerothermodynamics, this type of flow convects more heat to the surface of a vehicle than laminar flow at a given condition.

31. NASA referred to the first four Space Shuttle flights (STS-1 through 4) as orbital test flights.

32. *Pushover-pullup* is a flight maneuver that significantly changes the angle of attack of a flying vehicle. Pushover refers to pushing the angle of attack down and the pullup refers to bringing it back up.

33. *Boundary layer transition* is a complicated process where the flow transitions from a smooth laminar condition to a chaotic turbulent condition. The boundary layer is near the surface of an object. When a fluid flows over a surface, there is a layer of the fluid that needs to speed up from the surface of the object where the velocity is zero to the free stream flowing fluid speed. The boundary layer is significantly different for laminar (smooth) and turbulent (has mixing) flow, so when a vehicle transitions from laminar to turbulent flow, there is a change in vehicle performance (how it flies) and in the amount of heat going to the surface of the object in the flow.

34. Joe H. Engle commanded STS-2, the second Space Shuttle mission, one of four orbital test flights. Selected in 1966, Engle participated in the Approach and Landing Tests and went on to command two Shuttle missions. Learn more about his career by downloading his biographical datasheet available online at NASA's Astronaut Homepage. In 2004, Engle sat for a series of interviews with the JSC Oral History Project, which are available on the JSC History Portal.

35. STS-4, the final orbital test flight, flew in the summer of 1982. Learn more by visiting NASA's website http://www.nasa.gov/mission_pages/shuttle /shuttlemissions/archives/sts-4.html.

36. Rockwell, the prime contractor for the Orbiter, was located just outside of Los Angeles.

37. Glen Goodwin worked at the Ames Research Center in California, where he previously headed the Thermo and Gas Dynamics Division. Edward Teller, a theoretical physicist, contributed to the development of the atomic bomb during World War II and is known as the "father of the H-bomb." Hans Mark served as director of the Ames Research Center for eight years, from 1969 to 1977. Read more about Mark's career by downloading his oral history interview on the JSC History Portal.

38. A *French curve* is a drafting tool used to draw smooth, curved lines.

39. Vought Corporation in Dallas, Texas, manufactured the reinforced carbon-carbon on the Space Shuttle's nose cap and the leading edge of the wings.

40. The orbital maneuvering system provided thrust for the vehicle once in space.

41. NASA Headquarters, based in the nation's capital, guides and directs the agency.

42. The Orbiter, or more commonly called the Space Shuttle, had a wide variety of subsystems managers; JSC's Engineering Directorate supported Lee, who oversaw and helped to design, develop, and review the hardware. Spacecraft subsystems were numerous.

43. Robert C. Ried, Jr., began his career in New York at Grumman Aircraft Engineering Corporation; he later joined the Space Task Group and moved to Houston, where he worked until he retired in 1995 as chief engineer for Research and Development. An interview with Ried is available on the JSC History Portal.

44. Aaron Cohen oversaw the Orbiter Project, located at the Johnson Space Center, from 1972 to 1982, and Milton A. Silveira was his deputy. They were responsible for directing the design, development, production, and testing of the spacecraft. Interviews with Silveira and Cohen are available on the JSC History Portal.

45. Lee is referring to the STS-51L accident that occurred in January 1986. Learn more about that tragic flight by visiting NASA's mission archives at http://www.nasa.gov/mission_pages/shuttle/shuttlemissions/archives/sts-51L.html.

46. Michael J. "Mike" Smith was the pilot on STS-51L. The AIAA is the American Institute of Aeronautics and Astronautics.

47. Lee retired from NASA, according to her biographical datasheet, in 1987.

48. Bass Redd worked at NASA for twenty-six years on a variety of projects, including Apollo and Space Shuttle.

49. NASA leased temporary space when they first moved to Houston before the center opened in Clear Lake. To meet the president's goal, the government leased a former Canada Dry bottling plant and even apartments. Some employees worked in buildings off of Telephone Road or at Ellington Air Force Base.

50. The General Schedule is the pay scale for government employees. The scale, based on education and experience, starts at GS-1 and goes through GS-15.

51. Ivy F. Hooks participated in the JSC Oral History Project in March 2009.

52. Hooks graduated from the University of Houston in 1963 with a degree in mathematics.

53. Humboldt C. "Hum" Mandell, Jr., continued to develop cost estimates for other advanced programs like the Space Shuttle and Space Station. To learn more about this concept, read his interview on the JSC History Portal.

54. In brief, guidance calculates the changes needed to follow a trajectory or computes the data to reach a target. Spacecraft position, velocity, and orientation are provided by navigation systems. Reaction control engines are small thrusters used by spacecraft to maneuver in orbit.

55. Caldwell C. Johnson held only a high school degree but worked closely with Max Faget in Engineering and was one of the inventors of the Mercury spacecraft. He began working for NACA in 1937, when he was only eighteen, and worked for NASA until 1974 when he retired. Transcripts of his oral history interviews are available on the JSC History Portal.

56. That was the Flight Performance and Dynamics Branch.

57. Joe D. Gamble analyzed and simulated how well the Apollo launch escape system and entry capsule would perform in an emergency. Later, as NASA developed its first reusable spacecraft, he helped to develop and test the vehicle's entry flight control system and served as a subsystems manager for the Orbiter.

58. In hopes of landing on land, not water, NASA initially studied parasail landing systems but abandoned the concept.

59. Edward H. "Ed" White completed the first American spacewalk in June 1965 during Gemini IV.

60. The December 1968 Apollo 8 mission was the first to circle the Moon. For more details, see https://www.nasa.gov/mission_pages/apollo/missions/apollo8.html. Florida's Kennedy Space Center, or the "Cape" as some refer to the field center, launched human spaceflight missions.

61. John E. DeFife joined NASA in 1963 after graduating from the University of Pittsburgh. He spent most of his career in the Engineering and Development Directorate, leaving in 1990 to work for the Information Systems Directorate. The JSC Oral History Project interviewed John DeFife in 2000. Learn more by reading his transcript on the JSC History Portal.

62. Ops refers to the flight controllers working in Mission Control.

63. Betty M. Ensley worked as Max Faget's secretary at the space center from 1962 to 1977.

64. *Jane's All the World's Aircraft* was first published in 1909 and was the standard reference tool for aviation questions at the time engineers began working on the Shuttle design.

65. Burt Reynolds's nude centerfold in *Cosmopolitan* in April 1972 caused a sensation in the United States, not just NASA.

66. Hooks worked for the Engineering Analysis Division. From 1970 to 1979, Joyce H. Koplin served as secretary of the division.

67. At the time, Bruce Jackson headed the Engineering Analysis Division. He started with NACA as a PARD intern in 1955. Eventually he became part of the STG, moved to Houston, and remained at JSC until he retired in 1985. In 2009, he participated in the JSC Oral History Project. To learn more about his career, read his transcript on the JSC History Portal.

68. Hooks needed the boosters tilted or canted so that the particles would not hit the Orbiter.

69. Robert F. "Bob" Thompson served as the Space Shuttle program manager until 1981, when the Space Shuttle *Columbia* first flew into orbit. Read more about his career on the JSC History Portal.

70. The Tullahoma Arnold Engineering Development Center is located in Tennessee.

71. A Titan is a type of rocket developed in the 1950s and used by the agency to launch its Gemini spacecraft. NASA used the Titan II GLV, not the IIIC. The Department of Defense, specifically the US Air Force, was the primary user of the IIIC launch system. The problems described by Hooks were not experienced during Gemini because the early Titans did not have side motors.

72. Rick Barton headed the Entry Analysis Section of the Engineering Analysis Division in 1973.

73. Rockwell was based in Downey, California, where a good deal of the production on the Orbiter was performed.

74. The Shuttle carrier aircraft (SCA) crew included Fitzhugh L. Fulton, Thomas C. McMurtry, Arda J. Roy, Victor W. Horton, Thomas E. "Skip" Guidry, William R. Young, and Vincent A. Alvarez. The astronauts flying *Enterprise* consisted of two-man crews: Commander Fred W. Haise and Pilot C. Gordon Fullerton, and Commander Joe Engle and pilot Richard H. "Dick" Truly. To learn more about the *Enterprise* separation flights, read the astronauts' oral histories on the JSC History Portal.

75. NASA attached a tail cone to *Enterprise* for several of the Approach and Landing Tests to reduce drag.

76. Astronauts trained for missions in the Shuttle Mission Simulator (SMS) and verified flight software in the Shuttle Avionics Integration Laboratory (SAIL).

77. Christopher C. Kraft was JSC's second center director, a position he held until he retired in 1982. Just out of college he began a career with NACA in 1945, and in 1958, after NASA was established, became a member of the Space Task Group. He is best known for his role in the creation of spaceflight operations and serving as NASA's first flight director. Learn more about Kraft and his time at NASA by reading his oral histories at the JSC History Portal.

78. *The Little Red Hen* is a children's story in which a hen finds a grain of wheat. She asks for help in planting the seed, but no one agrees. When the wheat must be harvested, milled, and baked, the other animals refuse to help. Yet everyone willingly volunteers to eat the bread once the hard work is done.

79. The Space Shuttle main engines and its thermal protection system were pacing items for the Space Shuttle before its first mission.

80. Hooks retired from NASA in 1984.

Chapter 3

1. Eugene F. Kranz started working for the Space Task Group in 1960. He went on to become a flight director and led the Mission Operations Directorate from 1983 until his retirement in 1994. His book, *Failure Is Not an Option: Mission Control from Mercury to Apollo 13 and Beyond*, details his time in flight control. Read more about his lengthy career by downloading his oral history transcripts on the JSC History Portal.

2. Ginger Kerrick sat for this interview with the project in September 2012.

3. Duane L. Ross headed the Astronaut Selection Office for more than thirty-five

years. Learn more about his experience by reading his interview on the JSC History Portal website.

4. Building 4 South houses flight controllers.

5. The ECLSS systems make the ISS spacecraft habitable for astronauts. One function is to maintain cabin pressure; another provides life support for astronauts: oxygen and the removal of carbon dioxide from the atmosphere.

6. Expedition 1 was the first Space Station increment. Read more about that mission at: http://www.nasa.gov/mission_pages/station/expeditions/expedition01 /index.html.

7. William M. "Bill" Shepherd, a veteran of four spaceflights, was selected to command the first ISS increment in 1996, four years before they launched. He first flew in space onboard STS-27, a classified Department of Defense mission. Learn more by downloading his biographical datasheet on NASA's Astronaut Homepage.

8. Brock "Randy" Stone started working for the Landing and Recovery Division (LRD) at the Manned Spacecraft Center in 1967. Two years later, as part of his LRD duties, he joined the three astronauts of the Apollo 12 crew in the Mobile Quarantine Facility upon their return from the Moon, and stayed with them until NASA released them from quarantine. He later became a flight director, then retired in 2004 as JSC's deputy director. Read more about Randy Stone's career on the JSC History Portal.

9. More detailed information on the Functional Cargo Block (FGB) and Zvezda modules can be found on the NASA website at: http://www.nasa.gov/ mission_pages/station/structure/elements/fgb.html and http://www.nasa.gov /mission_pages/station/structure/elements/sm.html.

10. Node 1, known as Unity, linked American and Russian modules in space. Read more about this module at http://www.nasa.gov/mission_pages/station /structure/elements/node1.html and learn about the Laboratory Module, which is known as Destiny, by visiting http://www.nasa.gov/mission_pages/station/structure /elements/destiny.html.

11. Scott J. Kelly was then NASA's director of Operations in Star City, Russia. He spent 340 days onboard the ISS to enable researchers to learn more about space and its impact on the human body. He published a book about his year in orbit called *Endurance: A Year in Space, A Lifetime of Discovery*. Read his NASA biography on NASA's Astronaut Homepage.

12. The name "CapCom" is a holdover from the early space days when crews flew in capsules and that individual communicated with the crew from the Mission Control Center.

13. The third increment to the ISS, Expedition 3, included an American astronaut and two cosmonauts. Read more about their mission at http://www.nasa.gov /mission_pages/station/expeditions/expedition03/index.html.

14. Norman D. Knight started working at JSC in 1998 as a co-op and became chief of the Flight Director Office in 2012. In 2020, he served as deputy director of the Flight Operations Directorate.

15. Frank L. Culbertson, Jr., a naval aviator, commanded this mission. Selected as an astronaut in 1984, he flew STS-38 and STS-51 and went on to manage the Shuttle-Mir Program. Read more about his NASA career by downloading his biographical datasheet on NASA's Astronaut Homepage or by reading his oral history transcript on the JSC History Portal.

16. Vladimir Dezhurov was one of the cosmonauts assigned to this flight.

17. Flight controllers monitor spaceflight operations over communication loops. In the Control Center, flight controllers plug their headsets into these loops to hear conversations about mission-related topics. Controllers might listen to half a dozen or more at one time.

18. The IMMT includes managers who weigh in on key decisions. Overall, the team provides direction for the on-orbit operation of the Station.

19. Between September 2005 and April 2006, Expedition 12's two-man crew lived onboard the ISS. Michael Lopez-Alegria, Mikhail Tyurin, and Sunita L. Williams made up the crew of Expedition 14, which flew between September 2006 and April 2007. To learn more about the missions and crew of Expedition 14, visit http://www.nasa.gov/mission_pages/station/expeditions/expedition12/index.html and http://www.nasa.gov/mission_pages/station/expeditions/expedition14 /index.html. Read Lopez-Alegria's and Williams's biographies on NASA's Astronaut Homepage.

20. Kerrick spent time in Star City, Russia, as the Russian training integration instructor.

21. EVAs are spacewalks.

22. The Progress is an unmanned Russian spacecraft that brings hardware, space parts, repair gear, and supplies to the ISS.

23. Called a "home improvement" mission, STS-126 delivered a whole host of items to the International Space Station, including a water recycling system, sleeping quarters, a kitchen, and a new space toilet. Read more about the flight at http://www.nasa.gov/mission_pages/shuttle/shuttlemissions/sts126/main/index.html.

24. Christopher J. Ferguson flew three spaceflights, serving as commander for STS-126 and STS-135. Read more about his spaceflight experience by downloading his online biography on the NASA's Astronaut Homepage.

25. STS-120 (also known as 10A because it was an ISS assembly flight) featured the second female commander of a Space Shuttle, Pamela A. Melroy. She is featured in this book. Learn more about that mission by visiting http://www.nasa.gov /mission_pages/shuttle/shuttlemissions/sts120/.

26. Astronauts simulate spacewalks in the Sonny Carter Training Facility Neutral Buoyancy Laboratory, located near JSC in Houston.

27. Kerrick is referring to a loss of communication with the crew in orbit.

28. Kerrick is referring to the Expedition 32 spacewalk in August 2012. Suni Williams and Akihiko Hoshide encountered problems when installing the Main Bus Switching Unit, which is part of the station's power system, distributing power, such as from solar arrays. Just a few days later, they finished installing the unit.

29. Sarah L. Murray sat for this interview with the JSC Oral History Project in August 2012.

30. NASA selected John H. Casper in May 1984. He was heavily involved in ensuring the agency implemented the *Columbia* Accident Investigation Board recommendations following STS-107. Read more about his NASA career by downloading his biographical datasheet on NASA's Astronaut Homepage.

31. An instrumentation and communications officer, or INCO, is an officer in Mission Control that keeps an eye on in-flight communications and onboard instrumentation systems.

32. Gary Morse worked as the network director for the Space Shuttle at Goddard Space Flight Center from 1981 to 2000.

33. The tracking and data relay satellite system (TDRSS) allows NASA and

other agencies to communicate with satellites, the Space Shuttle, and International Space Station, replacing an earlier network of ground stations that dotted the globe.

34. Federal contractors work for companies like RSOC, which receive contracts with the federal government to perform a job or provide services or materials to the government. They are not government employees but work on government projects. Civil servants, by contrast, are employees of the federal government.

35. The Electrical, Environmental, and Consumables Manager (EECOM), a flight controller, oversees the thermal controls of the vehicle and monitors the cabin atmosphere, among other things.

36. Prop, a propulsion engineer, oversaw the orbital maneuvering and reaction control systems.

37. During liftoff, the booster, another controller, monitored the Shuttle's three main engines, solid rocket boosters, and external tank during ascent.

38. The Russian interface officer, or RIO, is the primary interface between the US and Russian control teams.

39. Mir was a Russian Space Station; NASA worked with Russia on the Shuttle-Mir Program, which paved the way for the International Space Station.

40. The blue Flight Control Room (FCR) handled Space Station missions.

41. The Space Shuttle *Columbia* broke up over East Texas and Louisiana upon reentry on February 1, 2003.

42. Gary C. Horlacher started out as a contractor and became a civil servant in 2006. By the time NASA selected Horlacher as a flight director in 2008, he had supported seventy-five Space Shuttle missions.

43. The Hotter'N Hell Hundred is an annual one-hundred-mile bicycle ride in Wichita Falls, Texas.

44. Douglas H. "Doug" Wheelock came to NASA in 1998 from the US Army. He was a rookie on STS-120 and then went on to serve as Flight Engineer for Expedition 24 and commanded Expedition 25. Read more about his experiences in his biography on the NASA Astronaut Homepage.

45. David W. Whittle headed the *Columbia* Recovery Office in 2003 and 2004. Read his oral history transcripts online on the JSC History Portal.

46. Senior Executive Service (SES) employees are basically executive managers for the federal government.

47. The beta gimbal assembly (BGA) is a mechanism used to maneuver a US solar array; the solar alpha rotary joint (SARJ) is a rotary joint that allows the solar arrays to track the sun. The remote power controller (RPC) controls the flow of electrical power to users.

48. STS-115 flew ISS assembly flight 12A, while STS-116 flew assembly mission 12A.1 More details can be found online at http://www.nasa.gov/mission_pages/station/structure/iss_assembly_12a.html.

49. United Space Alliance (USA) was an operations contractor.

50. Basically, Onboard Data Interfaces and Networks (ODIN) handled computer software and hardware, while the communications and tracking officer (CATO) oversaw on-orbit communication systems.

Chapter 4
1. Federal contractors work for companies who bid on and are awarded contracts with the federal government to perform a job or provide services or materials to

the government. They are not government employees but work for independent companies that receive funds for government projects. Civil servants, by contrast, are employees of the federal government.

2. S. Jean Alexander sat for this interview with the JSC Oral History Project in August 2014.

3. Before JSC opened, many NASA employees worked at Ellington.

4. Project Gemini was the second American human spaceflight program, following Project Mercury.

5. Safety, Reliability, and Quality Assurance (SR&QA) ensures the safety of all NASA missions, hardware, and its personnel.

6. William M. "Bill" Bland, Jr., began working for the National Advisory Committee for Aeronautics (NACA) in 1947. When NACA became NASA, he started working in the Systems Test Branch for the Space Task Group. From 1964 until his retirement in 1979, Bland developed and implemented reliability and quality assurance programs for the center.

7. Glen E. Brace served as head of the Awards Section at JSC from 1970 to 1985.

8. Until Chernobyl, the Three Mile Island nuclear accident, which occurred on March 28, 1979, was the worst nuclear power disaster in history. Bland was appointed to serve on the President's Commission on the Accident at Three Mile Island to determine the cause of the accident and how such future incidents could be handled. Upon completion of the investigation, Bland received the Exceptional Service Award from the commission.

9. Phyllis A. Morton spent her entire career—from 1967 to 1991—in the same division, the Crew Systems Division, which later became the Crew and Thermal Systems Division.

10. Walter W. "Walt" Guy oversaw the Crew Systems Division. Learn more about his career by reading his oral transcripts on the JSC History Portal.

11. The extravehicular mobility unit (EMU) is the suit that an astronaut wears for protection from the extreme temperatures of outer space.

12. At the time James O. "Jim" Schlosser managed the crew escape system.

13. Read more about the experiences of the other NASA suit techs, Ronald C. Woods, Alan M. Rochford, Troy M. Stewart, and Joseph W. Schmitt, on the JSC History Portal.

14. The suits used for the first four Space Shuttle missions were ejection escape suits, similar to those worn by flight crews in high-altitude reconnaissance airplanes.

15. Astronauts began wearing coveralls on STS-5 in November 1982. An image of Sally Ride wearing the outfit can be viewed at http://www.nasa.gov/content/president-obama-awards-presidential-medal-of-freedom-to-sally-ride/.

16. Building 9 housed the full-fuselage trainer and two crew compartment trainers on which Space Shuttle crews trained for missions. In 1983, Sally K. Ride was the first American woman to fly in space. Learn more about her career by reading her biographical information on NASA's Astronaut Homepage and her oral history interviews on the JSC History Portal.

17. After the *Challenger* accident, astronauts started wearing the bulky orange launch entry suit, which was a partial-pressure suit. If the crew cabin lost pressure during ascent or reentry, the suits provided protection.

18. Crews are secured into their seats by a team of folks on launch day; this is commonly called "strapping in" at NASA.

19. Troy Stewart and Al Rochford both retired in January 1998.

20. STS-110 was a Station assembly flight. Learn more about the mission's objectives at https://www.nasa.gov/mission_pages/shuttle/shuttlemissions/archives/sts-110.html.

21. STS-1 was the first Space Shuttle mission. Read a brief summary of this flight at http://www.nasa.gov/mission_pages/shuttle/shuttlemissions/archives/sts-1.html.

22. The Astrovan takes the crew out to the launch pad and transports them after the landing.

23. Suit techs traveled to Florida's Kennedy Space Center, or the Cape, as some refer to the field center, to support the launch.

24. The commander and pilot of each mission practiced landing the Orbiter on one of the STAs, a fleet of four Gulfstream II aircraft modified to simulate Shuttle landings.

25. Astronauts trained on the KC-135, an aircraft that, when flying a specific parabolic trajectory, gave flight crews and others onboard the opportunity to experience a few moments of weightlessness.

26. George D. "Pinky" Nelson was an astronaut selected in 1978 who flew three missions: STS-41C, STS-61C, and STS-26, the Return to Flight following the *Challenger* accident. To learn more about his NASA experience, read his online biography on NASA's Astronaut Homepage, and his oral history transcript on the JSC History Portal.

27. If an accident occurred at the launchpad and the crew were unconscious, the suit tech and astronaut support person (ASP) would have had to enter the vehicle and remove the astronauts from harm. If an astronaut were unconscious, the astronaut might have to be dragged, not carried, from the Orbiter to safety.

28. The ASP, or "Cape Crusader," also helped configure the vehicle's cockpit for launch; this included setting the switches and placing checklists, cue cards, and crew equipment in the cabin.

29. The VITT, just like the suit techs, were assigned to a crew. Astronauts describe them as their caretakers, the people who took care of them while they were at the Cape for prelaunch activities and then following landing.

30. Selected as an astronaut in 1978, Anna L. Fisher was the first mother to fly in space. She retired in 2017. Learn more about her historic career by reading her biography on NASA's Astronaut Homepage and her oral histories on the JSC History Portal.

31. Astronauts enter the spacecraft through the White Room and make any final preparations there before being strapped in. The "baskets" offered astronauts and personnel at the launchpad an escape route in case of an emergency.

32. Shannon W. Lucid, one of the first six women astronauts selected in 1978, first flew in 1985. She went on to set an impressive spaceflight endurance record onboard the Russian Space Station Mir. Read more about her accomplishments with the space agency on NASA's Astronaut Homepage. Her oral history detailing her increment on the Mir Space Station is available on the JSC History Portal.

33. Marsha S. Ivins became an astronaut candidate in 1984, achieving an important lifelong goal. She began working at JSC ten years earlier, first working as an engineer on the Orbiter and later as a flight engineer on the STA. Learn more about her missions by reading her online biography on NASA's Astronaut Homepage.

34. MS-2 was the flight engineer who assisted the Space Shuttle pilot and commander during launch.

35. STS-41D featured the first flight of the Space Shuttle *Discovery* and an industrial payload specialist. Learn more about this historic flight at http://www.nasa.gov/mission_pages/shuttle/shuttlemissions/archives/sts-41D.html.

36. Judith A. Resnik was the second American woman in space. She perished onboard the *Challenger* in 1986. Read her astronaut biography to learn more about her accomplishments on NASA's Astronaut Homepage.

37. Astronaut Michael J. "Mike" Smith was STS-51L's pilot and perished onboard the Space Shuttle *Challenger*. Read more about his naval career in his biography on NASA's Astronaut Homepage.

38. Kneeboards are essentially lap desks. They are also used onboard the T-38s to jot down clearances or hold cards in place.

39. Charles F. Bolden, Jr., was the NASA administrator from 2009 until 2017; he was selected as an astronaut in 1980. Read about his historic flights and NASA contributions in his oral history transcripts on the JSC History Portal and his biography on NASA's Astronaut Homepage.

40. The full-pressure suit was the advanced crew escape suit (ACES).

41. The Space Shuttle Program consisted of numerous subsystems, like its protective tiles (the thermal protection system) and reaction control system (small rockets that allowed the Shuttle to move in space).

42. Ellington Field is where NASA houses its planes, including the T-38s that astronauts fly.

43. Early in the Shuttle Program, astronauts trained in the Weightless Environment Training Facility (WETF), a large pool, for spacewalks.

44. Frederic Dawn designed, developed, tested, processed, and fabricated materials for spacesuits and the spacecraft.

45. Many of the astronauts who wore the launch entry suit complained that the pressure suit was too warm and uncomfortable. To cool themselves they came up with their own solution. Alexander remembers that they tied their lanyards to their Capilene underwear, which was a long underwear worn under the suit. Doing so increased ventilation in the suit and provided the astronauts greater comfort, but the solution brought significant risk that the neck dam, which fit tightly over the astronaut's head, could be damaged; that seal would be important if the crew had to bail out of the vehicle because it prevented water from entering the suit. Multiple studies explored the issue of ventilation and the LES in the years following the *Challenger* accident.

46. The United Space Alliance (USA), formed in 1995, was a spaceflight operations contractor. NASA signed a contract with USA to consolidate Shuttle contracts to reduce the costs of flying and processing the fleet.

47. Discrepancy reports (DR) identify known problems or errors that need to be fixed. DRs are given a number and remain open until the issue is resolved.

48. Terminal countdown demonstration tests (TCDT) are the final dress rehearsals for launch at Kennedy Space Center (KSC). More details about these events can be found online at http://www.nasa.gov/mission_pages/shuttle/flyout/tcdt.html.

49. Payload specialists were not professional astronauts. They could be representatives of other nations, companies, or, prior to the *Challenger* accident, even members of Congress. They typically monitored or managed a specific on-orbit experiment.

50. STS-53 was known as the Dog Crew. Learn more about the mission and crew by reading the flight summary at http://www.nasa.gov/mission_pages/shuttle/shuttlemissions/archives/sts-53.html.

51. STS-78 called themselves the Rat Crew. Read about the mission at http://www.nasa.gov/mission_pages/shuttle/shuttlemissions/archives/sts-78.html.

52. John W. Young and Robert L. "Bob" Crippen flew on the first Shuttle mission, while Joe H. Engle and Richard H. "Dick" Truly were on the second flight of the program. Young had flown on the Gemini Program and walked on the Moon. He was also the chief of the Astronaut Office at the time. Selected as an astronaut in 1966, Engle participated in the Approach and Landing Tests and commanded two Shuttle missions: STS-2 and STS-51I. Crippen and Truly were pilots for the first two Space Shuttle flights, STS-1 and STS-2. Both had been selected for the US Air Force's Manned Orbiting Laboratory, a military space station. When the program was cancelled they transferred to NASA, becoming part of the astronaut corps. Learn more about all of these Shuttle crewmembers by reading their biographies on NASA's Astronaut Homepage. Oral history interviews with Crippen, Truly, and Engle are available on the JSC History Portal.

53. George W. S. Abbey was head of Flight Operations at the time.

54. Crew quarters were in the Operations and Checkout (O&C) Building at KSC.

55. Sharon Caples McDougle sat for an interview with the JSC Oral History Project in July 2010.

56. Learn more about the flight that deployed the Gamma Ray Observatory in 1991 at http://www.nasa.gov/mission_pages/shuttle/shuttlemissions/archives/sts-37.html.

57. Linda M. Godwin joined NASA in 1980. Selected as an astronaut candidate in 1985, she flew four spaceflights between 1991 and 2001. Learn more about those missions and her experiences on NASA's Astronaut Homepage.

58. STS-47, a joint American and Japanese flight, featured the Spacelab and the first African American woman in space, Mae C. Jemison. Read more about the flight at http://www.nasa.gov/mission_pages/shuttle/shuttlemissions/archives/sts-47.html. Jemison became an astronaut candidate in 1987 after the *Challenger* accident. STS-47 was her first and only spaceflight. She left the agency in 1993 to pursue her passion in a new and exciting way. For more about her career, visit NASA's Astronaut Homepage.

59. The ACES suit is designed to protect astronauts during launch and re-entry.

60. STS-114 was the Return to Flight after the STS-107 accident. Read more about that historic mission at https://www.nasa.gov/mission_pages/shuttle/shuttlemissions/archives/sts-114.html.

61. Building 5 is the Jake Garn Mission Simulator and Training Facility, named for Utah Senator Edwin Jacob "Jake" Garn, who flew on STS-51D in 1985.

62. Building 9 is the Space Vehicle Mockup Facility.

63. During the Shuttle Program, crews participated in bailout training, where they donned their pressure suits (LES or ACES) and "bailed out" of the spacecraft.

64. Crews train for spacewalks at the Sonny Carter Training Facility Neutral Buoyancy Laboratory at JSC.

65. The Crew Transport Vehicle, first used with STS-40 in California in 1991, pulled up to the Orbiter after a flight. If needed, the astronauts were treated in the vehicle, and then taken to another location for additional medical treatment.

66. Joan E. Higginbotham is featured in this volume.

67. The G suit has two main purposes: to prevent a space flyer's blood from pooling in the legs and to provide support to an astronaut's abdomen during reentry.

68. The communication carrier assembly (CCA) caps fit under the astronaut's helmet and are equipped with a microphone and earphones.

69. Learn more about *Discovery*'s and *Endeavour*'s final flights, STS-133 and STS-134, at http://www.nasa.gov/mission_pages/shuttle/shuttlemissions/sts133/main/index.html and http://www.nasa.gov/mission_pages/shuttle/shuttlemissions/sts134/main/index.html.

70. STS-75 was a reflight of the Tethered Satellite System; read more about this system and the crew at http://www.nasa.gov/mission_pages/shuttle/shuttlemissions/archives/sts-75.html.

71. Learn more about the flight of STS-78 at http://www.nasa.gov/mission_pages/shuttle/shuttlemissions/archives/sts-78.html.

72. The LES and ACES suits are orange.

73. Lee Greenwood's popular song "God Bless the U.S.A." features the lyrics "And I'm proud to be an American."

Chapter 5

1. Dee O'Hara sat for this interview in 2002 with the JSC Oral History Project.

2. George M. Knauf was a staff surgeon at the Air Force Missile Test Center at Patrick Air Force Base. In 1962, he transferred to NASA Headquarters to become the deputy director of Aerospace Medicine in the Office of Manned Space Flight.

3. Cape Canaveral Air Force Station was the cradle of the US human space program.

4. Langley Research Center was home to the Space Task Group and the astronauts.

5. Alan B. Shepard and Virgil I. "Gus" Grissom were the first two Americans to fly in space in 1961. Learn more about their selection and Mercury flights by reading their online biographies on NASA's Astronaut Homepage. Alan Shepard sat down for an oral history interview in 1998; it is available on the JSC History Portal.

6. James E. Webb served as NASA administrator from February 1961 to October 1968.

7. John H. Glenn, Jr., was the first American to orbit Earth onboard Friendship 7 in February 1962 and later flew onboard the Space Shuttle *Discovery*. Read more about his career by visiting NASA's Astronaut Homepage and read his oral history on the JSC History Portal.

8. For his work in aerospace medicine, William K. Douglas was inducted into the International Space Hall of Fame in 1992.

9. A Snark is an Air Force missile.

10. Grissom nearly drowned when his spacecraft door opened prematurely. For more information, visit http://www.nasa.gov/mission_pages/mercury/missions/libertybell7.html.

11. The Apollo 1 fire happened at Cape Kennedy in 1967. A cabin fire during a launch pad test resulted in the death of the three astronauts in the vehicle.

12. NASA named Clifton C. Williams as an astronaut in 1963, and he died four years later. His biography is available on NASA's Astronaut Homepage and gives a brief overview of his pre-NASA career.

13. Because of an onboard explosion that crippled the spacecraft, the Apollo 13 crew forfeited their Moon landing. For more details on the harrowing events, read

the flight summary at https://www.nasa.gov/mission_pages/apollo/missions/apollo13.html.

14. The Lunar Landing Training Vehicle simulated the flying qualities of the lunar module. The astronauts trained on the vehicle at Ellington.

15. Vickie Kloeris spoke with the JSC Oral History Project in September 2012.

16. President Ronald Reagan called upon NASA to build a space station within a decade in his 1984 State of the Union address. In 1988, the station was named Freedom.

17. Skylab was America's first space station. Read about its three crews and experiments at http://www.nasa.gov/mission_pages/skylab.

18. During Bill Clinton's presidency, the Freedom Program transformed into the International Space Station.

19. Charles T. Bourland was manager of the Space Station food system at the time. For more information about his career with NASA, read his oral history transcript on the JSC History Portal.

20. Natick Research Center, Massachusetts, is the Army's research and development arm; their food laboratories supported NASA's early spaceflight programs.

21. Meals Ready to Eat are military rations provided to troops.

22. ISS Phase One was known more commonly as the Shuttle-Mir Program. Read more about the Shuttle-Mir Program and the oral histories of some of the people who worked on the program and lived on Mir on the JSC History Portal.

23. Helen W. Lane worked as a life scientist at JSC and conducted research on subjects such as metabolism and nutrition.

24. The Progress is a Russian ISS resupply vehicle. For more information, see http://www.nasa.gov/mission_pages/station/structure/elements/progress.html.

25. Norman E. Thagard was part of the 1978 class of astronauts. A medical doctor, he studied how zero gravity impacted the human body during his first spaceflight, STS-7. He flew three additional Shuttle missions before flying to Mir in 1995. Learn more about his career on NASA's Astronaut Homepage and read his oral history transcript on the JSC History Portal.

26. Nutritionist Scott M. Smith worked for Helen Lane in 1994.

27. Some buildings at JSC contain chambers where they test space hardware using the heat, cold, or pressures found in space. These tests can be crewed or uncrewed, depending on the requirements.

28. At the time, Nigel Packham worked for Lockheed Martin as a life support system scientist.

29. Space Center Houston is JSC's visitor center.

30. SLS-1 was a Spacelab flight dedicated to life sciences and the first to include three women on the crew. Learn more about their experiments at http://www.nasa.gov/mission_pages/shuttle/shuttlemissions/archives/sts-40.html.

31. TCDTs (terminal countdown demonstration tests) are the final dress rehearsals for launch at KSC. More details about these events can be found online at http://www.nasa.gov/mission_pages/shuttle/flyout/tcdt.html.

32. Among other things, the Cape Crusaders helped configure the vehicle's cockpit for launch; this included setting the switches and placing checklists, cue cards, and crew equipment in the cabin.

33. FEPC (Flight Equipment Processing Contract) was the name of the contract awarded by NASA to Boeing Aerospace Operations for crew and flight

equipment for the Space Shuttle Program. Items included spacesuits, food, and communications equipment.

34. The ATV (Automated Transfer vehicle) is an ISS resupply vehicle, supplied by the European Space Agency. Learn more by visiting http://www.nasa.gov /mission_pages/station/structure/atv.html.

35. Rita Rapp worked on food systems for the Apollo, Skylab, Apollo-Soyuz Test Project, and Space Shuttle flight crews.

36. That payload specialist was Rodolfo Neri Vela, who flew on STS-61B.

37. The Heights is a neighborhood in northwest-central Houston.

38. The bacteria *Clostridium botulinum* produces neurotoxins that can cause botulism.

39. SpaceX is NASA's first commercial supplier to the Space Station. Their spacecraft successfully docked with the ISS in May 2012.

40. Learn more about Expedition 33 and its crew by visiting http://www.nasa .gov/mission_pages/station/expeditions/expedition33/index.html.

41. That crew flew on Soyuz TMA-07M in December 2012. Chris Hadfield became the first Canadian to command an ISS increment, Expedition 35. Read more about his spaceflight experience by visiting http://www.asc-csa.gc.ca/eng /astronauts/biohadfield.asp. Roman Romanenko, a Russian cosmonaut, was one of the expedition's flight engineers. Learn more about his experience by visiting http://www.gctc.ru/main.php?id=204. Thomas H. "Tom" Marshburn joined NASA as a flight surgeon. In 2004, he was selected as an astronaut candidate and first flew on STS-127. He was also a flight engineer on the increment. Read his biography on NASA's Astronaut Homepage.

42. Read an interview with former ISS program manager Michael T. Suffredini on the JSC History Portal. He held that position for ten years, from 2005 to 2015.

43. HTV (H-II Transfer Vehicle) is Japan's ISS resupply vehicle. Information about HTV-4 can be found online at http://www.nasa.gov/content/htv-4-readies-for-departure-as-station-prepares-for-crew-exchange.

44. STS-135 was the final flight of the Space Shuttle Program. Read more about this historic mission at http://www.nasa.gov/mission_pages/shuttle/shuttlemissions /sts135/main/index.html.

45. NASA refers to MPLMs as "moving vans." To learn more about these modules, visit http://www.nasa.gov/mission_pages/station/structure/elements/mplm.html.

46. The Space Shuttle Program ended in July 2011.

47. That astronaut was Luca Parmitano. Read his online ESA biography for more information: http://www.esa.int/Our_Activities/Human_Spaceflight /Astronauts/Luca_Parmitano.

Chapter 6

1. Biographies are available online for all of the astronauts mentioned in this chapter on NASA's Astronaut Homepage. Some of the astronauts have oral histories available on the JSC History Portal.

2. Kathryn D. Sullivan sat for this interview with the JSC Oral History Project in May 2007.

3. In very general terms, mission specialists were scientists who flew on the Space Shuttle; pilots operated and flew the Orbiters in space and, as commanders, landed the vehicle at the end of a mission.

4. After completing doctoral work, some students accept postdoctoral research positions to further their research and gain additional training before they accept or pursue an academic or research post.

5. NASA selected eleven scientist-astronauts in 1967, who named themselves the XS-11 (pronounced "excess eleven"). None actually flew until the Shuttle Program became operational with STS-5 in 1982.

6. Frederick D. Gregory, one of Sullivan's close friends and one of the first African American astronauts, was selected in 1978.

7. Terry McGuire was a psychiatrist, and Joseph Atkinson was head of Equal Opportunity Programs at JSC.

8. On January 16, 1978, NASA released a press announcement about its newest class of astronauts. See release 78–03 at http://www.nasa.gov/centers/johnson/pdf/83130main_1978.pdf.

9. George W. S. Abbey was head of the selection committee and Flight Crew Operations at JSC.

10. Christopher C. Kraft served as center director from January 1972 to August 1982. Ivy F. Hooks and Carolyn L. Huntoon are included in this book; see chapters two and nine. Also, their oral histories are available on the JSC History Portal.

11. Judith A. Resnik was working as an engineer for the Xerox Corporation in California, Anna L. Fisher was an emergency room doctor, and Shannon W. Lucid was a research associate for the Oklahoma Medical Research Foundation.

12. Sally K. Ride, the first woman from the US in space, was then a doctoral student in physics at Stanford University.

13. Three African Americans were selected in 1978: Fred Gregory, Ronald E. McNair, and Guion S. Bluford. Ellison S. Onizuka was the first Asian American astronaut.

14. Margaret Rhea Seddon, a surgeon, was the only blonde of the group. In 2015, she published her memoir, *Go for Orbit*.

15. Resnik died onboard the *Challenger* in 1986.

16. C. Gordon Fullerton and Richard H. "Dick" Truly had been selected as astronauts after the cancellation of the Manned Orbiting Laboratory Program, a Department of Defense space station, and flew orbital test flights for the Space Shuttle Program.

17. Cape Crusaders traveled to Florida's Kennedy Space Center, or the Cape as some refer to the field center, to support the crew.

18. In 1978, test pilots Daniel C. Brandenstein and Loren J. Shriver were named astronauts. Brandenstein flew a total of four spaceflights; Shriver's missions included STS-51C, 31, and 46.

19. Donald E. Williams piloted STS-51D and commanded STS-34, which deployed the *Galileo* spacecraft.

20. In NASA-ese, the EVA suit is called an extravehicular mobility unit, or spacesuit.

21. Richard M. "Mike" Mullane wrote *Riding Rockets: The Outrageous Tales of a Space Shuttle Astronaut*.

22. Astronaut Offices were and still are located in Building 4.

23. Selected as an astronaut in 1966, Paul J. Weitz flew on Skylab 2 and STS-6.

24. Joan E. Higginbotham sat for this interview with the JSC Oral History Project in August 2012.

25. Higginbotham is referring to the *Challenger* accident, which occurred in January 1986, resulting in the loss of the vehicle and its crew. Reductions in force are a permanent reduction in the civil servant workforce.

26. Jay F. Honeycutt then served as the director of Shuttle Management and Operations at Kennedy Space Center. He had previously worked as George Abbey's right-hand man for the 1978 astronaut selection.

27. CapCom is the person in the Mission Control Center who speaks directly with the crew in orbit. The title is a holdover from the days when NASA flew crews in capsules.

28. Astronauts are required to fly in the T-38s for a certain number of hours each month, regardless of their background and training. Pilots flew the T-38s while non-pilots (mission and payload specialists) sat in the backseat.

29. The Glavni is the Russian CapCom in the Russian Mission Control Center, located in Korolev.

30. NASA selected Kent V. "Rommel" Rominger in 1992; he went on to fly five missions and retired from NASA in September 2006. STS-117 carried a pair of solar arrays and two truss segments to the ISS. Learn more about the flight by visiting https://www.nasa.gov/mission_pages/shuttle/shuttlemissions/sts117/main/index.html

31. The *Columbia* accident occurred on February 1, 2003, as the Shuttle was returning to Kennedy Space Center in Florida after a sixteen-day mission. Read more about STS-107 at http://www.nasa.gov/mission_pages/shuttle/shuttlemissions/archives/sts-107.html.

32. Richard A. "Rick" Mastracchio, a veteran of four spaceflights, moved to Houston in 1987 to work for Rockwell, a NASA contractor. Prior to becoming an astronaut candidate in 1996, he worked as a NASA engineer and flight controller.

33. Mark L. Polansky spent twenty years with the space agency, from 1992 to 2012, as an engineer, research pilot, and astronaut. Details of the STS-116 mission, a Space Station assembly flight, are available at http://www.nasa.gov/mission_pages/shuttle/shuttlemissions/sts116/main/.

34. A test pilot by training, Robert L. Curbeam was selected as a mission specialist in 1994. He went on to fly three missions: STS-85, 98, and 116, before retiring from NASA in 2007. Selected as an astronaut candidate in 1998, William A. Oefelein piloted STS-116, his first and only spaceflight. In 2007, he returned to the United States Navy. Nicholas J. M. Patrick earned his PhD in mechanical engineering from the Massachusetts Institute of Technology in 1996, two years before he was selected as an astronaut candidate. He flew on STS-116 and STS-130. Christer Fuglesang, a Swede, is an ESA astronaut. Read more about his background online on the ESA website: http://www.esa.int/Our_Activities/Human_Spaceflight/Astronauts/Christer_Fuglesang.

35. During the flight, Higginbotham operated the Space Station's robotic arm to construct the outpost.

36. The truss is the Space Station's backbone.

37. The astronauts practice spacewalking at the Sonny Carter Training Facility Neutral Buoyancy Laboratory.

38. Sunita L. Williams, a member of the Expedition 14 crew, was a passenger on STS-116, catching a lift to the ISS on the Space Shuttle.

39. RTLS was an abort where the crew flew downrange to burn as much fuel

as possible from the external tank feeding the main engines. Theoretically, the crew would turn around, jettison the tank, and land at KSC. Crews never used the RTLS option.

40. Higginbotham is referring to the gravitational forces (Gs) experienced by astronauts during launch.

41. About eight minutes after launch, the three Shuttle main engines shut down—called MECO—and jettisoned the external tank from the vehicle. The tank provided fuel and oxidizer to the engines.

42. The CBMs (common berthing mechanism) are used to mate pressurized modules on the ISS.

43. The P5 truss was nicknamed "Puny."

44. Higginbotham is referring to the contingency landing site, the White Sands Space Harbor in New Mexico, where the Shuttle landed only once: for STS-3 in 1982.

45. Most crewmembers swallowed salt tablets and drank water before returning to Earth as a countermeasure to feeling unsteady after landing, known as postflight orthostasis.

46. ASPs (astronaut support person) are also known as the Cape Crusaders.

47. Steven W. Lindsey served as chief of the Astronaut Office from September 2006 to October 2009.

48. STS-126 was another ISS assembly mission. For details visit https://www .nasa.gov/mission_pages/shuttle/shuttlemissions/sts126/main/index.html.

49. Eileen M. Collins sat for this interview in February 2013.

50. Those astronauts included Daniel C. Brandenstein, Michael L. Coats, Richard O. Covey, Steven A. Hawley, Jeffrey A. Hoffman, and Rhea Seddon.

51. Duane L. Ross headed the Astronaut Selection Office for more than thirty-five years. Read his oral history transcript on the JSC History Portal.

52. Teresa Gomez worked as the assistant manager of the Astronaut Selection Office for more than thirty years.

53. Donald R. "Don" Puddy began working at NASA in 1964 as a flight controller and went on to become the lead flight director for the Approach and Landing Tests. He had rotations at the Ames Research Center in California and at NASA Headquarters. Puddy retired in 1995 while assigned to the JSC Russian Project Office.

54. John W. Young flew on the Gemini Program, walked on the Moon, and commanded the first Space Shuttle mission. He was well-respected in the eyes of spacefarers.

55. Astronauts trained for spacewalks in the Weightless Environment Training Facility before the NBL was built.

56. The US Air Force used the T-37 as its primary pilot training vehicle.

57. STS-63 was Collins's first flight. Learn more about that mission by reading the online summary at http://www.nasa.gov/mission_pages/shuttle/shuttlemissions /archives/sts-63.html. To learn more about the Shuttle-Mir Program, visit the JSC History Portal.

58. STS-63 was James D. Wetherbee's second command; he went on to be the first American to command five spaceflights.

59. C. Michael "Mike" Foale started out as a payload officer in Mission Control and became an astronaut candidate in 1987. He went on to fly six space missions: four Space Shuttle flights, a Shuttle-Mir increment, and an ISS increment.

Bernard A. Harris, Jr., was the first African American to walk in space and was payload commander for STS-63. Janice E. Voss flew into space five times between 1993 and 2000. She worked as a co-op at JSC from 1973 to 1975, and fifteen years later was selected as an astronaut candidate. Vladimir Titov had been a cosmonaut since 1976 and had a great deal of spaceflight experience before the Shuttle-Mir Program.

60. The three main engines and the lines that connect the external tank are part of the main propulsion system; the APU power the vehicle's hydraulic systems; and the fuel cells provide the spacecraft's electrical power.

61. To learn about Victor D. Blagov, read his online oral history transcript conducted for the Shuttle-Mir Program on the JSC History Portal.

62. R1 upper refers to the Orbiter's reaction control system. The RCS are thrusters on the Orbiter that provide attitude control or help to change the vehicle's velocity.

63. Charles J. Precourt was an Air Force test pilot and became an astronaut candidate in January 1990. Selected as a pilot astronaut, he flew first as a mission specialist, then piloted STS-71. He went on to command two missions.

64. Vasili Tsibliev commanded the Mir-23 mission.

65. Read the mission highlights from STS-93 online at http://www.nasa.gov/mission_pages/shuttle/shuttlemissions/archives/sts-93.html.

66. For nearly a decade, from April 1992 through November 2001, Daniel S. Goldin served as NASA's administrator. Learn more about Goldin's contributions to the agency by reading his online biography at http://history.nasa.gov/dan_goldin.html. An oral history with the administrator is also available on the JSC History Portal.

67. Stephen Hawking was a cosmologist and theoretical physicist.

68. Sam Donaldson was a reporter for ABC News.

69. Eileen Hawley, then head of JSC's External Relations Office, worked for JSC's Public Affairs Office for fifteen years.

70. Steven A. Hawley was an astronomer completing a post-doc at Cerro Tololo Inter-American Observatory in La Serena, Chile, when he was selected as an astronaut candidate in 1978. STS-93 was his final Shuttle mission. A year before she was selected as an astronaut candidate, Catherine "Cady" Coleman graduated from the University of Massachusetts with a PhD in polymer science and engineering. She flew two Shuttle missions aboard the Space Shuttle *Columbia* and on Expedition 26/27. Michel Tognini, a Frenchman, was an ESA astronaut. Read more about his spaceflight experience by visiting http://www.esa.int/Our_Activities/Human_Spaceflight/Astronauts/Michel_Tognini. Jeffrey S. Ashby piloted two Shuttle missions, STS-93 and 100, and commanded STS-112.

71. Learn more about Lisa M. Reed, who is featured in this book, by reading her oral histories on the JSC History Portal.

72. David Brady began working at NASA in 1989. For more than twenty years he provided assistance with payloads as an operations engineer and flight controller.

73. For ten years, from 1993 to 2003, Bryan Austin was a Space Shuttle and ISS flight director. In 1982, after completing his degree in computer science, he started working for Rockwell Shuttle Operations as a trainer. He accepted a position with Boeing in 2003, and four years later went to work for Lockheed Martin.

74. To learn more about the telescope, visit http://chandra.si.edu/.

75. A short biography of Chandra's namesake is available online at http://chandra.harvard.edu/about/chandra.html.

76. Hubble's primary mirror was flawed, creating blurred images.

77. Todd Halvorson was a senior aerospace reporter for *USA Today*.

78. Danica Patrick is a NASCAR driver, and James Bond is a fictional British secret agent created by Ian Fleming.

79. STS-114 was the Return to Flight mission after the loss of *Columbia* and her crew.

80. CACOs, casualty assistance calls officers, provided assistance to the families of the *Columbia* disaster.

81. Andrew S. W. Thomas became an astronaut in 1993 and flew four times, spending more than 177 days in space. NASA selected helicopter pilot Wendy B. Lawrence as an astronaut candidate in 1992. After completing four spaceflights, she retired in 2006. Charles J. "Charlie" Camarda started working at the NASA Langley Research Center in 1974 as a research scientist. More than twenty years later, in 1996, he was selected as an astronaut candidate and flew once on STS-114, the first Return to Flight mission following the loss of *Columbia* and her crew.

82. Described as a backflip, the rendezvous pitch maneuver allowed astronauts onboard the ISS to photograph the Orbiter's heat shield to determine if there was any damage on liftoff. Based on the photos, management determined whether or not the crew could safely return home.

83. James M. Kelly was the pilot for STS-114. Since his selection as an astronaut candidate in 1996, he has piloted two Space Shuttle flights.

84. Stephen K. Robinson has a long history with the agency. He began working at the Ames Research Center in 1975 as a co-op and later at the Langley Research Center in Hampton, Virginia. In 1994, NASA selected Robinson as an astronaut. He flew four Space Shuttle missions and retired from NASA in 2012.

85. That flight was STS-121, the second Return to Flight mission after the *Columbia* accident. Information about that mission can be found at http://www.nasa.gov/mission_pages/shuttle/shuttlemissions/archives/sts-121.html.

86. Pamela A. Melroy sat for this interview in November 2011.

87. Melroy is a petite blonde who doesn't fit the stereotype of the tough, rugged masculine image of an astronaut. She is five feet four inches tall.

88. NASA selected Melroy and Susan Kilrain in 1994. Kilrain, a naval aviator, piloted two missions in 1997, STS-83 and STS-94.

89. The Russian Service Module, Zvezda, launched in July 2000, and Melroy's first mission flew later that year. Read more about that module at https://www.nasa.gov/mission_pages/station/structure/elements/sm.html.

90. STS-92, an ISS assembly flight, flew in October 2000. Read the mission highlights at http://www.nasa.gov/mission_pages/shuttle/shuttlemissions/archives/sts-92.html.

91. Selected as an astronaut candidate in 1985, Brian Duffy piloted STS-45 and 57 and commanded STS-72 and 92.

92. Learn more about the STS-112 mission at https://www.nasa.gov/mission_pages/shuttle/shuttlemissions/archives/sts-112.html.

93. NASA selected David A. Wolf as an astronaut in 1990. During his tenure in the corps, he flew on three Space Shuttle missions and completed cosmonaut training for his Shuttle-Mir increment.

94. Sandra Magnus became an astronaut candidate in 1996 and first flew in 2002. She served as flight engineer 2 and science officer on Expedition 18 and was one of four crewmembers on the final Space Shuttle flight, STS-135. NASA

also selected Piers Sellers in 1996. He flew in space three times and later worked at Goddard Space Flight Center in Maryland. Fyodor Yurchikhin is a cosmonaut. Download his biography to read more about his background and training prior to STS-112: http://www.gctc.ru/main.php?id=219.

95. The IVA (intravehicular activity) individual helped suit up the astronauts for their EVA; she did not go outside and conduct an EVA but instead gave directions to the crew during their time outside of the vehicle.

96. The STS-107 crew was lost in 2003 as the *Columbia* disintegrated over East Texas and Louisiana.

97. Read Gregory Kovacs's online biography at http://web.stanford.edu/group /kovacslab/cgi-bin/index.php?page=gregory-kovacs.

98. The integrated product team (IPT) includes the flight integration manager, flight manager, and contractors who develop a plan and fix problems as they occur.

99. Information about STS-120 can be found on the NASA website at http:// www.nasa.gov/mission_pages/shuttle/shuttlemissions/sts120/.

100. Daniel M. Tani, selected as an astronaut candidate in 1996, flew on STS-120 to the ISS, where he served as flight engineer on Expedition 16. He retired from NASA in August 2012. Douglas H. "Doug" Wheelock came to NASA in 1998 from the US Army. He was a rookie on STS-120 and then went on to serve as flight engineer for Expedition 24 and commanded Expedition 25.

101. Node 2 was also called Harmony, and P6 was part of the Station's integrated truss. For more information about Harmony, visit https://www.nasa.gov/mission _pages/station/structure/elements/node2.html. NASA maintains a page about the integrated truss and its parts at http://www.nasa.gov/mission_pages/station/structure /elements/its.html.

102. For the first time two women were commanding spaceflights at the same time. Peggy A. Whitson was commander of the International Space Station from October 2007 to April 2008, while Melroy commanded the crew of STS-120.

103. George D. Zamka, a test pilot, was selected as an astronaut candidate in 1998. He flew two Space Shuttle missions, STS-120 and 130.

104. To see the highlights of this historic flight, visit http://www.nasa.gov /mission_pages/shuttle/shuttlemissions/archives/sts-71.html.

105. Selected as an astronaut candidate in 1978, Robert L. "Hoot" Gibson went on to fly five successful missions.

106. After crews return from their missions, they visit their hometowns, NASA centers, and contractor facilities, where they give presentations about their flight called PRs (public relations). Sometimes they travel abroad if they flew onboard an international flight.

107. Dan Tani, Peggy Whitson, and Pam Melroy moved the truss with the arm of the Shuttle and Station and then on to a mobile transformer to a place where they could reconnect the truss later. They accomplished this on day seven, in between EVA 2 and EVA 3.

108. The solar alpha rotary joint (SARJ) allow the ISS solar panels, which power the Station, to continually track the sun.

109. STS-120 was Scott E. Parazynski's fifth and final Space Shuttle mission. NASA selected Parazynski as an astronaut candidate in 1992 after he completed his residency in emergency medicine.

110. Tani, Whitson, and Melroy were watching events unfold inside the International Space Station. One of the primary goals of STS-120 was to relocate

the P6 Photovoltaic Power Module (PVM) from its temporary location to its permanent position on the Station. One of four Station PVMs, the P6 converts sunlight into power; the module then stores, regulates, and distributes electricity to the orbital workshop.

111. A *beta angle* is the angle between a spacecraft's orbital plane and the Earth-to-sun line. Flight planners avoided scheduling missions with the Space Shuttle docked to the ISS when the beta angle exceeded 60 degrees because of high temperatures and the impact on the generation of solar power. NASA avoided flying Shuttle missions if the beta angle exceeded 65 degrees because the Orbiter might exceed its thermal limits. For more information, see "Mission Control Answers Your Questions," NASA, December 1, 2002, https://spaceflight.nasa.gov /feedback/expert/answer/mcc/sts-113/11_23_20_01_179.html (Web site decommissioned Feb. 25, 2021); "Mission Control Answers Your Questions," NASA, April 7, 2002, https://spaceflight.nasa.gov/feedback/expert/answer/mcc/sts-92/10_15_03_59_48.html. (Web site decommissioned).

112. In spite of the high beta angle, the Space Shuttle remained docked to the Space Station while the solar array was unfurled.

113. A solar array consists of two solar cell "blankets" attached to a mast. The solar arrays are divided into sections, or "bays"; the P6 arrays had thirty-one sections. The retracted solar array had to be unfurled, or deployed, once it was relocated. Prior to the launch of STS-120, NASA raised concerns about this process. Crewmembers received training on how to identify any warning signs in the event of a snag or damage to the arrays as they were deployed, and the importance of stopping the deployment if damage occurred.

114. Derek Hassmann was the lead flight director for STS-120 and STS-134. He worked at NASA for twenty-four years before moving on to Bigelow Aerospace, UTC Aerospace Systems, and Axiom Space. NASA selected Rick LaBrode as a flight director in 1998.

115. Kapton tape helped to insulate the spacewalkers against electrical currents.

116. After *Columbia*, crews used the Orbiter boom sensor system to inspect the vehicle's heat shield for damage.

117. Paolo Nespoli is an ESA astronaut. Read more about his career and interests on the European Space Agency's website at http://www.esa.int/Our_Activities /Human_Spaceflight/Astronauts/Paolo_Nespoli.

118. Wheelock was wearing his extravehicular mobility unit, or spacesuit, but he was in the shadow of the Space Station—where temperatures plunged to about 300 degrees Fahrenheit below zero—for more than seven hours. He later told the *Texas Standard* he was so cold that "I really almost had a little mini panic attack. I was so scared that I wasn't going to be able to close my hands to be able to grab onto things." Wheelock warmed his hands by raising them into a sliver of sunlight. Joy Diaz and Terri Langford, "NASA's Longest Work Commute Celebrates Twentieth Anniversary," *Texas Standard*, November 2, 2020, https://www.texasstandard .org/stories/nasas-longest-work-commute-celebrates-20th-anniversary/.

119. Stephanie D. Wilson came to JSC in 1996 from the Jet Propulsion Laboratory, where she worked on the *Galileo* spacecraft. She went on to fly three Space Shuttle missions.

120. "Inspection" refers to the crew's inspection of the vehicle's thermal protection system.

Chapter 7

1. Dispersion analyses are the study of deviations from a prescribed flight path. In this case, Anne Accola analyzed the errors that might occur during rocket burns for Apollo 9 and 11, the first lunar landing.

2. RIFs are a permanent reduction in the workforce.

3. The *Challenger* and her crew were lost on January 28, 1986.

4. Anne Accola sat for this interview with the JSC Oral History Project in March 2005.

5. MSC was the Manned Spacecraft Center, which became the Johnson Space Center, or JSC.

6. Eugene F. "Gene" Kranz was then chief of the Flight Control Division. The JSC Oral History Project has posted a series of interviews conducted with Kranz that are publicly available on the JSC History Portal.

7. Because of the way the federal hiring process works, Accola was the only candidate Kranz could select for this position. Normally, selection officials must hire a veteran if they made the final cut, but Accola does not recall there being any on the list.

8. Melvin F. Brooks was assistant chief for Systems, while Jones W. "Joe" Roach was assistant chief for Operations. Their transcripts are available on the JSC History Portal.

9. Kranz's memoir details how the speech came about and the basic ideas behind flight control. See *Failure Is Not an Option: Mission Control from Mercury to Apollo 13 and Beyond* (New York: Simon and Schuster, 2000), 203–204.

10. Richard A. "Dick" Hoover worked in flight control for many years. He was one of several men assigned to the Simulation Task Group for the Space Task Group, working on simulations for Project Mercury. In the early 1970s, he served as the chief of the Mission Simulation and Requirements Branch and its various iterations.

11. The command module simulator mimicked spacecraft performance in orbit, as did the lunar module simulator.

12. Betty Friedan's watershed book was published in 1963.

13. That woman was Frances "Poppy" Northcutt. Her transcript is available on the JSC History Portal.

14. Read more about the Apollo 15 mission at https://www.nasa.gov/mission_pages/apollo/missions/apollo15.html.

15. The lunar roving vehicle (LRV) allowed the crew to further explore the Moon by expanding the area they could visit. Apollo 15, 16, and 17 astronauts used the rover.

16. The lunar module was the spacecraft that landed on the Moon.

17. EVAs are known as spacewalks. The astro sims involved training people pretending to be astronauts completing the procedures.

18. Hiram G. Baxter worked on simulations from Apollo 9 through the end of the Space Shuttle Program. During the Apollo Program, geologist James W. Head trained the flight crews, analyzed lunar landing sites, and studied rocks and soil samples brought back by the astronauts. Read about Head's recollections of the Apollo Program in his oral history transcript on the JSC History Portal.

19. Read more about M. P. "Pete" Frank's operation experiences by downloading his oral history transcripts available on the JSC History Portal.

20. Learn more about one of the final lunar landing missions by visiting the

online mission highlights at https://www.nasa.gov/mission_pages/apollo/missions /apollo16.html.

21. Read more about Apollo 17, the final lunar landing flight, at http://www .nasa.gov/mission_pages/apollo/missions/apollo17.html.

22. The capsule communicator, known as CapCom, was the individual in Mission Control who spoke directly with the crew.

23. Harrison H. "Jack" Schmitt was the lunar module pilot for Apollo 17 and the first scientist-astronaut to fly in space. Learn more about his career by reading his online NASA biography on NASA's Astronaut Homepage and his oral histories on the JSC History Portal.

24. Skylab was America's first space station. Read more about that program at http://www.nasa.gov/mission_pages/skylab/.

25. The orbital workshop was the Space Station, a modified S-IVB rocket stage, where the crew lived and worked during their missions. The multiple docking adaptor and airlock were two of the major components of the Station, another being the Apollo telescope mount.

26. The simulation supervisor and the training team worked in the SCA just off the front room in Mission Control, writing scripts for mission simulations and monitoring training.

27. When the micrometeoroid shield and sunshade tore off during the launch of the Skylab workshop, temperatures inside the workshop rose quickly. Without a sunshield, Skylab was uninhabitable. MSC's Jack Kinzler proposed that the astronauts of Skylab 2, which was the first crewed flight of the program, should deploy a parasol to serve as a heat shield. After its deployment the temperatures dropped, allowing the Skylab crew to complete its twenty-eight-day mission. That three-man crew included Moon-walker Charles "Pete" Conrad, Jr., Paul J. Weitz, and Joseph P. Kerwin. Learn more about Conrad, Weitz, and Kerwin by reading their biographies on NASA's Astronaut Homepage; both Weitz's and Kerwin's oral history transcripts are available on the JSC History Portal.

28. Tracking stations were set up to monitor space vehicles and the crew. This enabled flight controllers in the Mission Operations Control Room, or MOCR, to receive real-time data (voice, radar, and telemetry) from the vehicle and astronauts.

29. At the time, Jack Knight was working for the Flight Control Division. Learn more about his career by reading his online transcripts on the JSC History Portal.

30. Flight controllers sat at consoles and monitored the spaceflights from the MOCR. It is commonly referred to as the Mission Control Room, or MCC, but the MOCR is the front room where they controlled the flights.

31. The Apollo-Soyuz Test Project (ASTP) was the first international spaceflight, a joint mission conducted by the United States and the Soviet Union. To learn more about this historic mission, visit http://science.ksc.nasa.gov/history/astp/astp .html.

32. The environmental control system made the Orbiter a habitable environment for crews, providing temperature, pressure, and humidity controls as well as heat and cooling.

33. Robert K. "Bob" Holkan headed up the Control Systems Section.

34. Robert L. "Bob" Crippen and Richard H. "Dick" Truly were pilots for the first two Space Shuttle flights, STS-1 and 2. Both had been selected for the US Air Force's Manned Orbiting Laboratory, a military space station. When that program was cancelled, they transferred to NASA, becoming part of the astronaut corps.

Learn more about the two by reading their online NASA biographies on NASA's Astronaut Homepage and their oral history transcripts on the JSC History Portal.

35. In 1977, NASA put the Orbiter prototype known as *Enterprise* through a series of tests designed to determine if the vehicle could safely land after a spaceflight. These series of flights were called the Approach and Landing Tests.

36. The first four flights of the Space Shuttle, STS-1 through 4, were part of the Orbital Flight Test Program, designed to prove the vehicle and its systems.

37. Astronauts fly the T-38s, a jet aircraft, and are required to fly a certain number of hours per month. Pilot astronauts fly the T-38s while mission specialists sit in the back seat.

38. Flight crews trained for their missions on the Shuttle Mission Simulator at the Johnson Space Center, which consisted of a fixed base and motion base simulator.

39. Accola, Jerry W. Mill, and John D. "Denny" Holt worked for the Flight Training Branch at the time. To learn more about Denny Holt's NASA career, read his online transcripts available on the JSC History Portal. The simulation supervisor, the "sim sup," is responsible for flight simulations and monitors the flight control and flight crew decisions.

40. John W. Young and Crippen were assigned to the first Space Shuttle flight, known as STS-1, while Joe H. Engle and Truly flew STS-2. Learn more about mission commanders Young and Engle by visiting NASA's Astronaut Homepage. In 2004, Engle sat for a series of interviews with the JSC Oral History Project, which are available on the JSC History Portal.

41. Discrepancy reports identify known problems or errors that need to be fixed. DRs are given a number and remain open until the issue is resolved.

42. Experts sat in the Mission Evaluation Room and helped flight controllers when they encountered malfunctions or problems with their systems.

43. In April 1981, NASA launched the first mission of the Space Shuttle Program, known as STS-1. Read more about that historic flight at http://history.nasa.gov/sts1/.

44. STS-2 was *Columbia's* second flight. Learn more about that mission by visiting http://science.ksc.nasa.gov/shuttle/missions/sts-2/mission-sts-2.html.

45. At the time, John Cox headed the Orbiter/Computer Section.

46. George W. S. Abbey was head of Flight Crew Operations Directorate at the time. Gene Kranz was then Abbey's deputy in Flight Crew Operations.

47. Cecil E. Dorsey joined MSC in 1967 and began working in Flight Operations. He later worked in Center Operations and retired after thirty years in 1997.

48. Spacelab was a laboratory module placed in the Orbiter's cargo bay where astronauts conducted experiments during flight. The remote manipulator system, called the "arm," captured satellites and maneuvered EVA crewmembers in orbit.

49. The fixed base simulator was a replica of the Orbiter's flight deck and middeck. The crews practiced mission activities in that simulator. The motion base simulator was a replica of the Orbiter's forward flight deck. It rotated and pitched, allowing the mission's commander, pilot, and flight engineer to practice ascent and entry procedures.

50. Astronauts often trained on the remote manipulator system (also known as the Canadarm) simulator at Spar Aerospace in Toronto.

51. Single-system training was a basic training session where an astronaut met with an instructor to practice on or learn about a single Orbiter system (electrical and propulsion systems, for instance) and to review the checklists.

52. John W. Young was then chief of the Astronaut Office.

53. Lisa Reed sat for this interview in June 1998 as part of the Shuttle-Mir Oral History Project.

54. STS-71, which flew in the summer of 1995, was the first docking mission between the Space Shuttle Orbiter and the Mir Space Station. To learn more about the flight, visit http://www.nasa.gov/mission_pages/shuttle/shuttlemissions /archives/sts-71.html.

55. Shuttle-Mir was a joint project between the United States and Russia. Between 1994 and 1998, NASA gained experience in developing and building a space station by docking with the Mir and learning how to work with international partners. By the end of the project, seven astronauts had lived and worked onboard the Russian Space Station. For more information about the program, visit the Shuttle-Mir section of the JSC History Portal.

56. Gregory J. Harbaugh, the flight engineer for STS-71, had flown two missions prior to the historic docking mission. Robert L. "Hoot" Gibson, a naval aviator, was a veteran flyer, with four spaceflights under his belt before this Shuttle-Mir mission. Read more about Gibson and Harbaugh by downloading their NASA biographies on NASA's Astronaut Homepage. To learn more about Hoot Gibson, read his oral history transcripts on the JSC History Portal.

57. Read more about the first flight of the Shuttle-Mir Program at http://www .nasa.gov/mission_pages/shuttle/shuttlemissions/archives/sts-60.html.

58. Learn more about Russian cosmonauts Sergei Krikalev and Vladimir Titov by reading their online biographies at http://www.gctc.ru/main.php?id=930 and http://www.gctc.ru/main.php?id=1098. Titov's oral history is available on the JSC History Portal.

59. STS-47 was a joint mission between NASA and the National Space Development Agency of Japan (which merged with two other space and aeronautics agencies in 2003, becoming JAXA); the flight featured a number of spaceflight firsts: the first African American woman in space, the first Japanese astronaut to fly on-board the Orbiter, and the first and only married couple to fly. To learn more, visit http://www.nasa.gov/mission_pages/shuttle/shuttlemissions/archives/sts-47.html.

60. Just a few months after his inauguration, President Bill Clinton and his Russian counterpart, Boris Yeltsin, agreed to expand the joint Shuttle-Mir Program, established under his predecessor, President George H. W. Bush.

61. The Space Shuttle Orbiter had three Auxiliary Power Units (APUs) that provided power to the vehicle's hydraulic systems during ascent and descent.

62. Henry Lampazzi eventually became a simulation supervisor.

63. Charles F. Bolden, Jr.'s, last command was STS-60 in 1994. Fifteen years later, in July 2009, he became NASA's administrator. Learn more about his NASA career by reading his online biography on NASA's Astronaut Homepage, and reading Bolden's oral history interviews on the JSC History Portal.

64. Mission specialists flew onboard the Space Shuttle and performed experiments onboard the spacecraft, operated the vehicle's arm, and conducted spacewalks.

65. STS-60 kicked off the Shuttle-Mir Program in 1994, and the landing of STS-91 ended the first phase of the International Space Station in 1998.

66. Learn more about these docking flights by visiting their mission pages at http://www.nasa.gov/mission_pages/shuttle/shuttlemissions/archives/sts-76.html and http://www.nasa.gov/mission_pages/shuttle/shuttlemissions/archives/sts-79.html.

67. STS-84, 89, and 86 were other docking missions to the Mir. Read the mission highlights at http://www.nasa.gov/mission_pages/shuttle/shuttlemissions

/archives/sts-84.html, http://www.nasa.gov/mission_pages/shuttle/shuttlemissions
/archives/sts-89.html, and http://www.nasa.gov/mission_pages/shuttle/
shuttlemissions/archives/sts-86.html.

68. In space, astronauts are not limited to placing objects on the floor, thanks to zero-G, or weightlessness.

69. The SPACEHAB was a module flown in the Shuttle's payload bay in which crews conducted scientific research. It was essentially a fully functional laboratory in space.

70. Charles J. Precourt, an Air Force test pilot, retired from NASA in March 2005. Read more about his spaceflights by downloading his online biography on NASA's Astronaut Homepage and also read Precourt's oral history interview on the JSC History Portal.

71. Eileen M. Collins is featured in chapter six.

72. Links to the STS-84 crew biographies can be found on NASA's Astronaut Homepage.

73. That mission was STS-93, which Eileen M. Collins commanded.

Chapter 8

1. Debra L. Johnson sat for this interview with the JSC Oral History Project in August 2012.

2. Building 419 was then known as the Logistical Service Offices.

3. Building 1 is JSC's administrative building, where the center's leadership resides.

4. JAXA is the Japan Aerospace Exploration Agency, an international partner with the United States.

5. Commercial Orbital Transportation Services was a partnership between the federal government and private aerospace companies to re-envision low-Earth orbit transportation. With NASA funding and assistance, SpaceX and Orbital Sciences Corporation developed the technology and hardware to deliver experiments and supplies to the International Space Station. Space Act Agreements provide NASA the flexibility to enter into a partnership to achieve its mission, but these acts are not subject to traditional procurement rules or regulations. Attorney Paul G. Dembling provided the agency this authority when he drafted the National Aeronautics and Space Act of 1958.

6. Commercial Crew is another partnership between NASA and aerospace companies to provide astronauts transportation to the ISS.

7. Aircraft Operations Division manages the planes at Ellington Field. The Flight Crew Operations Directorate included the Astronaut Office and Aircraft Ops.

8. Michael L. Coats served as center director from November 2005 to December 2012. Jefferson D. "Beak" Howell preceded him as center director, serving from March 2002 to November 2005. Read more about these JSC center directors in their NASA oral histories on the JSC History Portal.

9. Stephen J. "Steve" Altemus was then director of Engineering.

10. Benchmarking refers to the process of comparing NASA's practices and policies with those of other companies and federal agencies.

11. Astronauts practice for spacewalks in the Sonny Carter Training Facility Neutral Buoyancy Laboratory pool, which simulates the experience of working in a zero-G environment.

12. Source evaluation boards evaluate proposals submitted for government contracts.

13. NASA Headquarters, located in the nation's capital, provides the agency direction.

14. Award fee contracts allow federal contractors to increase their profits through incentives.

15. Robert D. "Bob" Cabana served as deputy director from 2004 to 2007 and has been KSC director since 2008. Read more about Cabana's career by downloading his online biography on NASA's Astronaut Homepage or reading his NASA oral history interviews on the JSC History Portal.

16. The center's leadership, including the director, sit on the ninth floor of Building 1.

17. SpaceX and Orbital are part of the public-private procurement model to provide cargo and payload launch services to the International Space Station.

18. The federal government pays employees based on education, experience, and time in service. The General Schedule pay scale starts out at GS-1 and goes up to GS-15.

19. Natalie V. Saiz sat for this interview in August 2012 with the JSC Oral History Project.

20. Carolyn L. Huntoon directed Space and Life Sciences from 1987 to 1994. Her oral histories are available on the JSC History Portal, and she is included in this volume.

21. Mission support includes positions with Human Resources, Public Affairs, and Legal. By contrast, NASA considers engineering and labs technical positions.

22. The inclusion and innovation concept takes different employee opinions and perspectives into account, allowing more voices to be heard. NASA leadership believes diversity of thought and opinions is beneficial, resulting in more creative thinking and innovation.

23. ERGs, employee resource groups, are a part of the inclusion and innovation concept implemented at JSC in 2011 and a recruitment tool to provide employee networks for the benefit of NASA's mission. Other companies like Texas Instruments and American Airlines have successfully used these groups to advance their goals.

24. The PPMD, Program and Project Management Development Program, was developed to prepare future managers for the long-term challenges facing the agency.

25. All of these men had long and illustrious careers with the agency; Arnold D. Aldrich as director of the National Space Transportation System, known as the Space Shuttle Program. Thomas W. "Tommy" Holloway, Glynn S. Lunney, and Jay H. Greene served as flight directors, among many other positions at NASA. You can read all of their interviews by visiting the JSC History Portal.

26. The Gilruth Center, named for Robert R. Gilruth, JSC's first center director, includes a fitness area and conference rooms.

27. Hurricane Ike hit the Gulf Coast of Texas in 2008 and caused widespread damage and flooding to the Houston/Galveston area.

28. President George W. Bush announced a new Vision for Space Exploration in January 2004. He declared that the Space Shuttle would stop flying in 2010 and that the International Space Station would be complete by that time. By 2020, he proclaimed, America would return to the Moon. NASA developed Constellation in response to President Bush's plan, but the Obama administration later cancelled the program.

29. Building 44 is the Communications and Tracking Development Lab, which tests spaceflight communication systems.

30. Patrick S. Pilola started working at NASA in 1986. In addition to working at JSC, he also worked at the Kennedy Space Center. In 2017 he gave a presentation entitled, "Workplace Violence: A Personal Look," which addressed the JSC tragedy Saiz discussed. In 2020, he was chief of the Project Management and Systems Engineering Division.

31. The Employee Assistance Program (EAP) assists employees and their immediate families with issues that affect employee health, the safety of the JSC workforce, one's career performance, attendance, and productivity, including psychological referrals.

32. Building 2 includes a large auditorium, named for Congressman Olin E. Teague, where meetings are routinely held.

33. Critical Incident Stress Management training offers employees the skills they need when dealing with crises like traumatic deaths. Jaqueline E. Reese has served as head of JSC's Employee Assistance Program for more than twenty years.

34. The incident involved a civil servant and a contractor; a member of the contract team shot a civil servant. NASA relies heavily on contractors for much of its work at the centers and at Headquarters. Each contracting company has its own HR professionals.

35. The Joint Leadership Team is made up of veteran leaders from the civil servant and contractor ranks to establish better communication and relationships between the two groups and explore solutions for the benefit of human spaceflight. Jacobs Engineering is the company Saiz is referring to.

36. Duane L. Ross headed the Astronaut Selection Office for more than thirty-five years. His oral history is available on the JSC History Portal.

37. Among its many responsibilities, the Information Resources Directorate (IRD) manages IT systems at the Johnson Space Center.

38. Center Operations supports center facilities and provides utilities for the site. In addition, they manage logistics, handle center security, and manage JSC's environmental programs.

39. John P. Shannon and N. Wayne Hale both served as program managers for the agency's longest running program. Hale ran the Space Shuttle Program from September 2005 to February 2008, and Shannon followed from February 2008 to August 2011. Their oral histories are available on the JSC History Portal. Suzanne "Sue" Leibert served as an HR representative for the Space Shuttle Program for more than twenty-five years.

40. Buyouts are offered by federal agencies to decrease the size of their workforce. Employees who volunteer to resign or retire receive a lump sum payment for doing so.

41. Reductions in force (also known as RIFs) are a permanent reduction of the workforce.

42. In 2011, NASA Administrator Charles F. Bolden, Jr., announced that the agency would use the MPCV for human space exploration beyond low-Earth orbit.

43. Learn more about Leonard S. Nicholson by reading his online transcript on the JSC History Portal.

44. Building 20, one of the newer buildings on site, is an energy-efficient facility built with some recycled materials. Building 12 was erected during the sixties boom and was renovated.

45. Peggy A. Whitson talked with the JSC Oral History Project in the summer of 2012.

46. Brent W. Jett, Jr., was then director of Flight Crew Operations, serving in this position from November 2007 to February 2011. To learn more about his NASA career, read his online biography on NASA's Astronaut Homepage. Jett's oral history is available on the JSC History Portal.

47. On Monday mornings the Astronaut Office holds a meeting with all the active-duty astronauts.

48. Whitson launched with the Expedition 5 crew on STS-111 in June 2002 and returned to Earth with the crew of STS-113 on December 7, 2002.

49. She commanded Expedition 16 from October 2007 to April 2008.

50. Capsule communicators, or in NASA-ese, CapComs, are the people who communicate directly with the flight crew from the Mission Control Center.

51. "Orion" is the name for the next generation spacecraft to take astronauts beyond low-Earth orbit.

52. STS-135 was the final Space Shuttle mission. To learn more about this historic flight, visit http://www.nasa.gov/mission_pages/shuttle/shuttlemissions/sts135/main/index.html.

53. Astronauts are the public representatives of NASA and spend a lot of time handling public relations requests from the media about the mission, hardware, and their careers.

54. Christopher J. Ferguson flew three spaceflights, serving as commander for two. Read more about his spaceflight experience by finding his online biography on NASA's Astronaut Homepage.

55. Biographies of the STS-135 astronauts can be found on NASA's Astronaut Homepage.

56. Kent V. Rominger's nickname is "Rommel."

57. In 2011, the NRC found that NASA's model for the astronaut corps in the post-Shuttle years was inadequate. The agency could not reliably predict whether or not the flight manifest for the ISS, expected to operate through 2020, could be maintained with the number of space flyers in the office.

58. White papers are generally reports researched and written by the government that provide a detailed look at complicated issues.

59. Astronauts train in the T-38 jet aircraft and are required to maintain their proficiency.

60. Whitson is referring to the Commercial Crew Program.

61. The Mission Operations Directorate handles mission planning, training, and flight operations support for NASA's human spaceflight programs.

62. In 2008, John A. McCullough became chief of the Flight Director Office. He began working at NASA in 1989 and became a flight director in 2000.

63. Michael T. Suffredini served as the ISS program manager for a decade, from 2005 to 2015. His oral history is available on the JSC History Portal.

64. Flight readiness reviews are an assessment of a spacecraft, its systems, mission operations, and support functions to successfully complete a mission.

65. Mark E. Kelly flew four NASA missions, including the second Return to Flight following the *Columbia* accident. To read more about his career, visit NASA's Astronaut Homepage.

66. The House of Representatives maintains a biography of Gabrielle "Gabby" Giffords on its history page at http://history.house.gov/People/Detail/14267.

67. Selected as an astronaut candidate in 1994, Frederick W. "C. J." Sturckow flew four missions before leaving NASA in 2013. Learn more about his spaceflights by downloading his NASA biography from NASA's Astronaut Homepage.

68. Timothy L. Kopra served as flight engineer for Expedition 20, flying to the outpost onboard STS-127 and returning on STS-128. He served as flight engineer for Expedition 46 and commanded Expedition 47. To read more about his spaceflight experience and current assignments, visit NASA's Astronaut Homepage.

69. That mission was STS-133. Read more about the mission by visiting https://www.nasa.gov/mission_pages/shuttle/shuttlemissions/sts133/main/index.html.

70. Stephen G. Bowen had previously flown on two missions, STS-126 and STS-132, and had conducted five EVAs before being assigned to STS-133. To learn more, read his online biography on NASA's Astronaut Homepage

71. The flight engineer sat on the Orbiter flight deck during launch and helped the pilot and commander during ascent. Mission specialists performed experiments in space, operated the Shuttle's arm, and conducted spacewalks.

72. NASA selected Benjamin Alvin Drew, Jr., as an astronaut candidate in 2000. He has spent more than 612 hours in space, having flown on board the Space Shuttle twice. Learn more about his career by reading his online biography on NASA's Astronaut Homepage.

73. Ellen Ochoa sat for this interview with the JSC Oral History Project in August 2012.

74. That mission was STS-110, which flew in 2002. Read more about that mission at https://www.nasa.gov/mission_pages/shuttle/shuttlemissions/archives/sts-110.html.

75. That tragedy occurred on February 1, 2003, when the Space Shuttle *Columbia* and the crew of STS-107 perished. The Orbiter broke apart returning from its mission.

76. Bob Cabana served as head of Flight Crew Ops from November 2002 to March 2004, while Kent Rominger was chief of the Astronaut Office for four years, from 2002 to 2006. To read more about Rominger's spaceflight and management experience, read his online astronaut biography on NASA's Astronaut Homepage.

77. The Mission Management Team (MMT) included senior managers and critical decision makers from all elements of the Space Shuttle Program—not just from JSC but also from other NASA centers—and contractors. The team reviewed significant issues that popped up during missions, specifically issues that flight controllers brought to management, and developed plans to resolve those concerns.

78. Launch Control oversees launches at the Kennedy Space Center in Florida.

79. Rick D. Husband commanded STS-107. Read his astronaut biography on NASA's Astronaut Homepage to learn more about his spaceflight experience.

80. Husband and Ochoa flew together on STS-96.

81. The report issued by the *Columbia* Accident Investigation Board is available online at http://www.nasa.gov/columbia/home/CAIB_Vol1.html.

82. Phil Engelauf and John Shannon were flight directors at the time, but not serving in that capacity for this mission. Oral histories with both men are available on the JSC History Portal.

83. At the time, the deputy chief of the Astronaut Office (known by organization code CB) was Andrew S. W. Thomas. Learn more about his career by reading his biography on NASA's Astronaut Homepage and his oral history interview on the JSC History Portal.

84. That was the Expedition 6 crew.

85. Jefferson Davis Howell's nickname is "Beak."

86. Kenneth D. Bowersox was in orbit with Donald Pettit and Nikolai Budarin. They remained in orbit for nearly six months, from November 2002 to May 2003.

Links to the increment summary are available at http://www.nasa.gov
/mission_pages/station/expeditions/expedition06/index.html. Biographies of Bowersox and Pettit can be found on NASA's Astronaut Homepage. Read an interview
with Pettit on the JSC History Portal.

87. Ron Dittemore served as program manager for four years, from April 1999 to
July 2003. To learn more about his NASA career, read his online oral history
transcript on the JSC History Portal.

88. Linda Ham was the first woman to serve as a flight director; she became
lead flight director for STS-58, a Spacelab flight. Following the *Columbia*
accident, she received an appointment to the National Renewable Energy
Laboratory in Colorado.

89. Jerry L. Ross flew seven times onboard the Orbiter, a record held by only
one other astronaut, Franklin Chang-Diaz. To read more about his career, read
Ross' online biography on NASA's Astronaut Homepage as well as his oral history
transcripts on the JSC History Portal.

90. Because the Orbiter disintegrated over East Texas, NASA sent teams to the
area to retrieve the debris and crew.

91. When Dominic L. Pudwill Gorie retired from NASA in 2010, he had flown
in space four times. Learn more about his career and spaceflights by downloading
his astronaut biography from NASA's Astronaut Homepage.

92. Andrew S. W. Thomas became an astronaut in 1993 and flew four times,
spending more than 177 days in space. To learn more about his NASA career, visit
NASA's Astronaut Homepage and read his oral history on the JSC History Portal.

93. President George W. Bush spoke at the memorial in Houston.

94. Sylvia S. Stottlemyer started working for the Astronaut Office after graduating from high school. She was the crew secretary for Apollo 17, the Skylab 3 and 4
flights, and STS-1 mission. In 2003, when *Columbia* and her crew was lost, she was
working for FCOD.

95. Michael J. Bloomfield, a veteran of three spaceflights, works for an aerospace
contractor, Oceaneering Space Systems. To learn more about his missions, read his
online biography on NASA's Astronaut Homepage.

96. NASA lost two spaceflight crews before the *Columbia* accident, one in
1967—called Apollo 1—during a launch pad test at Cape Kennedy. In the second,
the crew of *Challenger* perished in January 1986 moments after liftoff.

97. The T-38s flew a missing man formation in memory of the astronauts lost
onboard STS-107.

98. David A. King was then the Marshall Space Flight Center's director.

99. Jerry L. Ross, a former astronaut, was chief of the Vehicle Integration Test
Office at that time. James D. "Jim" Wetherbee, a veteran of six spaceflights,
commanded a record five flights between 1992 and 2002. Learn more about their
NASA careers by reading their online biographies from NASA's Astronaut
Homepage. Their oral history transcripts are available on the JSC History Portal.

100. Two died and three were injured when a helicopter searching for *Columbia*
debris crashed in East Texas.

101. The Progress spacecraft delivers supplies to the ISS.

102. Control boards are governing boards with oversight over a certain task or
project.

103. The Shuttle carrier aircraft ferried the Orbiter across the country, and the
Shuttle commander and pilot simulated landing in the Shuttle training aircraft.

Chapter 9

1. Carolyn L. Huntoon sat for this interview in June 2002.

2. For nearly a decade, from April 1992 through November 2001, Daniel S. Goldin served as NASA's administrator. Learn more about Goldin's contributions to the agency by reading his online biography at http://history.nasa.gov/dan_goldin .html. An oral history with Goldin can be found on the JSC History Portal.

3. Christopher C. Kraft, Aaron Cohen, and Gerald D. Griffin were former center directors, while Henry E. "Pete" Clements had been JSC associate director. All of their NASA oral histories are available on the JSC History Portal.

4. Huntoon is referring to work with the Houston city government as well as Congress.

5. The Shuttle-Mir Program was a joint American-Russian space program and the first phase of the International Space Station.

6. The Technology and Transfer Commercialization Office, established in 1994, required NASA to share its engineering and scientific technologies with the private sector.

7. NASA Headquarters, based in the nation's capital, guides and directs the agency.

8. Astronauts train for spacewalks in the NBL.

9. The first Water Immersion Facility opened in 1966. Early in the Shuttle Program, astronauts trained in the Weightless Environment Training Facility (WETF), a large pool, for spacewalks.

10. EVAs are extravehicular activities, NASA-ese for spacewalks.

11. STS-61 was the first Hubble Space Telescope repair mission. Read more about the Hubble and the missions to repair it at https://www.nasa.gov/mission _pages/hubble/main/index.html.

12. The Man-Systems Division handled crew support functions like flight crew equipment (clothing and hygiene systems) and food systems, as well as neutral buoyancy simulation. Huntoon was director of Space and Life Sciences from 1987 to 1994.

13. The Marshall Space Flight Center, in Huntsville, Alabama, had the Neutral Buoyancy Simulator decommissioned in 1997.

14. President Ronald Reagan approved Space Station Freedom, which evolved into the International Space Station. Ellington Field houses the Aircraft Operations Division; it manages JSC aircraft including the T-38s, which the astronauts fly as part of their training.

15. Established in 1997, the Space Biomedical Research Institute studies the impact of long-duration missions on the human body and proposes countermeasures to the physical and health-related issues encountered by crews.

16. The National Institutes of Health is America's medical research agency.

17. Established in 1976, the Office of Science and Technology Policy advises the president about the impact of science, engineering, and technology issues on domestic policy as well as foreign affairs.

Index